New World Monkeys

NEW WORLD MONKEYS

The Evolutionary Odyssey

Alfred L. Rosenberger

PRINCETON UNIVERSITY PRESS
PRINCETON AND OXFORD

Published by Princeton University Press
41 William Street, Princeton, New Jersey 08540
6 Oxford Street, Woodstock, Oxfordshire OX20 1TR

press.princeton.edu

ISBN: 978-0-691-14364-4
ISBN (e-book): 978-0-691-18951-2

British Library Cataloging-in-Publication Data is available

Editorial: Alison Kalett and Abigail Johnson
Production Editorial: Ellen Foos
Production: Jacquie Poirier
Publicity: Matthew Taylor and Julia Hall
Copyeditor: Lucinda Treadwell

Figures 1.1, 2.1, 2.6, 2.11, and 4.4 are copyright 2013 Stephen D. Nash /
IUCN SSC Primate Specialist Group

Jacket image courtesy of Luciano Candisani / Instagram: @lucianocandisani

This book has been composed in Adobe Text and Gotham

Printed on acid-free paper. ∞

Printed in the United States of America

10 9 8 7 6 5 4 3 2 1

This book is dedicated to Rivka Rosenberger,
my SuzieQ,
my muse,
twice my wife,
forever the love of my life.

CONTENTS

ILLUSTRATIONS

Tables

Today, New World monkeys may be the most intensely studied group of primates. When I first became intrigued by the question of platyrrhine evolution as a college student in the 1970s, few people were interested in New World monkey fossils, morphology, ecology, or behavior. The focus then was on apes, Old World monkeys, and lemurs. It was my great good fortune that primatology grew rapidly as a discipline at the same time that I matured to become a professional primatologist, working in the lab, in the field, and in the museum. The methods of study developed, new technologies were introduced, and the scope of research interests broadened, thus transforming our knowledge of platyrrhine evolution. This made possible the modern study of what Charles Darwin described as "descent with modification," meaning how a lineage or a taxonomic group of organisms becomes altered over generational time while adjusting to the circumstances of an ever-changing environment. We call this phylogeny and adaptation. That's what this book is about.

Other key evolutionary concepts that form the basis of our understanding of the harmonious coexistence of the long-lived New World monkeys and their diversification, each group in its ecological niche, will become clear in the story of their odyssey.

My interest in platyrrhines and primatology had many beginnings. Words and names and a charismatic professor, Warren Kinzey, drew me in when I was a sophomore at City College of New York. The words that attracted me were the invented, Latinized taxonomic names of these animals written in italics to offset them from everyday vocabulary. They were often exotic compound terms—*Para | pithecus, Australo | pithecus, Calli | cebus*—which when deciphered, might reveal an author's hypothesis about where a particular fossil or modern primate fit into the scheme of evolution, a geographic place of origin, or what physical trait set one species apart from others. Professor Kinzey, whom I soon knew as Warren, though I was half his age, tossed these names about in lectures as he explained ideas, challenging his students to learn in what seemed like speaking in code. He prodded us to think visually as he used old-fashioned chalk to annotate an anatomical image on the blackboard, sometimes drawing with both hands simultaneously. The scientists, his contemporaries, about whom he talked in lectures, he called by first or nicknames, and he mentioned their academic affiliations to humanize them even more: Cliff was Clifford Jolly at NYU; Sherry was Sherwood Washburn at Berkeley. Warren used surnames to note the pantheon of greats cited in our textbook, the only advanced book on primate evolution at the time, *The Antecedents of Man* by Sir Wilfrid E. Le Gros Clark, which remains a classic. I could see that becoming a physical anthropologist would allow me to combine my attraction to a small branch of science for which I seemed to have an aptitude, with my desire to write.

As a first-year graduate student at the City University of New York (CUNY) Graduate Center, my first real taste of professional-level study occurred in a primate anatomy course, one of only two offered in the entire country, during the weekly evening appointments I had with a dead gelada baboon in a high-rise office building on 42nd Street, off Broadway, in midtown Manhattan in the heart of New York City. I dissected it in the physical anthropology lab. It was an invaluable learning experience, to open the animal up, isolate the individual muscles, follow their ribbonlike paths to see how the muscles spanned the joints, and learn the routes of coursing nerves and blood vessels.

My years at CUNY consolidated my interests in form, function, behavior, theory and practice, and New World monkeys. During two summers I did fieldwork in the Peruvian Amazon, "chasing monkeys" as it was called, learning what monkey life was like and how little we knew about them. I learned how important it was to study morphology and behavior together in order to reveal how these primates evolved. The results of this work in the field were published in several coauthored articles, and this was my initiation into the world of professional scientists.

Meanwhile, my sanctum became the study collections of skeletonized and preserved primates in the American Museum of Natural History in New York. There, evolutionary puzzles and anatomical questions could be probed by studying skulls, teeth, limbs, and pelts. I was encouraged by my mentors to do problem-based research, to express my opinions and publish them when I had something original to say.

At that time at the Museum, graduate students could freely roam the stacks of one of the world's best research libraries. There I spent hours, days, months, poring over reprint collections that had once belonged to, and bore the signatures of, the intellectual giants in the field. They included the Museum's curator of vertebrate paleontology, comparative anatomy and ichthyology—three departments all at once—William King Gregory, who was the finest primate evolutionist of the first half of the 20th century. I was even able to peruse Linnaeus's 1758 edition of *Systema Naturae*, written in Latin, which became the starting point of modern taxonomy.

In the stacks, everything seemed there for the touching and reading; rare books, defunct foreign journals, mimeographed and photocopied personal logs, dog-eared separates gifted to Museum scientists by far-flung colleagues, the collected writings of the famous and the forgotten systematic biologists and paleontologists whose works were collected into individual boxes and books bound for convenience. In the stacks, I could turn the pages of rare, oversized, leather-bound volumes dating back 200 years and more, with hand-colored illustrations describing wild primates never before seen by Europeans and with gorgeous lithographs depicting the minute anatomical details of newly discovered skulls and tiny fossil teeth. Examples of this brilliant artistry from the naturalist Blainville, published between 1839 and 1864, are presented in chapter 2. The library was the resource vault behind the taxonomic names I found so intriguing. Therein lay the historical evidence that primates have long claimed the imaginations not only of scholars, but also of adventurers, explorers, artists, amateurs, expatriates, courtiers, monarchs and Jesuit priests, writing in English, French, Spanish, Italian, German, Dutch, Latin, and more. It was a group whose ranks I was eager to join and whose legacy I would be privileged to carry onward.

In the 1970s the American Museum was a hotbed of the cladistic revolution, a newly refined method of reconstructing phylogeny, taxonomic genealogies, that spread from Europe. It was a truly tumultuous period in evolutionary biology, with a public, often hostile battle of ideas and egos among scholars forcing a rethinking of a fundamental scientific approach, and a rewriting of the evidence for primate evolution. What was being replaced was the rather loose way scientists had approached drawing the genealogy of organisms, the Tree of Life, from fossils and living species ever since Darwin's *On the Origin of Species* in 1859 convincingly argued that all of life was interconnected by a web of ancestry and descent.

At that time, two core members of the evolutionary primatology faculty at CUNY were carrying out a thorough review of primate evolution. They were my graduate school mentors who published prolifically for decades. Eric Delson went on to cofound, and is, as of this writing, the head of the New York Consortium of Evolutionary Primatology (NYCEP), and Frederic Szalay is now retired after having had a tremendous impact on the field. While they were frequently on opposite sides of the cladistics debate, there was between them a mutual respect and appreciation for each other's opinions and expertise. They spent several years jointly writing *Evolutionary History of the Primates*, an indispensible resource which immediately became a classic when published in 1979.

As they cowrote their magnum opus I became involved as an avid listener, occasional sounding board, and also an enthusiastic contributor. It became clear that there was a topical gap in their coverage. While Eric and Fred commanded an unparalleled knowledge of the evolution of Old World anthropoid primates, and of Madagascar's lemurs and the early primates, it was not so for New World monkeys. Little was known then about platyrrhine evolutionary history. At that time I was writing my doctoral thesis, *Phylogeny, Evolution and Classification of New World Monkeys (Platyrrhini, Primates)*, and I was given the opportunity to present some new ideas about their phylogeny, how they should be extensively reclassified, how their adaptive diversity unfolded, and how long-lineages played an important role in platyrrhine evolutionary history. I was privileged to be heard and trusted as Eric and Fred filled in what was missing from the platyrrhine chapter of their book, based on the research for which I earned a PhD in the same year. For that, I will always be grateful to them.

I went on to publish papers and edited volumes with Eric, Fred, and 100 other collaborators—men, women, professionals, peers, and students. My inner circle of colleagues with whom I have continued to work is narrow, but our combined network is broad, spanning five continents and fields as diverse as geology, ethology, conservation, behavioral ecology, software engineering, 3-D computer graphics, scuba diving, and more. This is the nature of scientific research and discovery. It is only because of the work of this dedicated network of professionals, and many others, that the story of the New World monkeys' odyssey can be told.

Fifty years ago, when I began to study primatology, the world was a different place and I could never have imagined that the rainforest would be disappearing and the primates living there would be facing extinction. I am grateful for the opportunity to study these magnificent animals, to learn from them, and to write this book. As a small token to aid in the efforts to save the platyrrhines and other primates, I feel it is only right to donate a portion of the proceeds of the sale of this book to primate conservation efforts.

CHAPTER 1

WHAT IS A NEW WORLD MONKEY?

This book is about the evolutionary odyssey of New World monkeys, the South and Middle American platyrrhines, though it is mostly about their evolution in South America where most of platyrrhine history was played out. Their odyssey appears to have begun 45–50 million years ago when an ancestral population of monkeys arrived in South America to found one of the most diverse and colorful adaptive radiations produced by the Order Primates. A robust view of what platyrrhines have become and how can be gleaned from the living animals today and the fossil record, which, though still limited, documents the major features of New World monkey evolution during roughly the last 25 to 35 or 40 million years of their existence, although the record is exceedingly sparse for periods older than 20 million years. Unlike other major primate groups, the history of New World monkeys is one in which the separate lines of descent leading to many of the 16 extant genera recognized herein can be traced back in time for millions of years by fossils and by molecules. This long-lineage pattern is what gives the structure of platyrrhine evolution its distinctive shape, and it is a centerpiece of this book. It also serves as a poignant point of reflection in thinking about the platyrrhines' future. Fourteen of the 16 living genera include species that are now classified, according to the International Union for Conservation of Nature (IUCN), as Critically Endangered or Vulnerable.

There is an abundant record of fossil South American mammals that dates back nearly to the beginning of the Age of Mammals, about 66 million years ago. However, the oldest New World monkeys we know of date back only 36–40 million years. Given that South America was an island continent for most of the last 66 million years, as the world's living mammals began to flourish and before Isthmus of Panama emerged to firmly connect North and South America 3 million years ago, the questions arise: Where did their ancestors come from, and how did they get there? Whether primates originally came from Africa by rafting across the Atlantic Ocean on a floating mat of vegetation, or mostly overland from North America, two scenarios detailed in chapter 10, they arrived as pioneers in a landscape where monkeys had never existed before.

The ways in which these animals evolved and thrived on the isolated continent, always in the trees, is a history of radical change and enduring stasis, novel adaptive solutions and predictable transformations. It is a story of giants, dwarfs, brainy predaceous tool users, dim vegetarians, fungus feeders, and bark-gnawing gum eaters. It is an account of cautious quadrupeds, acrobatic arm-and-tail swingers, quiet nocturnal denizens, and roaring diurnal howlers. Their mating strategies include codominant monogamists, and alpha males and alpha females living in large social groups. In some species females use scent to control the breeding success of their daughters; in another, males queue up on big branches waiting their turn to copulate with one female. By inhabiting a range of niches so varied in ecological and anatomical solutions to feeding and locomotion, or in social arrangements for group living, mating, and rearing offspring, platyrrhines have produced one of the most diverse adaptive radiations among the primates.

How did this happen? The present is key to understanding the past. There are two intertwined models describing how platyrrhine evolution has unfolded, the Long-Lineage Hypothesis and the Ecophylogenetic Hypothesis. What this means is that the many kinds of monkeys we see today have been around for millions of years and that some have existed for at least 20 million years with little change in their ecological situation, to the extent that their adaptations are documented in the fossil record. Furthermore, at another level, genetically related subgroups of New World monkeys, clusters of genera linked by their shared phylogenetic histories, have found success in various ecological niches defined by the particular sets of characteristics inherited from their remote common ancestors. Today, more than a dozen extant platyrrhine species belonging to all the six major subgroups can be found packed into a single rainforest locality, forming a harmonious monkey community. The fossil record suggests that this phylogenetic and ecological framework may have been in place for the entirety of the modern platyrrhines' long-lived existence, setting the stage for the evolution of more refined divisions of niches by the procession of the living genera and species.

As further discussed below, I use the term lineage to mean a genus-level line of descent, an evolutionary stream carried in DNA that is embodied in a species, or a collection of intimately related species, and is manifested as a distinct ecological lifestyle. When examining an entire radiation such as the platyrrhines, the taxonomic level of genus, not species, is the most appropriate perspective. Genera exemplify and define the combinations of anatomical

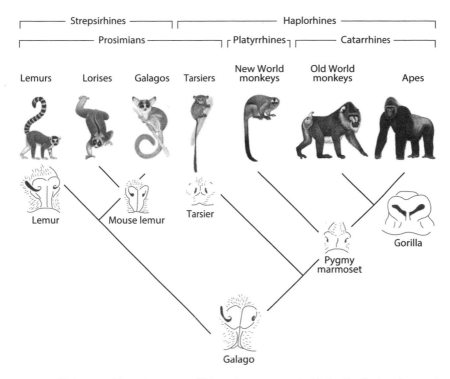

FIG. 1.1. Cladogram of the major groups of living primates mapped with the distribution of external nose shapes. Primate images courtesy of Stephen Nash.

and behavioral characteristics that are of particular ecological relevance, and that separate all the significant lines of descent that compose an adaptive array.

What is a monkey?

We regularly call platyrrhines monkeys, but the word monkey has no scientific significance. There are two groups of primates commonly called monkeys, the New World monkeys and the Old World monkeys. However, they are not grouped together in formal taxonomic language because they lack the evolutionary connection that is the main reason animals are classified jointly in particular groups: a genetic, or phylogenetic, relationship. The two groups we call monkeys are less closely related than the use of the word monkey suggests. In fact, the primates we call Old World monkeys, such as olive baboons and the rhesus macaques, are more closely related to apes than they are to New World monkeys (fig. 1.1). New World monkeys are a separate group entirely, an offshoot of the primate family tree that appeared about 25 million years before the earliest appearance of today's Old World monkeys and apes documented in the fossil record. The sameness implied by the word monkey is an anachronism that may date back to the 14th century, according to the Oxford English Dictionary, an old-fashioned word based on an equally old, pre-evolutionary idea about the natural world. It was meant to distinguish these animals from apes and the other nonhuman primates, the lemurs, lorises, galagos, and tarsiers of Africa and Asia. They are all very different from monkeys and apes in many ways, including the structure of their skulls, their dentition and skeletons, sensory systems, and behavior, reflecting separate evolutionary histories.

Taxonomic groups that are formally recognized and named as units in classifications, such as species, genus, family, and order, are called taxa, the plural form of the word taxon. The term taxonomy, which means arrangement, is derived from the words taxon and taxa. The groups mentioned thus far—primates, platyrrhines and New World monkeys, Old World monkeys, apes, tarsiers, lemurs, lorises and galagos—are all taxa that have formal names in classifications as well as these common names. But monkey is not a taxon and has not been thought of in that way since Darwin introduced us to evolution and phylogeny, and reinforced the notion that classification should be based on relatedness, which previously was only a vague idea. The word is applied to two different groups of taxa that are actually not each other's closest relatives.

Some labels for primate groups are like nicknames and have no scientific standing. Sometimes they are holdovers from the pre-Darwinian period when natural history was not a secular enterprise and scholars used such terms to express their ideas about how far a group was stationed along an imagined trajectory, a ladder of ascent, reflecting the Scale of Nature or the Great Chain of Being that emanated from Creation. Humans were considered the pinnacle of creation and all other animals were said to occupy standings below that high point, as lower grades or stages in the procession of life. The early naturalists arranged their classifications accordingly and their informal language sometimes expressed those views. Thus the term monkey referred to the group of primates grouped with the apes as "higher primates" and gradistically situated between apes and the "lower primates," the tarsiers, lemurs, lorises, and galagos. The latter were called prosimians, meaning near monkeys and apes. Eventually, Darwin made it quite clear that the two great groups of monkeys were distinct: Old World monkeys are the closest living relatives of apes and New World monkeys are a separate line of evolution within the monophyletic group—the unique descendants of a common ancestor—we call Anthropoidea, informally anthropoids, the taxonomic equivalent of "higher primates," composed of New World monkeys, Old World monkeys, apes, and humans.

Even in the Darwinian era grade-thinking persevered throughout biology, and particularly when it came to discussing nonhuman primates as human relatives. Darwin's most effective scientific ally, Thomas Henry Huxley, wrote of primate diversity and evolution in 1863, in *Man's Place in Nature*, four years after *On the Origin of Species* was published. He said, "Perhaps no order of mammals presents us with so extraordinary a series of gradations as this—leading us insensibly from the crown and summit of the animal creation down to creatures, from which there is but a step, as it seems, to the lowest, smallest, and least intelligent of the placental Mammalia." In the next 100 years the gradistic mindset faded from research practice but it still endures in our everyday language as a convenience, hence the word monkey. As a way of viewing the world, however, gradistics failed with the onset of a methodological revolution known as cladistics that occurred in the 1960s, which sought to organize and classify groups according to their placement on the appropriate branch, or clade, of the phylogenetic Tree of Life, as will be fully discussed later. That failure had important consequences in spurring a wholesale re-thinking of platyrrhine evolution.

The geographic modifier in the name New World monkey is also an anachronism. Since the Age of Discovery, in the 15th century, European writers have referred to the Western Hemisphere as the New World, ostensibly discovered by Columbus, in contrast to the Old World, comprising Eurasia and Africa. Similarly, platyrrhines are also often called Neotropical primates, meaning primates of the New World tropics. In an ecological sense, that term may conjure up a misunderstanding about the habitats where platyrrhines live, and what the relevant environments of South America in particular look like. It delimits the wide swath of South and Central America straddling the equator, the tropical zone, where the climate is moist, warm or hot all year round and supports dense, evergreen, jungle vegetation. But that landscape is not all continuous rainforest, and platyrrhines are not strictly jungle dwellers.

South America is a vast continent that is two-and-a-half times the size of the Amazonian rainforest, where most platyrrhines are found. Another vitally important tropical and subtropical region, the Atlantic Forest of southeastern Brazil, supports a smaller, unique ensemble of monkeys including several endemic forms, meaning they are found nowhere else in the world (fig. 1.2). Most of them are presently endangered as a result of the wholesale decimation of the Atlantic Forest that occurred during the last 500 years which, as discussed in chapter 11, has reduced their habitat to disconnected, relict forest fragments

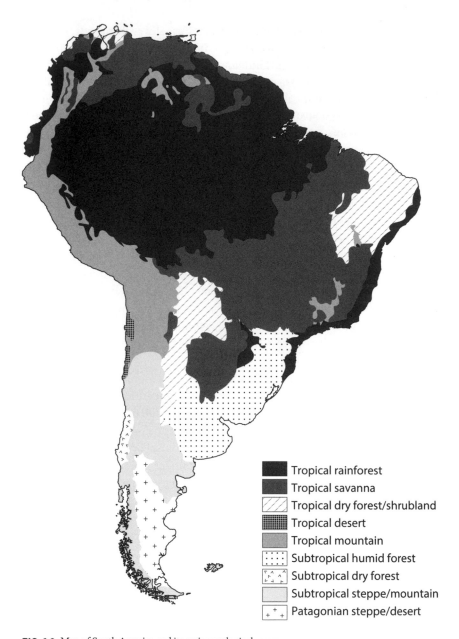

Tropical rainforest
Tropical savanna
Tropical dry forest/shrubland
Tropical desert
Tropical mountain
Subtropical humid forest
Subtropical dry forest
Subtropical steppe/mountain
Patagonian steppe/desert

FIG. 1.2. Map of South America and its major ecological zones.

about one-tenth the size it was when European colonists first arrived in Brazil half a millennium ago.

The full geographic range encompassed by monkeys in South America extends from the northern edge rimming the Caribbean Sea and the Atlantic and Pacific Oceans, to northern Argentina in the distant south. The habitats mapped out in this enormous expanse are predominantly evergreen rainforests, semideciduous forests where trees lose their leaves seasonally, and open-country savannas, grasslands, and shrublands. Primates can be found in all these areas, though the greatest concentration of species and the most densely packed communities of platyrrhine species occur in the rainforests. In drier, more sparsely vegetated zones, only a few generalist species of monkeys, or those with a special set of adaptations to procure food from a limited, local supply, manage to get by. There they are often found in narrow strips of forest situated alongside water courses. Of all things, New World monkeys need trees no matter where they live.

Why is this so? Comparing the vegetation map of South America with the distribution maps of the living species highlights an intensely interesting question about platyrrhine evolution: Why are there no terrestrial species? In Africa, another enormous continent with a similarly varied distribution of habitats, Old World monkeys have evolved an impressive array of terrestrial and arboterrestrial species, living in forests and even extending into bone-dry, near-desert areas. In contrast, while platyrrhines are obviously an exclusively arboreal radiation, there is nothing about the design of their bodies or their dietary needs that makes it impossible for a New World monkey to habitually visit the ground and benefit from it. Actually, some species do so occasionally in order to cross large gaps in the forest or obtain drinking water in drier places when the forest does not provide them with enough because watery fruits are in short supply.

Juvenile monkeys sometimes play on the ground. Clever capuchin monkeys living in swampy areas have even learned to collect clams on the ground when the tide recedes. Yet, no living platyrrhines have evolved terrestrial adaptations or a terrestrial lifestyle. Given their long evolutionary history, however, and knowing that South American forests have waxed and waned over the entire continent, it may be that the fossil record will at some point turn up a ground-dwelling New World monkey. In fact, there is already a hint of this in the few remains of an extinct Caribbean platyrrhine, *Paralouatta*, to be discussed in a later chapter. With all that biologically built-in ecological flexibility and a vast area of the continent as potentially exploitable habitat, under the forest canopy and beyond, the absence of living terrestrial platyrrhines seems quite the mystery.

What is a platyrrhine?

The technical name for New World monkeys is Platyrrhini; platyrrhines, colloquially. It means flat- or wide-nosed. The name was given to them in 1812 by the French naturalist Étienne Geoffroy Saint-Hilaire, who was then sorting and cataloging specimens of mammals held in the collections of the Muséum National d'Histoire Naturelle in Paris. He found that the shape of the nose turned out to be a useful way to identify several groups of primates. In platyrrhines the nostrils are widely spaced and laterally facing, separated by a broad fleshy strip between the openings (fig. 1.1). In some, such as the Saki Monkey, the expression of this characteristic is rather extreme. A contrasting pattern occurs among Old World monkeys and apes, which have nostrils that are closely spaced and separated by a thin band of flesh. They are classified as Catarrhini; catarrhines, informally, meaning downwardly facing nose.

These distinctions, like many others used in identifying and classifying primates, are exhibited consistently among platyrrhines, but not universally. To see an exception, one has only to look at the gorilla-like face and nose of the largest living platyrrhine, the Muriqui, with its adjacent nostrils. The usefulness of employing these names, terms stemming from the same Greek root word for nose, *rhine*, is that they are physically descriptive and they bind together a naturally paired, phylogenetic set of primates. Platyrrhines and catarrhines are the two branches of the extant anthropoid primates, the taxonomic group consisting of New and Old World monkeys, apes, and humans that arose monophyletically from an exclusive common ancestor.

Because the nose is made of flesh, which under nearly all circumstances does not fossilize, paleontology is limited in what it can tell us about the evolution of the platyrrhine nose, and the contrasting catarrhine pattern as well; but is there is a way to reconstruct their morphological histories by examining the living animals? If so, what would the nose have looked like in the last common ancestor of anthropoids? Would its shape have been platyrrhine, catarrhine, or something else? In other words, what nose shape is the primitive form in anthropoids?

In fact, we have good reason to infer that in the first anthropoids the nose was platyrrhine-like. To arrive at that interpretation, we use information on the comparative soft anatomy of extant animals in order to envision the past, as a hypothesis, with an assist from fossil evidence. This method, called character analysis, involves examining the similarities and differences of inherited traits—or presumably inherited, since links between genes and anatomy are still difficult to establish—in closely related forms, with the aim of tracing the sequence in which the details of those features evolved. The approach applies to any observable trait and it is important for understanding how and why evolutionary changes happened functionally, although it does not always lead to adaptive insight because we often do not know the benefit of one pattern or another even when they are linked historically.

In the case of noses, character analysis entails invoking the primate clado-gram, a simplified family tree, as a map that guides us toward the common morphological denominators shared between the animals in question and their nearest relatives: platyrrhine and catarrhine noses are compared with the nose of the tarsiers of Southeast Asia (fig. 1.1). Tarsiers are small, giant-eyed, noctur-nal predators, and they have an external nose that is a close match for a platyr-rhine's even though most of the animal's other features look almost nothing like a platyrrhine or any anthropoid. Since its broad, laterally facing nostrils and pug nose conform to the New World monkey pattern, we can infer that the ancestral anthropoids also shared that morphology, perhaps comparable to a pygmy marmoset's.

The scientific logic behind this conclusion is that it is the most parsimo-nious, or efficient, explanation of the taxonomic distribution of nose shape among the three groups. Reasoning this way implies that New World monkeys inherited a tarsier-like pattern with little change from the original condition, and that catarrhines later evolved the newer, derived shape. An alternative inference would hypothesize that the catarrhine shape was ancestral in an-thropoids. But that means we would have to explain why the same wide-nosed morphology evolved twice in this one monophyletic group, once in the line leading to tarsiers and a second time in the ancestors of New World monkeys. Minimizing such parallelisms, which means minimizing the number of hy-pothesized evolutionary changes required to satisfy existing morphological and taxonomic conditions when there is no reason to think otherwise, is basic to the protocol of the character analysis strategy. That's what is meant by par-simony, and explanatory efficiency. Regarding the evolution of the two nasal shape patterns in this exercise, we still have no sound explanations concerning functional significance, but we do have possible explanations for some of the more oddly shaped, superwide external noses found in a few living platyr-rhines, such as the Saki Monkey, as we shall see below.

Focusing on the nose to identify a primate or other mammal, and formal-izing it descriptively in the structure of a taxonomic name, is a common prac-tice in mammalogy. The rhinoceros, formally the genus *Rhinoceros*, meaning horn-nosed in Greek, is a familiar example. Among catarrhine primates, there is the Proboscis Monkey, *Nasalis*, meaning of or pertaining to the nose in Latin, a genus in which females have a striking, projecting nose and males have an extremely large, pendulous nose.

It may seem odd or even trivial that scientists continue to sort major, higher taxonomic groups of primates such as the platyrrhines and catarrhines by the shapes of their noses because of a tradition dating back to the early 1800s, particularly if we have few ideas about any adaptive significance or benefit to the different morphologies. True, nose shape once served as nothing more than a convenient descriptor and identifier for early naturalists who had limited knowledge of the deeper anatomy, or the actual lives, of the animals whose remains they studied. But as understanding of anatomy and behavior accumulated, this approach began to yield important clues about primate evolution.

The Order Primates is divided into two major extant groups (fig. 1.1), called Strepsirhini (strepsirhines) and Haplorhini (haplorhines). The extant strepsirhines include lemurs, lorises, and galagos. They have wet noses with slitlike, comma-shaped nostrils: strepsirhine, from the Greek *streph*, means twisted nose, a reference to the shape of the nostril's opening. The haplorhines are tarsiers, New World monkeys, and Old World monkeys, apes, and humans. They have dry noses with rounded nostrils. *Hapl*, also Greek, means simple, an illusion to the rounded nares.

We now understand that these names represent profoundly different biological systems. They are only parts of a larger anatomical complex that is functionally and behaviorally important in regulating communication and even how these animals tend to perceive the world, how the two groups gather fundamental information about their surroundings. While all primates are highly competent visual animals, the strepsirhine primates, which are mostly nocturnal and live in low-light conditions, favor olfaction over vision as sensory input. Their acute sense of smell is tied to the structure of their noses. Haplorhine primates, who are mostly diurnal, favor visual input over olfactory information. Consequently, they are less dependent on the anatomy of the nose, and the snout has evolved in another direction.

The outward, easily seen differences in nostril shape, traits that are still without a good adaptive interpretation, are accompanied by other, functionally significant features. A slit or rounded nostril is one piece of a more important whole, the nose itself. Strepsirhines have a bulbous external nose, much like a dog's, covered in a perpetually moist, textured skin. Situated at the very tip of the bony snout, the nose extends as a broad flap directly into the mouth, and splits the hairless upper lip in the middle. As a result, the mouth is not ringed by muscle, and no lemur, loris, or galago is able to control the contour of their lips to shape the mouth to produce facial expressions—no smiling, grimacing, or pouting.

The textured surface of the wet nose is designed to collect molecules of scent from the air they breathe and shunt them down a strip of skin toward a chemosensory organ, the vomeronasal, or Jacobson's organ, situated in the mouth behind the upper incisors. It is part of the secondary olfactory system that is the seat of pheromonal communication, a scent-based adaptation that is especially important in the exchange of sexual signals between males and females. The primary olfactory system, which has sensors located in the nose itself, is concerned with the broad range of environmental smells. The processing centers of the strepsirhine brain, of course, are coordinated, and they emphasize the olfactory areas rather than the visual ones. As one example of this pattern, the forebrain has a conspicuously large olfactory bulb in strepsirhines, while the area responsible for processing visual information in the back of the cerebrum, the occipital lobe, is not emphasized.

Haplorhines have dry, non-textured, untethered external noses, separated from the mouth by a continuous, fleshy, mobile upper lip and a patch of furry skin. A secondary olfactory system still exists in some haplorhines, but it is greatly reduced. Bands of muscle encircling the mouth are buried in the upper and lower lips of haplorhines, giving them varying levels of freedom to shape the mouth in communication. The occipital lobe important to visual processing is well developed, while the olfactory lobe is reduced compared with strepsirhines. With a haplorhine-based potential for elaborating the mobility and importance of the lips, among platyrrhines the capuchin monkeys have evolved well-differentiated oral musculature, which makes it possible for this monkey to produce grins, grimaces, smiles, frowns, puckers, and a host of other visual gestures and sounds to support its sophisticated forms of communication.

There are other important structural features of the cranium, and the eyes, that relate to the differences between the strepsirhine and haplorhine primates, and the trade-offs each of these groups has evolved in supporting what we generally think of as a smell-dependent or sight-dependent lifestyle. For example, the eyes of strepsirhines, which are designed for night vision, are set wide apart. They are separated by the structure of the cranium, by the space where the large olfactory bulb is situated, and by the rear end of the capacious chamber that makes up the bony nose inside the rostrum, which houses an impressive array of scroll-like bones covered in smell-sensitive epithelial tissue. The eyes of haplorhines are set closer together. They are designed for daylight, and the hollow that forms the bony nose is much smaller in volume, with a much smaller complement of olfactory scrolls.

As far as spelling goes, if there appears to be an inconsistency in forming these *rhine*-based, compound, informal taxonomic names—strepsirhine, haplorhine, platyrrhine, and catarrhine—it's not a typo. In spite of a recent push for uniformity, to employ the comparable double-*rr*, platyrrhine-catarrhine spelling when writing the strepsirhine and haplorhine terms, it was decided in this volume to maintain these single-*r* spellings because it adheres to common, published usage established over more than a century, thus preserving continuity of language. Doing so does not conflict with the ethos of taxonomy or zoological nomenclature. There are no naming rules for categories above what we refer to as the family level, meaning terms given to formal taxonomic classes like Superfamily, Family, and Subfamily. At the same time, a major tenet of the rules of nomenclature to which zoologists abide stresses the conservation of names to maintain clarity.

Platyrrhines and catarrhines

As the field of comparative anatomy grew in the 1800s, a variety of cranial and dental features were discovered to distinguish extant platyrrhines and catarrhines in addition to their nasal morphologies (fig. 1.3). For example, the sidewall of the cranium in the region where the braincase joins the face on the backside of the orbit is composed of several bones that fit together like puzzle pieces to form what is called the postorbital mosaic. In platyrrhines, the mosaic is completed horizontally by a suturing of the zygomatic and parietal bones. The postorbital mosaic of catarrhines is closed vertically, by a suturing of the frontal and sphenoid bones. As with the platyrrhine vs. catarrhine external noses, the differences are consistent yet there are exceptions. What these contrasting suture patterns mean functionally, if anything, has not yet been determined.

A third trait distinguishes New World monkeys from Old Word monkeys and apes. It is the shape of the ectotympanic bone that supports the eardrum, or tympanic membrane, by encircling it (fig. 1.3). The vibration of the eardrum initiates hearing when it is struck by sound waves traveling through the air. In platyrrhines, this thin bone is an open, ringlike or U-shaped loop that sits flatly against another bone that houses the hearing mechanism, thus producing a prominent "hole" in the ear region. In catarrhines, the ectotympanic bone is shaped like a tube, more or less horizontally disposed. Its medial (inner) end holds the tympanic membrane in place and its lateral (outer) end opens to the auditory environment. Because the tubular ectotympanic bone narrows laterally, catarrhines tend to have a relatively smaller opening that can be seen on the side of the cranium.

A fourth trait that distinguishes platyrrhines from catarrhines is the dental formula, or tooth count (fig. 1.3). By convention, the dental formula enumerates the teeth in each functional tooth group—incisors, canines, premolars, and molars—in the four quadrants of the mouth—right, left, upper, and lower.

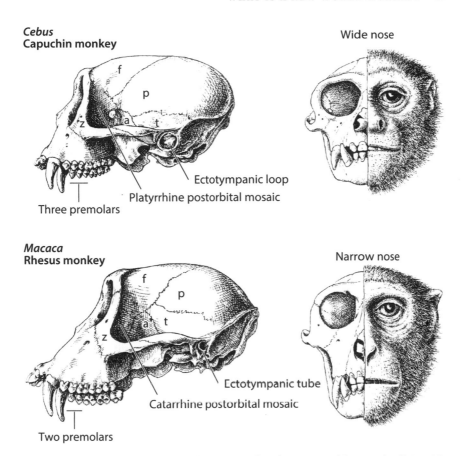

Cebus
Capuchin monkey

Wide nose

Ectotympanic loop
Platyrrhine postorbital mosaic
Three premolars

Macaca
Rhesus monkey

Narrow nose

Ectotympanic tube
Catarrhine postorbital mosaic
Two premolars

FIG. 1.3. Skull and face of a capuchin and a rhesus monkey showing cranial features that distinguish modern platyrrhines and catarrhines. Abbreviations of bone names: a, alisphenoid; f, frontal; p, parietal; t, temporal; z, zygomatic. Adapted from Schultz (1969).

When the count is the same in the upper and lower jaws, the pattern can be described using a simple string of numbers. Living platyrrhines have two dental formulae because the numbers of molars differ among genera, but they always have three premolars. Their dental formulae are: 2-1-3-3 and 2-1-3-2. Among the extinct platyrrhines, an unusual genus from the Caribbean also has a reduced count of two molars. The contrasting formula among living catarrhines is 2-1-2-3. One premolar has been lost. This shortens the non-molar, front end of the toothrow, a shift that indicates an emphasis on chewing food with the molars, whereas New World monkeys have maintained an emphasis on the premolar battery for biting, a processing step that precedes molar-mastication.

The hard-anatomy differences between platyrrhines and catarrhines, which had been diagnostic for more than a century, came to be revised in the mid-1960s when Elwyn Simons, the great American primate paleontologist, discovered the fossil *Aegyptopithecus zeuxis* and other 30-million-year-old anthropoids in the Fayum Depression, a geological basin south of Cairo, Egypt. The Fayum has produced an extraordinary trove of material that exponentially increased the fossil record of early Old World primates and other mammals of this period. It led to the discovery that the archaic Old World anthropoids of that age resembled living platyrrhine morphology rather than the extant catarrhines in two of the four diagnostic features, in having a three-premolar dental formula and a non-tubular ectotympanic bone. As far as the other two distinguishing features discussed above, the earliest Egyptian fossil crania are ambiguous as to the morphology of the postorbital mosaic, and, of course, none of them inform us about nasal shape.

There is, however, an opportunity to discover more about the olfactory behavior of Fayum primates by examining the bony anatomy inside the nasal opening. It can provide clues about the secondary olfactory system that, as mentioned, is well developed in strepsirhines and plays a role in communication via scent, especially in connection with reproduction. The nerve that joins Jacobson's organ to the brain runs in a midline groove that is observable in some well-preserved fossil crania. The width of the groove corresponds to the thickness of the nerve. A study of *Aegyptopithecus* crania reveals that the groove resembles the reduced thickness of modern platyrrhines. This provides fossil corroboration of the hypothesis originally based on living species, that the last common ancestor shared by platyrrhines and catarrhines was already less reliant on the sense of smell than a strepsirhine primate.

These critical fossil finds have demonstrated that platyrrhines are the more primitive of the two lines of extant anthropoids in some traits. It suggests that living platyrrhines, rather than the Old World monkeys or apes, should be used to model the behavior and adaptations of these early Old World anthropoids. *Aegyptopithecus zeuxis* is a good example in its postcranial traits as well as in the cranial morphologies mentioned. Its elbow morphology and limb proportions do not resemble any Old World monkey or ape, but very closely resemble a platyrrhine, the Howler Monkey. This indicates that in life *Aegyptopithecus* engaged in a style of locomotion that was very different from that of any living Old World anthropoid, but would have resembled the deliberate form of quadrupedalism seen in howlers. Another example is the skeleton of the small Egyptian fossil *Apidium phiomense*. It closely resembles the Squirrel Monkey rather than any of the Old World monkeys, indicating it used leaping in its locomotor repertoire. Various other examples involve similarities between early

African forms and platyrrhines in the functional morphology of the dentition. The modern platyrrhines are thus a living laboratory for testing hypotheses about the nature of early anthropoid ecology, behavior, and evolution.

Platyrrhine taxonomy

The taxonomy of platyrrhine genera and species, their identification and arrangement in classification, remains a subject of some debate among scholars. The 16 living genera recognized and discussed in this book are based on the work of myself and many others, involving intensive study of the morphology of all the living platyrrhines at the genus level, and studies of the taxonomy, behavior, and ecology of species contained in each genus. This count has been a relatively conventional and stable figure since about 1925; however, there has been an accelerating trend since 2000 to re-taxonomize platyrrhine genera and species based almost exclusively on molecular studies, and now more than 20 genera are recognized by some.

Even more controversial is the number of platyrrhine species. CITES, the Convention on International Trade in Endangered Species of Wild Flora and Fauna, an authoritative organization that tracks biodiversity, listed 146 living platyrrhine species in 2018. *The Handbook of Mammals of the World. 3. Primates*, a 2013 landmark treatment of primate biology written by active field biologists and conservationists, identified 156 species. In contrast, *Mammal Species of the World*, a comprehensive text organized by the Smithsonian Institution and written by experts in the taxonomy of each mammalian order, recognized 85 platyrrhine species in 1993, and fewer than 50 species were presented in 1976 by P. H. Napier, a pioneering primate specialist who was then writing catalogs covering all the primates housed in the research collections of the British Museum.

The progression from roughly 50 to more than 150 species did not occur because we discovered more than 100 new species between 1976 and 2018 that had been hidden in the jungle; perhaps there were a handful. It happened because different approaches were being employed by the scientists working on the taxonomy of species and genera, in the evidence used, and in the conceptual models they applied to species and genera, which will be discussed in later chapters. An example of how this new methodology changes things is the taxonomic status of Titi Monkeys. Over a 60-year period ending in 2016, three separate scientific revisions of the classification of titis variously concluded that there are 3 species, 13 species, or 34 species. Though it was long accepted that all titi monkeys constitute a single genus, *Callicebus*, the authors of a 2016 study felt the need to organize the species into three genera instead of one.

There are consequences to this strategy, which has been called taxonomic inflation, an artificial increase in the perceived number of species and genera in nature. It begins with a question of credibility, because none of the research done with this approach has ever reduced the number of species in a multispecies genus, as might be expected when powerful DNA methods are applied to sort out any taxonomy involving many populations. Instead, the taxonomic standing of monkey populations previously classified as subspecies has been elevated to the rank of species, which changes the biological significance of their names but does not actually alter our knowledge of their existence, as if they had not been previously discovered in nature. As to the significance of such changes to a research program, flattening the species confounds a very basic theoretical tenet of evolution, that variation *within* species is what provides the material basis for potential species change. Taxonomic inflation has the effect of homogenizing the perceived variability within species by eliminating the geographically distinct subspecies divisions whose smaller size and spatial distribution can encourage genetic isolation, for example, an early step in the evolution of fresh traits that can transform populations and generate new species.

Another repercussion of the taxonomic inflation trend is that different, incompatible methods are being applied to document biodiversity and classify living and fossil primates. This is not only a matter of theoretical interest. Lack of a consistent method of recognizing and classifying living and extinct species undermines the fundamental way we inventory biodiversity. Such difficulties extend to the challenge of reconstructing what happened during the course of evolution, too. They make it virtually impossible to investigate the possibility that nominally extinct species evolved into extant species.

One reason for this radical taxonomic shift since 2000 is that the concept of species has always been difficult to define scientifically, and while it has changed over time, it is likely to remain problematic because of ambiguity. In *The Origin*, Darwin wrote, "No one definition [of species] has yet satisfied all naturalists, yet every naturalist knows vaguely what he means when he speaks of species. . . . Nor shall I here discuss the various definitions which have been given of the term species."

Since Darwin's time, we have tried to develop what we call an operational definition of species applicable to living and extinct forms by identifying natural, universal biological properties. Seen through the prism of evolution, the aim is to apply a formula that integrates biological knowledge about the extraordinarily varied lives and circumstances of organisms like animals in a replicable, yet elastic, way as species are formally recognized by science.

By the 1930s, scientists understood that the species is a fundamental unit of evolution and it was proposed that the fundamental property of a species is exclusivity of reproduction. The biological species concept became the dominant paradigm. Its most widely accepted definition was given by the eminent 20th-century evolutionary biologist and ornithologist Ernst Mayr, who in 1942 explained that species are groups of populations in nature whose members mate with their own kind, act accordingly, and are thus isolated from other such groups.

However, it is very difficult—impossible for the vast majority of cases—to actually test for interfertility between two potentially distinct living species, even more so for the extinct ones. Many have seized on this methodological dilemma, making it a principal reason for discarding the biological species concept and replacing it with the idea that species are lineages, which is a phylogenetic concept typically applied to higher taxonomic groups. Therefore, in order to operationalize the biological species concept, researchers understood that species are, in effect, distributed networks of reproductively compatible individuals having unique combinations of genes that are likely to be manifest or mirrored in morphology or behavior, as a design.

Subspecies can be thought of as a spatial array of nodes that are connected via the network. That means we can recognize species by finding specific morphological and behavioral patterns that are known or thought to be genetically based, and sufficiently distinct so as to inhibit crossbreeding with another species at any of the subspecies nodes. The indirect evidence that interbreeding is unlikely to happen may come from genes, body proportions, craniodental anatomy, coat color, mating rituals, vocalizations, and more, any combination of important traits that sets two species-like entities apart in a statistical sense and, when observable, in nature. When it comes to comparing fossils that may belong to two distinct species, we apply empirically developed observations of living relatives as a yardstick to delimit interfertility, theoretically.

As mentioned, the taxonomy of platyrrhine genera is also a matter of debate. In some cases this reflects different views of the genus concept, which is not the same as the dispute over the meaning of species. It is generally agreed that species are real entities in nature, each with a unique genetic template and each one being an individual, direct product of evolution. The genus, in contrast, is not a real thing in nature. There is no natural process that produces a genus per se. It is a construct utilized by scientists to aggregate species that are identified by a uniquely shared phylogenetic and adaptive origin that establishes a unique ecological position for the collective. There are no direct or indirect tests, as there might be for species no matter the difficulty of applying them. That is why classifying at the genus level is a subjective process. In cases where a genus comprises only a single species, the factors determining its taxonomic status as a species are the same as those identifying it as a genus. Two such examples are presented in the next chapter, concerning the Pygmy Marmoset and Goeldi's Monkey. They reveal another practical difference between classifying at the species level and the genus level. There is no unifying criterion that determines their taxonomic status, like the breeding standard. Different details are used to define each genus because each one is adapted differently; that is, body size may be construed as a primary genus-level character in one instance and craniodental morphology may be the defining character in another. Another example involves the current controversy regarding the number of genera representing capuchin monkeys, also discussed below.

Why does the actual number, or the best scientific estimate, of genera and species of New World monkeys matter? Because these classifications tell us different things. To study the fine points of evolution is to study species. Natural selection, the universal process by which traits benefiting reproductive success are preserved over generational time, among other factors, acts on individuals, and their genetic contribution to a larger population, to the species, is what determines what features will change or remain the same. Thus it matters greatly to be able to properly identify species. To study the structure of an adaptive radiation is to study genera, what constitutes each genus and how many genera there are. The genus is the taxonomic level at which we can trace the distinctive pattern of platyrrhine evolution, which comprises many multimillion-year lineages of genera and monophyletic collections of genera.

The formal taxonomic names for the 16 living platyrrhine genera used in this book are the established ones employed for many decades, and are italicized according to nomenclatural rules. The informal names are not subject to the same conventions and have varied over the years, but they are capitalized as the name of a genus. Therefore, as an example, the name Squirrel Monkey is capitalized when it refers to the genus *Saimiri*; the lowercase squirrel monkey is used as a generalization.

20 million years of evolution

16 genera of extant platyrrhine primates

Genus-level descriptions of each of the living platyrrhine primates are presented in the following chapter. The genera recognized in this book have been identified as such for decades, although there have been a few cases where a species has been moved from one genus and placed into a different one. The Pygmy Marmoset is an example. It is generally agreed now that this one living species is different enough from all other platyrrhines to warrant placement in a genus of its own, *Cebuella*. In the past, however, some, including the present author, preferred to place that single species elsewhere, classifying it with other types of monkeys in the genus *Callithrix*, a group of monkey species with overlapping adaptations.

In studying how living primate genera are situated ecologically, the most important characteristics are body size, diet, locomotion, the brain, activity cycles, reproductive patterns and behaviors associated with social organization, and mating strategies. Some of these features can also be examined in the fossil record in various ways, which amplifies the importance of understanding them. They are introduced in chapter 2, and other details concerning the evolution of these traits as adaptations are further discussed in subsequent chapters. In most respects the characteristics that provide the basis for recognizing platyrrhine genera are the same kinds of traits that delineate genera in the larger world of mammalogy.

Body size, diet, locomotion, cognition, and social behavior are examples of adaptive complexes that are all linked biologically at several levels. Still, in the analysis of what makes an animal successful, even a single trait or complex can be highly informative. It may set a genus apart from its relatives for purposes of identification and also serve as a primary correlate or building block with respect to other traits that support a given lifestyle. For example, in pygmy marmosets a tiny body size—adults rarely weigh more than 120 g, roughly 4 oz—enables the animals to subsist on an unusual diet that includes large amounts of natural gum that exudes from trees. Locally, this diet reduces feeding competition with other platyrrhines and it also minimizes a pygmy marmoset's daily energy output by saving it the expense of searching widely for other foods.

Specialized incisor and canine teeth enable these very small monkeys to access gums by scraping away patches of tree bark. The tree responds by forming a dribble of gum to heal the wound. Coupled with these features are postural adaptations of the skeleton and especially the fingers and toes that allow the monkeys to position themselves on trees so gouging can be done effectively. A practical benefit of the constellation of adaptations is that an entire family unit of pygmy marmosets may be able to subsist for long periods of time by feeding on a single tree that is rich with gum, as long as the tree can survive the onslaught of daily hole-gouging to stimulate the production of gum globules. When life revolves around a single tree, a limit is placed on home range and social group size, and a premium may be placed on territorial behaviors in defense of one, highly valued food resource.

A tiny body size is the adaptive cornerstone of the pygmy marmoset's existence. The coordination of adaptive systems involving food, movement, and interpersonal and intergroup behaviors with body size in *Cebuella* is comparable to the adaptive paradigms seen in every other platyrrhine genus, making each one unique. Body size is more than a descriptor. It is a fundamental design element governing an animal's lifestyle and evolutionary history, and it is strongly influenced by natural selection. As we shall see, there are platyrrhines 100 times larger than the Pygmy Marmoset, such as the largest Spider Monkeys and the Muriquis, and their body size plays a similar role in defining their lifestyles. In historical terms, this extensive range of body sizes is not a continuum. Rather, in reconstructing the evolution of platyrrhines it becomes apparent that different clades and genera have experienced different trajectories of body-size evolution. Some have gotten smaller and some have gotten larger over time. Even though it is difficult to accurately infer the magnitudes of these adaptive shifts, it is evident that, comparatively, some forms are phyletic dwarfs and others are phyletic giants.

In reconstructing the evolutionary history of platyrrhines it also becomes clear that the radiation of New World monkeys as a group is characterized by a preponderance of long-lived individual genera, generic lineages, and clades. A generic lineage can be thought of as a line of descent or a stream of genes effectively evolving in a column that produces a coherent set of characteristics that determine the unique ecological lifestyle shared by all its descendants. The genetic column may involve a fossil species that bears the same genus name as a living genus, as with a 12–14-million-year-old fossil Owl Monkey and its living counterpart, both named *Aotus*. Or, a generic lineage may involve two differently named genera that are separated by a significant amount of geological time, but they are monophyletically related and fall within the same lifestyle boundaries. In other words, while the anatomical evidence may make it too much of a stretch to hypothesize that the species of the older genus is a direct ancestor of a species belonging to the younger genus, the former is considered directly in or near the ancestry of the latter because the morphologies align, their temporal ages are consistent with the idea, and that hypothesis is not discounted by relevant evidence. We can infer the longevity of genus-level lineages through the fossil record and by using molecular methods which help us reconstruct how genera are linked up with one another cladistically, and when the splits between and among the branches of the platyrrhine Tree of Life occurred.

A half-dozen or more of the 16 living platyrrhine genera can be traced back to fossils, as genera or generic lineages, that date between 7 and 20 million years. The implication is that these living genera have remained much the same as they were millions of years ago, in the body parts that have been discovered in fossils. Furthermore, insight about how modern platyrrhines are organized locally tells us that these genera evolved in connection with one another, enabling them to coexist in harmony within the same community by occupying unique niches within an ecosystem. Having such a high proportion of genus-level lineages representing most of the major phylogenetic clades of living New World monkeys over such a long time interval reveals that platyrrhine history has proceeded in a pattern, as a unified radiation rather than an evolutionary venture that produced a chaotic ensemble of primates. There is plenty of unpredictability in the evolutionary process, but the manner in which the modern platyrrhine radiation unfolded was anything but random.

CHAPTER 2

DIVERSE LIFESTYLES

Every genus has a story to tell, written in its anatomy and behavior and contextualized by its ecology and history. In this chapter, the platyrrhine genera are arranged in family and subfamily groupings according to the classification used in this book, along with summary characterizations of each of the major groups (table 2.1). This is a cladistic approach. By using formal language to denote two-layered, hierarchical positions within the classification, with the level of family being above and bracketing the lower level of subfamily, it organizes the animals in a way that reflects their phylogenetic relationships. The informal names relating these groups, which are given for ease of discussion, are traditional simplifications of the formalized Linnaean terms. For example, the informal name for Family Pitheciidae is pitheciids; for Subfamily Pitheciinae it is pitheciines.

The proper scientific terms are based on explicit rules followed in zoology, the parent discipline of mammalogy, primatology, physical anthropology, etc. They are constructed to connote a particular category, or rank, within the Linnaean system, and they also provide an indication of their generic content. In the present example, this is done by beginning with a root that specifies that the Saki Monkey, genus *Pithecia*, is included in the suprageneric groups, and then combining it with alternative suffixes that stand for the two given ranks: *-idae* always refers to the official family and *-inae* refers to the subfamily. Thus Subfamily Pitheciinae and a second group of equal taxonomic status, Subfamily Homunculinae, are each composed of a monophyletic collection of genera, and they are classified together under the overarching Family Pitheciidae to indicate the subfamilies are sister taxa, two clades stemming from a unique ancestor. We commonly say that the hypothetical ancestor of these subordinated subfamilies was a pitheciid, and all its descendants are also called pitheciids.

There is one exception to this family and subfamily arrangement in which a taxonomic level between the subfamily and genus is used. The formal name for this rank is Tribe, identified by affixing the suffix *-ini*, and it is widely used in more detailed classifications of the platyrrhines and catarrhines. Here, the tribe becomes useful in discussing a subgroup of clawed New World monkeys, four genera that are offset from another genus that belongs to the same subfamily, the Callitrichinae.

Colloquial names for the genera are different. There is no international, disciplinary basis that governs them. They are often descriptive terms or based on native names, without any connection to the formal genus name. Their selection is a matter of personal preference. For example, the Brazilian Muriqui is the common name now used for the genus *Brachyteles*, but for decades before that became fashionable it was known as the Woolly Spider Monkey in English, and locally as *mono carvoeiro* in Portuguese, loosely meaning charcoal, black-faced, monkey. A person who burns wood to produce charcoal is a *carvoeiro*; *carvao* means coal.

In this book, common names for genera follow current convention in all but two cases. The genus *Leontopithecus* is called the Lion Marmoset, not the Lion Tamarin—except in the last chapter for reasons that will become evident—to suggest it is closely affiliated with other marmosets, as will be discussed later. The genus *Callimico* is called Goeldi's Monkey, not the Callimico, which may be confusing. The terms Lion Marmoset and Goeldi's Monkey were once standard common names, with a long, familiar history in the primate literature.

The genera are presented taxonomically, not listed alphabetically, in order to encourage comparisons among them and extract evolutionary lessons among close relatives. The gridlike summary in table 2.1 arranges several major elements used in the genus profiles according to their taxonomic distributions. What is immediately obvious and most important is that there is a patterned structure to the listing. Closely related monkeys tend to have similar body sizes, the sexes are either comparably similar or different in body size to the same degrees, and related groups share similar diets and locomotor behaviors. Yet, these characteristics alone do not determine how the animals are classified. Nevertheless, it becomes clear that the classification accurately reflects phylogeny, the genetic relationships among the animals. Genetic relatedness is the fundamental evolutionary condition that produces similarity.

While each of the taxonomic units—families, subfamilies, tribes, and genera—of platyrrhines are believed to be monophyletic groups, the descendants of unique common ancestors they shared unto themselves and not with any other New World monkeys, there is one outstanding case involving the Owl Monkey where other interpretations have been proposed based on different readings of the morphological and molecular evidence, and this will be discussed at several points in the following chapters. This is a deeply significant, big-picture debate with potential implications about the methods we use to reconstruct phylogeny, the basic historical outlines of two of the

TABLE 2.1. Summary of critical adaptations and behaviors of New World primates based on the information reported in this book, including measurements of behavior, raw field observations reported in the literature, and interpretation of traits crucial to each genus

	Body Size[a]	Sexual Dimorphism[b]	M/F Weight Ratio[b]	Diet[c]	Locomotion[d]	Social Organization[e]
Family Cebidae						
Subfamily Cebinae						
Cebus	M	M > F	1.3	F, I, P, N	Q, L, CC, T-AP	MM/MF
Saimiri	M	M > F	1.3	F, I, P	Q, L	MM/MF
Subfamily Callitrichinae						
Callimico	S	M = F		FN, F, I, G	L, Q, CL, VC	FF/MM
Saguinus	S	M = F	1	F, I, G, P	Q, L, CL	FF/MM
Leontopithecus	S	M = F	1.1	F, I, P	Q, L, CL	FF/MM
Callithrix	S	M = F	1.1	G, F, I, P	CL, Q, L, VC	FF/MM
Cebuella	XS	M = F	1	G, I	CL, Q, L, VC	FF/MM
Family Pitheciidae						
Subfamily Homunculinae						
Callicebus	M	M = F	1	F, HFH, S, I, L	Q, L, CC	P
Aotus	M	M = F	1	F, I, L	Q, L	P
Subfamily Pitheciinae						
Pitheciinae	M					
Pithecia	M	M >> F	1.4	F, S, HFH, L, I	L, Q, CC, VC	P(MM/MF)
Chiropotes	M	M > F	1.2	F, S, HFH, I	Q, L, CC	MM/FF
Cacajao	M	M > F	1.2	F, S, HFH	Q, L, CC	MM/FF

Family Atelidae

Subfamily Alouattinae

Alouatta	L	M>>F	1.4	Q, CC, S, T-AP	L, F	1M(MM/MF)

Subfamily Atelinae

Lagothrix	L-XL	M>>F	1.5	Q, CC, S, T-AP	RF, L	MM/FF
Ateles	L-XL	M=F	1.1	S, Q, CC, T-AP, T-AB	RF, L, S	MM/FF
Brachyteles	XL	M>F	1.2	S, Q, CC, T-AP, T-AB	F, L	MM/FF

Sources: DiFiore et al. 2011, Rosenberger 1992, Ford and Davis 1992, Fernandez-Duque et al. 2012.

[a] The five main classes reflect the average body mass of each genus. Weights can vary because of several factors, including measuring techniques, species differences, local variation within species, uneven sampling of males and females, time of year when data were collected, the reproductive status of females, and differences between weights measured in wild and captive individuals. XS, Extra Small, 100–120 g (3.5–4.0 oz). S, Small, 250–600 g (8.8–20.8 oz). M, Medium, 0.7–4 kg (1.5–8.8 lb). L, Large, 5–8 kg (11–17.6 lb). XL, Extra Large, 10 kg (22 lb) or more.

[b] Based on the average genus-level differences between males and females across species in body mass, expressed as the ratio of male:female weight. M=F, 1.0–1.1. M>F, 1.2–1.4. M>>F, 1.4–1.5. The categories prioritize the tendencies within genera irrespective of some overlapping values between the classes.

[c] Based on field observations and morphological adaptations. Several categories of fruit eating are used to specify particular specializations on fruit types, or use of special feeding methods, when appropriate. F, fruit eating (frugivory). HFH, hard-fruit harvesting. RF, ripe fruit. FN, fungus. G, gum eating (gumivory). I, insect

eating (insectivory). L, leaf eating (folivory or semifolivory). N, nuts. P, vertebrate prey. S, seeds.

[d] Q, quadrupedalism, above-branch walking, running, or bounding using all four limbs. L, leaping. CC, climbing, using all four limbs to ascend or descend tree trunks or large, inclined branches, and clambering, a form of pronograde quadrupedalism using all four limbs as handholds in moving awkwardly or clumsily through unstable branches. S, suspensory locomotion, below-branch movement using the forelimbs. SCC, suspensory climbing and clambering, in an orthograde manner using the forelimbs to power movement and the hindlimbs for stability while moving awkwardly or clumsily through branches. T-AB, tail-assisted brachiation, using the prehensile tail as well as the forelimbs in swinging through the canopy. T-AP, tail-assisted postures. VC, vertical clinging. CL, clawed locomotion.

[e] Categories reflect the numbers of adult males and females that form the core of different social groups and how the sexes interact on a gross level. MM/MF, male oriented multi-male/multi-female units. FF/MM, female oriented multi-male/multi-female units. P, pair-bonded, a monogamous male/female group. 1M, one-male/multi-female group.

three major platyrrhine clades, and the phenomenon of parallel evolution, its breadth across adaptive systems and its frequency of occurrence among the platyrrhines.

A second controversy involves Goeldi's Monkey, which also has intriguing implications for interpreting critical aspects of the evolution of the callitrichine subfamily that also includes the marmosets and tamarins, here classified in Tribe Callitrichini, informally called callitrichins. And, there is a less consequential controversy about the cladistic position of the Muriqui. Such differences are expected in this area of science, as the history of research shows that no single source of evolutionary information is able to generate universally accepted hypotheses or incontrovertible support for alternative interpretations of a large and complex adaptive radiation, even those based on sound analyses. One of the stark and exciting lessons we have learned about evolution is that its consequences always hold surprises.

Predatory frugivores: Family Cebidae

There are seven extant cebid genera, composed of two subfamilies, the cebines and callitrichines (fig. 2.1). No other families of primates today are so disparate anatomically and behaviorally; nevertheless these genera share a baseline of critical commonalities in their adaptive profiles. The cebids include all the smallest-bodied New World monkeys and two larger genera that span the middle size range of the radiation. They are fruit eaters (frugivores), insect eaters (insectivores), and gum eaters (gumivores). Most, if not all, species also eat small vertebrates, such as frogs, lizards, and birds' eggs; they could easily be called faunivores. Though centered on the same sets of foods, the largest genus of the family, the Capuchin Monkey, *Cebus*, has a unique array of adaptations that make it one of the most remarkable primates in many respects, as we shall see repeatedly in later chapters. There is solid evidence based on morphological and molecular studies that this family grouping is monophyletic, descended from a single common ancestor.

A major, driving force behind the diversity within cebids is body-size evolution (fig. 2.2). There is a 35-fold difference in body mass between the smallest and largest cebid genera. The disparity is exceptional because it occurs in a closely knit group, in phylogenetic and adaptive terms. Body size is also a primary distinction between the two subfamilies. There is a twofold size difference between the largest callitrichine and the smallest cebine. At these scales of divergence, the biological ramifications are profound.

Though fruit is central to their diet, as it is for all platyrrhines, cebids behave like predaceous foragers, always looking for an opportunistic dis-

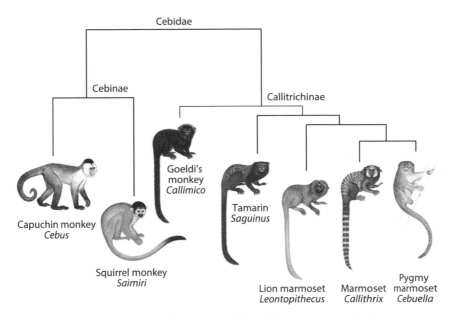

FIG. 2.1. Cladistic relationships and classification of living cebid genera, not to scale. Images courtesy of Stephen Nash.

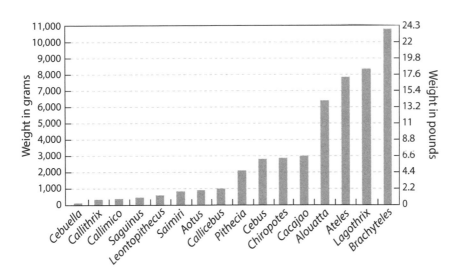

FIG. 2.2. Average body weights of the 16 living platyrrhine genera.

covery, or engaged in a purposeful hunt that results in an edible morsel of non-vegetable food. Some glean exposed invertebrates such as insects and arthropods (e.g., spiders, scorpions, centipedes) from leaves, branches, tree trunks and vines, or lianas. Others extract concealed vertebrate prey from birds' nests, bunched-up leaves, natural fissures and pockets in vegetation, or under the bark they strip from trees. These cebids are characterized formally as frugivore-insectivores to highlight their combination diet and the importance of invertebrate prey; the pursuit of small vertebrates dominates the anatomy and lifestyle of one genus, the Lion Marmoset, whose long arms and hands are highly specialized anatomically to forage for insects and small vertebrates that are deeply concealed in crevices, and certain plants that have a distinct morphology.

As is typical for small primates, cebids are generally quadrupedal and often engage in leaping as a means to cross substrate gaps in the trees or below the canopy when branches are out of their limited reach. They have oval or rounded braincases, flat or rounded foreheads, short faces, relatively large canine teeth, and, as an adaptation to biting large insects, the postcanine teeth tend to be dominated by pointy premolars—except, again, for Capuchin Monkeys, to be discussed below. Their third molars are either highly reduced in size as a primitive cebid trait, as in cebines and Goeldi's Monkey, or missing altogether as in the four callitrichin genera, in part to augment the biomechanical efficiency of pointy premolars for biting tough insect exoskeletons.

Subfamily Cebinae

Cebus, Capuchin Monkeys; *Saimiri,* Squirrel Monkeys

Squirrel Monkeys and Capuchin Monkeys are the largest monkeys of the rather predaceous cebid family. Though different in size—squirrel monkeys weigh less than 1 kg, 2.2 pounds, while adult male capuchins can weigh three to four times more—their activities and demeanor are similar. Linda Fedigan and Sue Boinski, who studied these monkeys in the wild for many years, describe them as "busily moving about poking, prying, peeling, and scraping substrates looking for tasty bits, and bustling about fruit sources, bumping and jostling each other like a litter of puppies at a food bowl." Capuchins and squirrel monkeys are highly vocal as they forage, especially the squirrel monkeys, which live in large troops of dozens of individuals that can become widely separated while roaming about. A constant chatter of contact calls enables them to stay in touch with one another and maintain vigilance.

Cebines share a compelling host of clear-cut, specialized resemblances. Because their brains and braincases are proportionately large they have rounded skulls. Tucked in below the orbits, their faces look small. They have close-set eyes. The canine teeth are large and sexually dimorphic, a phenomenon in which adult males and females differ from one another in appearance. In species in which dimorphism has been under significant natural selection pressure, the canine teeth are usually a target organ where intersexual differences are exaggerated. As discussed further below, canine dimorphism in cebines evolved as a correlate to living in multi-male/multi-female social groups where there is aggressive competition among males for access to females. The postcanine teeth of cebines exhibit an emphasis on the premolars, which are uncommonly broad, while the molars are proportionately small, especially the final, third molars. This arrangement is consistent with a biting strategy rather than a chewing strategy, the initial phase of food breakdown—or prey-killing—as opposed to the masticating phase, when a mouthful is reduced to a size than can be swallowed. Overall, Squirrel Monkey cheek teeth have more pointed cusps and sharper edges, better designed for piercing insect bodies, while Capuchin Monkey cusps and ridges are blunter, better able to resist damage and natural wear when biting into and feeding on a wide range of materials, including very tough vegetation. The Capuchin Monkey configuration is a derived one, a unique dental specialization that is part of the anatomical and behavioral syndrome that evolved as these monkeys transformed and extended their ancestral frugivorous-insectivorous feeding niche.

The cebines are good with their hands and good with their tails, capuchins more so than squirrel monkeys. Capuchins evolved a version of the opposable thumb and a semiprehensile tail, which squirrel monkeys did not. Larger and stronger than the squirrels, they have larger, bluntly cusped, and more wear-resistant cheek teeth covered in thick enamel, an adaptation for eating hard foods and breaking open and biting tough branches to get at the insects lodged inside.

The Capuchin Monkey's big brain, semiprehensile tail and grasping hands are an integrated adaptive package that supports its predatory technique of extractive insect-foraging. Their prey is often social insects, such as ants and termites that live embedded in colonies, and the capuchins must learn where to find them and how to test for their presence in a likely spot before investing time and energy in pursuit. They check for hidden ant nests by tapping dead branches with their fingers and then listening and looking for additional signs. Capuchins are strong enough to snap open the hard, thick fronds of palm trees to get at the nutritious, soft tissue inside, called pith. They also use rocks as tools to hammer hard nuts that cannot be opened by biting. They can prop themselves up against the branches of a tree, cantilevered in a tripod stance— two feet braced against a trunk and the tail wrapped around another support to anchor the body—as they use a two-handed, overhead pounding maneuver to smash the hardest kinds of fruit, seeds, or nuts against a solid branch. They will stand upright on the ground, with legs extended and the tail stiffened against the ground, adding gravity and body mass to muscular force as they deftly pound a hard palm nut with a heavy rock held between two hands. In drier habitats, thirsty capuchins use stones as tools to dig into the ground for water.

The squirrel monkeys scour the exposed surfaces of trees in search of insects, and they find others by manipulating and unrolling masses of curled-up leaves, including dried, dead clumps that get caught up in the canopy. Soft caterpillars and the pupae of butterflies and moths are favored prey. Eating caterpillars is a complex proposition, requiring squirrel monkeys to first remove the caterpillars' protective armor by rubbing it off against a branch or their own tail fur. The next steps are even more intricate, as described by Charles Janson and Sue Boinski, two leading experts on cebine natural history: "Once any spines or hairs have been dealt with, a squirrel monkey usually will eviscerate any caterpillar longer than 2 cm. It does this by carefully grabbing the head capsule in its teeth, gently severing the head capsule and adjacent body segment from the rest of the body, leaving the gut tract attached to the head capsule. The body is pulled away with the hands and the part in the mouth discarded before consuming the body." A similar process is employed to reach the nutritious, soft innards when squirrel monkeys feed on insects with sturdy exoskeletons.

The Squirrel Monkey tail does not have the same degree of manipulability as the Capuchin Monkey's tail, but it is relatively longer and perpetually deployed in muscular and tactile ways. Baby squirrel monkeys wrap their tails around the mother's abdomen when being carried on her back, to keep from falling off. Adults wrap their tails around their own bodies when resting, or across the body of another group member when socializing. Tails are commonly draped around an adjacent branch while sitting or moving, and used as a brace against the substrate, similar to the manner described above for capuchins.

These two, closely related cebine genera, both relatively well studied behaviorally in the wild and anatomically in the lab, present an unusual opportunity to examine how and why evolution has come to produce the different forms they now present, most notably their contrasting body types. Although we cannot yet determine how large their last common ancestor was, the most reasonable working hypothesis to explain their extreme body-size difference is that Squirrel Monkeys are a kind of dwarf and Capuchin Monkeys are giants relative to the ancestral, or primitive, cebine body size. The *why* in one sense is evident: to separate these genera ecologically and allow them to occupy distinctive, minimally overlapping niches among a crowded field of platyrrhine monkeys. Yet the *why* also has direct physical consequences in the anatomical and behavioral correlates associated with sheer size, some that have already been mentioned. *How* this happened is another matter.

One way that natural selection affects anatomical change is by altering the schedules of growth and development, of the body as a whole and of particular organs that selection may target. Selection may alter the timing of development—how long it lasts, and/or its rate—how quickly it unfolds. Both processes appear to have been factors in evolving the differences between these cebines.

Squirrel monkeys have a fast fetal growth rate relative to adult body mass, which means they have a developmental head start by the time they are born. They arrive as relatively large, precocious neonates that can rapidly attain motor, behavioral, and foraging skills, and they wean early. One way squirrel monkeys attain this advanced condition is by achieving 57% of their full, adult brain size at birth, which is remarkable. In contrast, the capuchins, with absolutely and relatively much larger brains, are born with brains that are only 40% of adult size. As neonates they also weigh proportionately less relative to maternal weight than the squirrel monkeys, making them far less advanced developmentally at that stage. This is all the more remarkable because squirrel and capuchin monkeys each gestate for the same length of time, nearly 170 days, which means squirrel monkeys have a prolonged gestation period for their size, allowing for more prenatal brain growth. Yet squirrel monkeys end up being much smaller as adults than capuchin monkeys, though they both follow the same postnatal growth trajectory. In the squirrel monkeys, natural selection favored small adult size and precociousness at birth and they have become miniaturized through a process of arrested development, by truncating a capuchin-like growth pattern.

Squirrel monkeys first reproduce, that is, become adults, at roughly three years, which is the expected age given their body size. In contrast, a capuchin's initial reproduction does not occur until about five-and-a-half years of age, and they wean infants later than usual for a platyrrhine of that size class. The extra time allows for continued brain growth to occur which, as noted above, lags behind that of the squirrel monkey's *in utero* rate and results in a relatively smaller size at birth. More time spent as a non-adult also allows additional time for capuchin juveniles and adolescents to learn, not only because procuring food may be challenging cognitively. It may also provide a vital educational opportunity to a young individual that belongs to a species in which adults live in complex and often dangerous contexts, as we will see below when discussing social organization. Since brain size is associated with body mass in primates, simply growing to a relatively larger size is an easy way to produce a large brain, although in the case of capuchins, adult brain size is well above that expected for a middle-sized platyrrhine monkey. So, part of the explanation as to *why* capuchins are larger in body mass than squirrel monkeys devolves to a method that naturally produces a species with a relatively large brain, among other ecological benefits large size offers their particular lifestyle.

The very large brain of the Squirrel Monkey and the exceedingly large size of the Capuchin Monkey brain are accommodated by a voluminous braincase, which grows faster than the facial part of the cranium and winds up dominating the morphology of the skull throughout the growth period and into adulthood. This presents other interesting aspects regarding their similarities and differences. At birth both genera exhibit a rounded forehead and braincase, and a centrally located foramen magnum, the large opening at the base of the head through which the spinal cord passes, which is not an unusual pattern among primates. What is unusual is that adult squirrel monkeys retain these features when they are fully grown. With the skulls of both developing along the same trajectory, at a point when the squirrel monkey head reaches adult size and ceases to grow it strongly resembles that of a juvenile capuchin monkey rather than the fully-grown cranium of other platyrrhines of similar body size, such as the Owl Monkey and Titi Monkey. In other words, growing along the same vector as a capuchin, the adult shape difference in squirrel monkeys results largely from a body size difference. The head stops growing when maturity is reached, and that happens at a younger age in squirrels than capuchins.

It is clear from this that natural selection operates during all the developmental phases of life. Having large babies may be a special advantage for squirrel monkeys, at birth and soon afterward, and it is also a pattern that translates into benefits reaped by adults that live most of their lives at a body size that advantages their ecological position. Having slow-growing juveniles may confer other types of advantages for capuchin monkeys, during non-adult stages and when they are fully grown and must assume very different social roles.

Cebus—Cebus is a relatively well-studied primate, behaviorally and anatomically (plate 1). Capuchins have penetrated pop culture in many instances, as pets, as the panhandling organ grinder's monkey seen on city streets during the early 1900s, and in television programs and films—sometimes misplaced geographically in an Old World context, like the monkey that ate poisoned fruit meant for film character Indiana Jones, who was hunting for archaeological treasures in Egyptian tombs. They are the smartest, most curious and resourceful New World monkeys, and one of the most highly intelligent non-human primates.

The Capuchin Monkey is a remarkable hyperinquisitive product of evolution, always watchful in demeanor, persistent, and often cooing and twittering to convey information to group members. With a small face set against a large head, and hands busily working at something, their close-set eyes convey a sense of intelligence, borne out by decades of observing and testing the skills and cognition of capuchins. These monkeys exhibit a set of exceptional evolutionary adaptations to achieve ecological versatility, driven and supported by the mind and the hand. They are tool users.

To paraphrase the classical definition of tool use proposed by Benjamin Beck, a longtime leader in the study of animal cognition, a tool is an object taken from the environment and purposefully applied to another object, or to the user, with a goal in mind. The first report of capuchins using tools in the wild was made in 1988 by one of the leading field researchers on squirrel monkeys. Sue Boinski had been following her focal group of squirrel monkeys when a capuchin troop approached and noticed a large venomous snake on the ground below them, a fer-de-lance. An adult male capuchin went to the ground with a 2-foot-long dead tree branch and began to club the snake with overhead blows, at least 55 times during a period of about 15 minutes. Another monkey joined in, hitting the snake with another branch. After 20 minutes had passed, Boinski confirmed that the *Bothrops* snake was dead, its head smashed.

Cebus eat virtually everything, plant and animal, and they go nearly everywhere, even to the ground in parched environments when moisture-laden fruits are scarce and they need to find drinking water, or places where a prized food, such as clams, can be dug out of the mud when swamp water drains away. Versatility and curiosity have translated into geographic success. As ecological pioneers, capuchins have spread out to inhabit the second-largest geographic distribution of any extant New World monkey. They reside in the most varied types of forested habitats between Argentina and Honduras.

The Capuchin Monkey is a robustly built, quadrupedal monkey with long limbs relative to the length of the trunk, and hindlimbs that are somewhat longer than the forelimbs (fig. 2.3; table 2.2), which makes it an effective leaper. It has a furry, semiprehensile tail of moderate length with respect to the trunk, but rather short compared with body weight. It has a good-sized head and a significant degree of sexual dimorphism (table 2.1) in body size and canine size. In capuchins, males can weigh 30% more than females. The caliber of male upper canines may be 35% larger than female canines. The number of capuchin species and genera is a matter of some debate. Some primatologists divide capuchins into two genera and as many as 11 species, whereas the traditional count is one genus and four or five species. We return to this topic in the next chapter on taxonomy and classification and, again, in dealing with the dietary adaptations of capuchins. Diet is the driving source of the morphological differences underlying the proposal that two genera are involved.

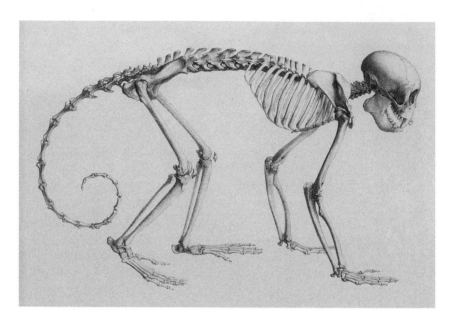

FIG. 2.3. Skeleton of *Cebus*, the Capuchin Monkey. From Blainville (1839).

TABLE 2.2. Comparative measures of living platyrrhine body proportions calculated as percentages against trunk length, except for IMI, the Intermembral Index ([humerus + radius length]/[femur + tibia length]) × 100. Outstanding proportions of various taxa are highlighted to emphasize adaptive specializations.

	Head size	Chest size	Tail length	Hindlimb length	Forelimb length	IMI
Saguinus	24	107	239	96	98	102
Saimiri	23	85	207	98	103	105
Cebus	26	105	196	118	125	106
Aotus	22	102	197	105	105	100
Pithecia	29	94	176	115	118	103
Cacajao	32	91	65	123	139	113
Alouatta	22	119	221	107	139	130
Lagothrix	27	117	264	130	162	124
Ateles	23	126	248	132	180	137

Sources: Schultz 1956, Fleagle 2013.

The capuchin has become an important model in anthropological studies pertaining to the evolution of high intelligence and dietary adaptations in early hominins. In the latter case, this is because several of its craniodental characteristics are associated with feeding on hard materials. The oldest fossils attributed directly to the long-lived *Cebus* lineage include some 11-million-year-old broken teeth from a Peruvian site that have been classified in the same genus. Another fossil capuchin from Brazil is 7–11 million years old. It is classified in a different genus because of its large size. *Acrecebus* is estimated to be about three times larger than an extant capuchin. There is evidence from the fossil record and molecular studies that the capuchin lineage originated at least 20 million years ago.

Saimiri—The Squirrel Monkey is slender and round-headed, with an instantly recognizable face (plate 2). A circular patch of dark skin precisely outlines the nose and mouth, intersecting the line of the lips and extending below to the area of the chin. A whitish dual eye patch frames the upper face and starkly highlights the eyes. In the wild, they are hyperactive, sometimes moving all day without rest, noisily scurrying about quadrupedally in large social groups, taking short leaps across gaps in the trees, making chirping contact calls, hands always at work searching for something to eat, especially prey. They are widely distributed in the Amazon Basin and are found as far north as Costa Rica in Central America. The number of species is between three and six, with some disagreement among scholars.

One of the more interesting aspects of Squirrel Monkey biology is the fatted-male syndrome, though it is not seen in all species. It involves mating, reproduction, and an annual transformation of male appearance and behavior during the breeding season. Though squirrel monkeys generally live in large social groups, sometimes comprising more than 60 individuals, males and females tend to self-segregate until the three-month mating season begins. When that happens, competition among males is fierce and their size differences become important in displays that aim to attract mates. Adult males undergo a hormonally induced physiological makeover. They rapidly bulk up in size, adding another 20% to their body weight, and they no longer sport the lean physique shared with females at other times of the year. Males develop more heft around the shoulders, upper arms, and torso by substantially increasing water retention and the amount of subcutaneous fat. The fur of fatted males also becomes fluffier, giving the appearance of added size without incurring additional metabolic or energetic costs.

The *Saimiri* cranium is highly unusual anatomically in ways that make it appear almost infantilized, which is in keeping with the hypothesis that a process of arrested growth and development was the driving mechanism

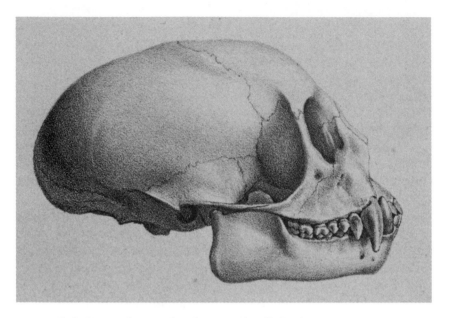

FIG. 2.4. Skull of *Saimiri*, the Squirrel Monkey. From Blainville (1839).

behind the evolution of dwarfism in the Squirrel Monkey lineage. It has a distinctive, oblong shape and is rounded in front and back, with a domed forehead and an inflated occipital region in the rear, contours that reflect the relatively large brain (fig. 2.4). The braincase is also very narrow, a characteristic mirrored in the face, where the orbits are set very close together. The overall design accomplishes several things. Chiefly, it envelops a capacious brain in the form of a narrow package. This is part of a pattern of obstetrical advantages because, as also mentioned, *Saimiri* neonates have exceedingly large brains at birth, and the skeletal framework of an adult squirrel monkey is very slender to begin with. In fact, females are known to have obstetrical difficulties at term, passing a large-headed baby through a constricted birth canal.

Another feature of the cranium is a most unusual orbit, a morphology seen nowhere else among mammals. The socket is typical of anthropoids in being a cone-shaped recess that encases the eye, but the inner (medial) side has a large, more or less oval gap where there is no bone. Because the right and left sockets are actually in contact in that area, effectively squeezed together in the narrow cranium, the presence of this window, the interorbital fenestra, means that in dry skulls the two bony orbits are continuous through the void. In life, however, there is a membrane that seals the space. Interestingly, this odd construction develops as the squirrel monkey grows to adulthood, as discussed in chapter 9, where the paleontological implications are considered. The fenestra is not present in newborns.

Here, too, the growth process offers insight. The Squirrel Monkey eyeball is quite large and it fits snuggly within the socket. Eye growth and brain growth are intimately tied in primates and, like the brain, the eye of an infant squirrel monkey is exceptionally large, as determined by Tim Smith and Valerie De-Leon, foremost experts in the growth and development of the head of nonhuman primates. As it grows in this confined space, the eye can potentially place pressure on surrounding bone and suffer from friction unless it is insulated at likely points of contact. The membrane covering the interorbital fenestra is a smooth, soft spot that gives the eyeball some "play" while the bones of the cranium harden as growth proceeds.

One additional feature of adult squirrel monkeys echoes the morphology of neonatal anthropoid crania and is a sign of infantilization via arrested development. On the underside of the braincase, the large opening that transmits the spinal cord, the foramen magnum, is centrally located at birth and it remains that way into adulthood in squirrel monkeys, even though the surrounding bone expands in size and the cranium grows in length accordingly. The foramen magnum of other quadrupedal primates tends to be located more toward the back of the cranium, where the skull is cantilevered in a semihorizontal orientation against the vertebral column.

DNA comparisons among squirrel monkeys suggest that the living species differentiated around 1.5 million years ago. This is interesting in light of the existence of fossils securely attributed to the Squirrel Monkey lineage that lived in the greater Amazon Basin 12–14 million years ago in Colombia. The teeth, jaws, and postcranial bones of the extinct *Neosaimiri*, meaning "new" or "young" squirrel monkey, are so similar to the extant forms that it has been suggested the fossil could be classified in the same genus. As further discussed in chapter 9, another fossil, *Dolichocebus*, from Argentina, is also a member of the Squirrel Monkey lineage. Its oldest remains are 20 million years old.

Subfamily Callitrichinae

Callimico, Goeldi's Monkey; *Saguinus*, Tamarin; *Leontopithecus*, Lion Marmoset; *Callithrix*, Marmoset; *Cebuella*, Pygmy Marmoset

The callitrichines are all about being small (fig. 2.2). They are phyletic dwarfs, though the process that drove their reduction in size via natural selection was different from the process of infantilization and arrested development in the evolution of the Squirrel Monkey. Callitrichines are a taxonomically diverse radiation consisting of five living genera, all physically smaller than the smallest monkeys of any other platyrrhine clade by a significant measure, and also far smaller than any living Old World anthropoid genus, the smallest of which, the Talapoin Monkey, *Miopithecus*, is two to three times the size of the largest callitrichine.

Callitrichines uniquely share many less visible features that are pinned to their small physique. Since so many of these highly specialized adaptations have evolved to define them ecologically and phyletically as an adaptive complex, one wonders whether any derivative, giant callitrichines with the same ensemble of traits have ever evolved, or could possibly evolve. Is there a miniature monkey threshold effect that determines callitrichine evolvability? There is at least one other primate radiation, comprising the tarsier and its fossil relatives, that seems to have been canalized, or channeled, this way. They diversified taxonomically but remained similarly dependent on small stature as the crux of their adaptations to a novel, markedly predaceous lifestyle. This evidently became a constraint. With very few exceptions, the tarsier group basically never escaped being diminutive primates weighing between less than 60 and 500 g, 2–17 oz, in about 55 million years of evolution. On the other hand, there are the Capuchin Monkeys. They are essentially giant-sized cebines, yet the fossil record indicates that the size of modern *Cebus* does not represent a body mass limit for this ecologically versatile group. The fossil *Acrecebus* was three or more times the size of a living capuchin. Among platyrrhines, and other primates—think of gorillas and orangutans—a heavy reliance on vegetation does not impose limits on upward size evolution in the manner that a heavily insect-oriented diet places limits on becoming larger.

Many of the peculiar aspects of callitrichine biology are further addressed in various chapters in this book. Characterizing the nature of these long-tailed monkeys means accounting for novelties pertaining to many systems found in four of the genera, and some that are present in a fifth. Goeldi's Monkey, *Callimico*, is the standout exception in several respects. But all five callitrichine genera share the same feature set: miniature body size, claw-based locomotion, a diet that includes tree gums seasonally or perennially, and a propensity to exploit both the canopy and subcanopy microhabitats for food. Four of them, including the Tamarin, *Saguinus*, and the three marmosets—the Lion Marmoset, *Leontopthecus*, Marmoset, *Callithrix*, and Pygmy Marmoset, *Cebuella*—always give birth to twins (rarely triplets), while *Callimico* gives birth only to singletons. The same four have a reduced dental formula in which the third molar has been lost; it is retained in Goeldi's Monkey.

Among the platyrrhines, this distribution of traits highlights an important break in adaptive continuity as well as a transition in callitrichine history. For convenience, this distinction is recognized here by sorting the animals into two groups (fig. 2.1). The two-molared, twinning genera are referred to as callitrichins (spelled without the e). This is an informal version of Callitrichini, which corresponds with the Linnaean category of Tribe, which is a level below the subfamily rank of the hierarchy. Goeldi's Monkey is set apart from them in a complementary tribe but, with only one genus living, there is no need to employ the term (Callimiconi, or callimiconins) in the present context. This arrangement is contested by molecular phylogeneticists, who maintain that *Callimico* is part of the callitrichin group, cladistically situated deeply *within* that clade, as the sister genus of the marmosets. Morphological analysis, however, suggests its three molars and single births are among the indications that Goeldi's Monkey is a primitive callitrichine, a branch outside the callitrichin clade that split off from the remote callitrichine ancestral stock before the reduced dental formula and twinning reproductive pattern evolved as derived, signature features of the callitrichins.

Among the twinning callitrichins, fathers and siblings are intensely involved in caring for young offspring, and breeding females have the capacity to hormonally suppress the reproduction of daughters that live in the group by emitting scent signals. This represents another adaptive dimension to the derived small-body complex, social adjustments to the high reproductive output of breeding groups. The degree to which the combination of unique forms of social organization and reproductive patterns underlies the origins, essence, and success of the callitrichins is rarely matched by any other major extant primate adaptive radiation. To explain these novelties, there are hardly any parallelisms exhibited in other groups to draw upon for reference.

All callitrichines are monomorphic, meaning adult females and males are the same size and virtually indistinguishable from one another in outward appearance. Two genera are monotypic, consisting of only a single species, *Callimico goeldii* and *Cebuella pygmaea*. The others comprise numerous species according to most taxonomies, ranging from perhaps three to more than a dozen. The potential to evolve a relatively large number of species is not surprising. Taxonomically, it has been shown that there is a relationship between body size and species biodiversity in non-marine animals. There are more species alive in the smaller size classes, possibly because they have shorter intervals between generations than larger animals. They are also less susceptible to extinction aggravated by large-scale environmental fluxes, which can quickly deplete the big food reservoirs required by larger animals, giving them less time and opportunity to adapt. As a correlate to their small size, reduced count and small size of the molar teeth, and relatively small brains, callitrichines have a very distinctive appearance, with short faces dominated by the orbits, lightly built mandibles, and flattened foreheads.

Within callitrichines, there is a 6.5-fold spread between the largest and smallest forms. This does not represent a single trajectory. Body size in callitrichines has evolved in both directions, toward smaller and larger sizes, as it did in cebines, as there may be benefits to being either large or small in essentially all ecological situations that fall within the phylogenetically defined limits of a lineage or clade. This is evident in the disparate sizes of species within callitrichin genera and among the genera as well. Body mass is constantly under natural selection and eminently evolvable because of inherited size variations that are ever-present among individuals within a species. The Marmoset, *Callithrix*, is a dwarf, 20%–40% smaller than some tamarins, and the Marmoset's closest relative, the Pygmy Marmoset, *Cebuella*, is extra small, a super-dwarf two-and-a-half times smaller than *Callithrix*. The largest genus, *Leontopithecus*, the Lion Marmoset, is a giant, an oversized callitrichin. It has become larger since splitting off as a separate lineage from the common ancestor it jointly shared with the Marmoset and Pygmy Marmoset at the other end of the size spectrum.

A body-size shift was instrumental in the origins of callitrichines. The last common ancestor of this subfamily had experienced a transformation downward in body size. How much of a size reduction was involved in the transition is open to speculation, but a decrease of 50% in mass is not unreasonable in the context of the radiation given that there are three other living platyrrhine genera weighing roughly 1 kg, or 2.2 lb, including the closely related Squirrel Monkey. But none of them share any of the critical physical or behavioral traits that have been selected in callitrichines. Within the space of their niche, which is centered on a frugivorous and faunivorous diet, there would always be new ecological separations and opportunities that could be had by further specializing. This is shown by the callitrichin genera that have evolved either as dwarfs or as giants within the clade. In the case of the Pygmy Marmoset, tiny body size augmented the efficiency of gum eating. For the Lion Marmoset, evolving to a larger size enhanced their ability to secure prey from large, structurally complex forms of vegetation, and to exploit an available niche in a unique community of platyrrhines in the Atlantic Forest composed of an unusual combination and a smaller number of genera than the communities in the Amazonian rainforest.

There are costs and benefits to being small. Among the costs are some related to feeding. There is a more limited range of items that can be ingested with a small mouth, and smaller muscles produce less force, making hard items difficult to bite into and chew. At the same time, smaller animals need less space, and less food to survive. Another benefit is that gravity poses less of a problem. Less muscular strength, thus less bulk, is needed to propel the body through the trees, and less food energy is needed to hang on in static postures. Anatomical tissues, such as those shaping claws, which interface with postural and locomotor surfaces like tree bark, do not have to be overly strong to resist gravity's pull. Claws that evolved in heavier arboreal mammals such as tree sloths are very long, thick, massive, hooklike structures, which are more expensive metabolically to grow and maintain.

Like other small mammals, callitrichines may also produce more offspring by having larger litters, and by producing them more frequently during the course of a reproductive lifetime. On the downside, being small may make one susceptible to predation by a wider set of larger-bodied animals, including avian raptors, snakes, and mammalian carnivores, such as the ocelot and other small native South American cats. But being small also means one can more easily hide to escape them. Nevertheless, with family groups simultaneously managing multiple offspring deriving from successive litters, this risk requires an aptitude for vigilance and cooperation among family members. Observations of their behavior patterns in the field show that callitrichines live under significant predatory pressure. They have well-established antipredator strategies to foil attacks from the air, alarm calling and then quietly descending en masse into dense vegetation in the understory, where they freeze until the danger passes. Preparing for the night, the members of a callitrichine group typically gather to sleep in a tangle of concealing vegetation. The Lion Marmosets, as a rule, spend the night sleeping in tree holes for protection.

The most important trait correlated with small size in all callitrichines is having clawed hands and feet, which is tied to their postural behaviors and locomotion that has been described as squirrel-like. This is more fully discussed in chapter 6, on locomotion. All the digits on the hands and feet, except for the nailed large toe, have narrow, downwardly curved, and pointedly tipped claws. The combination of small size and claw-based positional behaviors, meaning postures and locomotion, allows callitrichines to efficiently range up, down, and horizontally through the forest structure, including the subcanopy sector and forest-edge microhabitats. The subcanopy presents its own challenges. There, small trees and saplings tend to be more widely spaced, requiring more leaping. This is where light weight is an asset because leaping can be done efficiently. The callitrichine body type is also an advantage in dynamic microhabitats such as the forest fringes or openings produced by fallen trees, where new growth is dominated by shrubs, bushes, thin supports, and young saplings.

In contrast, large-diameter tree trunks pose difficulties, because many are wider than the grasping span of the hands and feet and even the wingspan of extended arms. This is where the claws come in. They are useful because they act like tiny grappling hooks, enabling the animals to scamper up and down or position themselves in a resting posture by clinging. A callitrichine does not require deeply embedded anchors to maintain traction on a supporting surface during locomotion. At their body weight, the claws need only a light touch to penetrate ever so slightly into tree bark, working in conjunction with the velvet-soft friction skin on the palms and soles to prevent them from slipping.

The unique reproductive patterns, mating strategies, and social adaptations of the two-molared callitrichins are central to their success. When environmental conditions are good, a callitrichin can give birth to twins twice a year. This high reproductive output is a fundamental Darwinian advantage at the individual level. It reflects the principle of "differential reproductive success," which means that some individuals are expected to have more offspring than others in their cohort if their features are better adapted to local conditions, or simply due to chance. In this case, the environmental context is in some sense merely a background factor. If the last common ancestor of callitrichins had genes that steered the reproductive system to produce twins, that feature alone would be strongly selected as long as the environment yielded enough food to sustain the population and parents did their jobs adequately. Twinning would then become fixed in the genome and perpetuated by descendants.

These are among the explanations for the unusual anatomy and behavior of callitrichines, the reasons for their abundance, evolutionary success, and lower threat levels for extinction now and in the past. For them, the selective benefits of being a small member of the primate community have outweighed the potential drawbacks.

Callimico—There is only one species of the monomorphic, jet-black (sometimes slightly tinged with brown), helmet-coiffed *Callimico* (plate 3). It was little known until the late 1990s, but this reclusive species was first described early in the century, in 1904. It was the last truly distinct genus-level platyrrhine to actually be discovered in the jungle, though it was not until 1911 that Goeldi's Monkey was formally recognized as a separate genus and given a genus name. Ever since, *Callimico* has been a conundrum to morphologists because its anatomy mixes traits that are typical of callitrichins with some that resemble all the other platyrrhines. We return to this matter, and its significance in unraveling the evolution of callitrichines, in later chapters. As to behavior and ecology, Goeldi's Monkey proved difficult to study long-term until Leila Porter began her work in 1998. She confirmed that these monkeys are quiet, shy, and have large home ranges, occur in low population densities, and usually live in groups of four or five individuals, a pattern that has contributed to their rare sighting. As described by Porter in 2007, *Callimico* is an understory/lower canopy specialist. They frequent the undergrowth below the canopy where light levels are low—and their blackness makes them inconspicuous—but where *Callimico* find and consume an unusual food source that grows advantageously in that microhabitat.

As in other callitrichines, fruit and fauna are important food resources for Goeldi's Monkey, but their dietary specialty is fungus. Rather than the familiar mushroom variety of fungus, with the fleshy caplike tops that tend to grow above the soil, these monkeys eat a different class of jellylike fungi that grow as blobs on rotting wood. Fungus is a low-quality food that requires a significant amount of time to digest, and the Goeldi's exhibit behaviors that are appropriate when consuming hard-to-digest foods, searching for and eating fungus late in the day before retiring, which gives the gut ample time to do its work. The fungus they eat is also widely dispersed in the forest, which is why they have large home ranges.

Among all the extant primates, *Callimico* are the most committed fungus feeders, although some marmosets have also been observed doing the same. Fungus makes up a fairly large proportion of their diet, more than 50% during several months of the year. They harvest fungi where it grows on decaying or dead wood. Moving through the subcanopy where small diameter supports are common, the relatively long-legged and long-footed Goeldi's Monkeys frequently use clinging resting postures and leaping moves, in addition to quadrupedalism. They inhabit the far western side of the Amazon basin in primary and secondary forests as well as thickets of bamboo, where fungus is common. Fossil remains of a genus from Colombia that may be part of the *Callimico* lineage, the 12–14-million-year-old jaw of *Mohanamico*, suggests that Goeldi's Monkeys are a long-lived lineage.

Saguinus—Tamarins are the most diverse and widely distributed callitrichines, with well over a dozen well-defined species that range throughout the Amazon basin and into Panama, in Central America. They are well studied. They occupy a wide variety of habitats, including primary and secondary forest and forest fringes, operating in the continuous canopy and understory layers. While ripe fruit, insects, tree gum, and nectar make up most of their diet, *Saguinus* also specialize in feeding on large winged insects such as crickets and grasshoppers that are available throughout the year.

Unlike the monochromatic Goeldi's Monkey, or the evenly dappled, grey-brownish Pygmy Marmoset, Tamarin bodies are often multi-hued, and the species are distinguished by many different colors and fur patterns arranged in various combinations on arms, shoulders, legs, hips, tails, backs, and bellies (plate 4). Their faces may be evenly tinted or mottled, smooth-skinned or hairy, and the head may be furry, bald, or sprouting a wedge of long white hair. In some species with dark body fur there is a white muzzle on the face. In others, a thin V-shaped blaze is drawn across the forehead above the eyes. Some have long whiskers extending out from the lips like an upside-down handlebar mustache. Others have a patch of white fur that stretches horizontally across the flap between the upper lip and nose. These patterns are each species-specific, and present in both males and females because the Tamarin is monomorphic.

The genus is genetically predisposed to evolving a colorful appearance. Several factors may have played a role in promoting such variations. They may comprise a set of visual cues to facilitate recognition of close kin, as well as species identity, and some aspects may also be associated with gestural communication or ritualized mating behaviors. The diverse evolution of coat colors and pelage (fur) patterns within the genus may be promoted by rapid evolution, a result of the callitrichine high reproductive rate, and varying local environments. Widely spread out in Amazonia, tamarin populations are prone to division and isolation by the large network of rivers that are subject to dramatic shifts, even annually, by the massive flooding of the Amazon River and its tributaries, events that would reshuffle the distributions of species and populations over time. These conditions may influence evolutionary changes in external traits by chance, through an evolutionary phenomenon called character displacement. Random mutations affecting appearance accumulate when a population becomes divided and separated into two. These traits become accentuated as species identification markers to prevent intermixing of the gene pools if the animals once again come into geographic proximity.

Paul Garber and Leila Porter have contributed much to the understanding of the genus *Saguinus* by surveying the ecology and behavior of many species. They found that the feeding behavior of the frugivorous-insectivorous tamarins includes sizable amounts of tree gum collected serendipitously as the animals forage widely. It is evidence that callitrichins may have inherited a fundamental predilection for gumivory as a dietary component from their last common ancestor, as one of the correlates to being small-sized. That is because they engage in this behavior without having any dental specializations to enable them to acquire gum independently, as do other callitrichins in which this resource is pivotal. The Marmoset, *Callithrix*, and Pygmy Marmoset, *Cebuella*, evolved a suite of dental features that make it possible to harvest gum by scraping away bark and gouging a tree to stimulate its production, thus assuring a staple food supply.

Monogamy was once thought to be at the core of *Saguinus* social systems, but Garber and colleagues, in an extensive 2016 report on several species, including detailed paternity studies within and between social groups, have challenged this notion. They showed that social polyandry, a system which centers on one breeding female having two or more mates at a time, is a common pattern and monogamy is not. Furthermore, paternity tests of captured (then released) monkeys performed on cells taken from hair follicles show that groups are not simple extended families. Resident males may not be the fathers of the breeding female's offspring, and two adult females that may be living in the same group may not have a mother-daughter relationship.

Overall, the behavior and ecology of *Saguinus* provides an important perspective in considering potentially primitive patterns in the clade of two-molared, twinning callitrichins, and probably for callitrichines as well. None of the fossil platyrrhines discovered thus far appear to have a specific phylogenetic link to the genus.

Leontopithecus—The largest of the callitrichines, and the one with a most impressive mane, *Leontopithecus* is here informally called a Lion Marmoset, as it was until the late 1970s (plate 5). My preference for this colloquialism, rather than "Lion Tamarin," reflects the hypothesis that *Leontopithecus* is more closely related to the Marmoset and Pygmy Marmoset, *Callithrix* and *Cebuella*, respectively, than to the Tamarin, *Saguinus*. The recent move to apply the familiar term tamarin was based on an old, incorrect anatomical distinction that divided callitrichins into "long-tusked" tamarins and "short-tusked" marmosets; however, this is both a descriptive misnomer, and quasi-cladistic, based on faulty evolutionary reasoning.

The lower canines of marmosets and pygmy marmosets only seem to be short-tusked, with low crowns that do not project much, but this is because they are set adjacent to the incisors, which evolved into very tall, tree-gouging teeth. This is a derived condition, a characteristic that has become modified evolutionarily over time from an earlier, more primitive state in which the incisors are not elevated in height. Primitive conditions, such as the short-tusked condition shared by the Lion Marmoset, Tamarin, and Goeldi's Monkey, and most other platyrrhines, are not useful phylogenetic indicators because they do not signify a uniquely shared ancestry by the taxa that exhibit them. Primitive traits can be widespread among taxa, some closely related and some that are distantly related. Derived conditions, on the other hand, are historical markers that indicate the animals sharing them are descendants of a unique common ancestry, which makes them powerful cladistic markers.

The Lion Marmoset, of which there are at least three highly endangered species, is a giant when compared with its closest relatives. Lion marmosets are twice the weight of the Marmoset and five times as heavy as the Pygmy Marmoset. They are endemic to the Atlantic Forest, which after 500 years of decimation, exists only in small fragments in little more than a narrow arc of land along the southern seaboard of Brazil.

Like the other callitrichines, lion marmosets occasionally feed on plant gum encountered in the forest, and fruits, but they are also specialized faunivores, feeding on small vertebrates especially when fruits are scarce. They use a different strategy from the approach typically used by tamarins when hunting. Rather than stalking any flying orthopterans as most tamarin species do, *Leontopithecus* uses its long forearms and hands to find and secure concealed prey by reaching deeply into large, tree-dwelling plants called tank bromeliads, an important element of the Atlantic Forest flora. The leaves of these bromeliads grow around a central bowl that stores water for the plant and tends to harbor a collection of relatively immobile insects and small frogs and reptiles that live in the pool of water that accumulates in the tank.

As reported by Anthony Rylands, a pioneering field biologist and conservationist who has contributed much to our knowledge of the ecology and behavior of lion marmosets and other callitrichines, and to preservation of the Atlantic Forest where monkeys are severely threatened by habitat loss and degradation, these monkeys also extract prey by foraging in rotten or loose bark, and in natural crevices and holes in trees. They are unlike the visually oriented tamarins and marmosets that generally glean exposed insects from surfaces; they specialize in searching out hidden prey by touch. In this sense, they are manipulators like capuchins and squirrel monkeys, though without comparably adroit hands or an evidently nimble mind.

There is one species, the White-mouthed Tamarin, *Saguinus nigricollis*, that has a similar foraging style and also has relatively longer hands and fingers than other tamarins. The evolution of a comparable anatomy and behavior, which is an example of the parallel evolution phenomenon, corroborates the link between long hands and a foraging style that involves manual probing. It is another illustration of how the evolution of postcranial features, normally seen as the province of positional behavior, is closely tied to feeding, especially when there is a strong selective rationale. This emphasis on predation and the methods employed explains why the Lion Marmoset has evolved to become the largest callitrichine, and it aids in understanding of the uniquely elongated forearm and hand.

Six primate species now exist in the Atlantic Forest. This is a much smaller ensemble than in the communities of the Amazonian rainforest, where the cebid family tends to be represented by the full-size spectrum of species, in gradations from about 100 g to more than 3 kg, 3.5 oz to 6.5 lb, and all operating within a broadly based ecological niche predicated on a frugivorous and faunivorous diet. The middle-sized genus in this taxonomic and ecological spectrum is *Saimiri*, the Squirrel Monkey, at about 1 kg, 2.2 lb. In the Atlantic Forest, where squirrel monkeys do not exist, and the small-to-large body size spread of monkeys is arrayed in large steps rather than a gradient, the Lion Marmosets are the middle-sized taxon of the three-genus cebid array; the Marmoset and Capuchin Monkey represent the extremes. In other words, in broad ecological terms the Lion Marmoset occupies a niche that resembles the Squirrel Monkey niche elsewhere on the continent. Because ecological niche is largely defined by body size, that middle position was filled in the Atlantic Forest as ancestral lion marmosets increased in size via natural selection and began specializing in a unique brand of faunivory.

The long hands and long feet of lion marmosets are also assets when clinging to large-diameter tree trunks in the Atlantic Forest and when climbing, but they influence locomotion in ways that can make the animals look awkward. Callitrichines do not have grasping hands and feet that rely on muscular opposition of the thumb or large toe, but the efficacy of grasping during quadrupedal locomotion in Lion Marmosets is further compromised by the proportions of the long palms and long midfoot. When walking quadrupedally on slender branches, lion marmosets tend to place the hands and feet diagonally on the supports, in parallel, facing one side and then the other as they move forward, rather than placing right and left hands and feet, respectively, on opposite sides of a branch in a clasping position. Alternating hand and foot placements in this way makes for an atypical, sideways gait as they move along a branch. They also exhibit an overstriding footfall pattern, in which a hindfoot advances ahead of a handhold, which has the effect of shifting the position of the rump from side to side as the animal progresses. This locomotor style is a lesson in the give and take of evolutionary change in which selection for highly specialized foraging hands required locomotor accommodations as a trade-off.

The unusual reproductive strategy of lion marmosets in their natural environment is highly seasonal, with conception able to occur only during three months of the year. Females may or may not ovulate at other times but they do not conceive, indicating that the seasonal cycle is not entirely mediated by female hormonal control. Male behavior seems to be an important, proactive factor. There is no significant difference in the degree to which females solicit mounting from males during the breeding and nonbreeding seasons, but the male patterns are different. During the nonbreeding season they mount females infrequently; they increase mounting attempts during the breeding season. The hormonal or neurobiological mechanism behind this pattern is not yet known.

Callithrix—Callithrix is the second-smallest platyrrhine, weighing roughly 225–450 g, 0.5–1 pound. Like the Tamarin *Saguinus*, the Marmoset has evolved into numerous species; estimates range from 5 to 21. Though the animals are less colorful in appearance than the tamarins, they include species whose faces are often embellished by furry ear tassels of different sizes and shapes (plate 6). The species are widely distributed across many environments in the middle of South America south of the Amazon River. The habitats of the versatile marmoset species may be humid, lush, or dry; forested woodland or treed grassland; or narrow gallery forests that grow in strips alongside rivers. The ways marmosets use these varied habits is constant in at least one genus-specific way. They prefer forest fringes and areas of secondary growth, where desirable fruit trees are concentrated, rather than large swaths of forest in which they are more scattered. This is where the sheltering vegetation that marmosets need is dense and insect life is abundant.

Marmosets are specialized gum eaters, or gumivores. This is a critical adaptation, particularly in environments where fruits are seasonally or annually scarce. It is the evolutionary reason behind many of the characteristics that make up the profile of the genus. In one study, more than 70% of all the identified plant parts the marmosets ate during the course of a year consisted of tree gum, while fruit, nectar, flowers, and seeds contributed 16%. As we have seen, tamarins and lion marmosets occasionally also feed on gum when they come across it exuding naturally from trees as they forage, and they lap it up. But neither of these genera have incisor and canine teeth suitable for scraping and gouging a tree to induce gum to flow (fig. 2.5). Such teeth have evolved only in the Marmoset genus and its sister genus, the Pygmy Marmoset, monkeys whose last common ancestor evolved a variety of anatomical and behavioral characteristics that enable them to procure globules or drippy gum intentionally, by scraping off patches of bark. The scraping stimulates the production of gum by the tree as a form of repair, to seal the hole against further damage. This harvesting adaptation may have first evolved in drier habitats at the edges of, or outside, the rainforest where fruit trees are less productive, as in the vast areas that exist now between Amazonia and the Atlantic Forest, or in highly seasonal montane forests.

The intrinsic ability of a marmoset group to select a limited set of gum trees and exploit them for food is a major strategic advantage over the manner of Tamarin and Lion Marmoset gum eating that relies on extrinsic factors leading to the production of tree gum within their ranges. Marmosets have the anatomical tool set to do this, which is lacking in the others. By producing and cultivating feeding sites as needed, *Callithrix* can maintain themselves in a smaller home range and do not have to travel so far each day in search of a

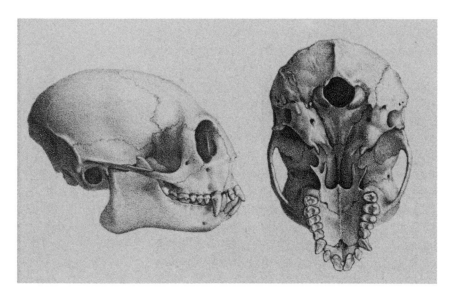

FIG. 2.5. Skull of *Callithrix*, the Marmoset. From Blainville (1839).

primary food staple, thus minimizing the energy expended in foraging. For example, the Common Marmoset, *Callithrix jacchus*, lives in thorn scrub habitats that may be quite dry and fruit-poor. There the monkeys live in home ranges that are 5–15 acres in size, roughly equal to 7–20 football fields. In contrast, the Saddle-backed Tamarin, *Saguinus fuscicollis*, which weighs 320–420 g, 11–15 oz, and is one of the smallest tamarin species, forages over a home range of about 74 acres. This is not to say that gums are always the food resource that is key to survival of marmosets in such arid regions. Insects are also known to be important in these extreme habitats, which some marmosets have probably come to exploit secondarily.

Cebuella—*Cebuella*, the Pygmy Marmoset, is a super-dwarf. It is the smallest living anthropoid, about 5.5 inches long from nose to tail, weighing roughly 110 g, about 4 oz. *Cebuella* is a monotypic genus with only one species, *Cebuella pygmaea*. It lives as a habitat specialist in low population densities in the western Amazonian Basin, in or at the margins of floodplain forests alongside the many watercourses that wind their way through the region.

Due to their extra small body size, few markings, and dull grey-speckled, or orange and tawny coat color, *Cebuella* is well camouflaged (plate 7). This is advantageous when feeding, because this species is an extreme gumivore and its harvesting behavior makes the Pygmy Marmoset quite vulnerable to predation. Getting gum involves being exposed on large tree trunks for prolonged bark-scraping sessions in a stationary position. They are perpetually at risk because they eat gum throughout the year and rarely eat any other plant part that may place them in the denser canopy, although pygmy marmosets also stalk and eat arthropods like insects and spiders. *Cebuella* use the same hole-gouging methods employed by *Callithrix*. They have the same set of dental adaptations, though the pygmy has smaller jaws, teeth, and muscles as a consequence of being less than half the size of the Common Marmoset, for example, so it takes more time and effort for the Pygmy Marmoset to gouge a comparable gum site. An interwoven set of novel traits like these in form, function, adaptation, and behavior are evidence that these genera are very closely related. How long has *Cebuella* occupied this gum-eating niche in western Amazonia? Tiny fossils found in Peru, 11 million years old, have been allocated to the same genus. Among the fossil specimens are front teeth with the same pattern of tooth wear damage seen in the tree-gouging Pygmy Marmoset, demonstrating that this lineage and its gum-eating adaptation has endured for at least 11 million years.

Fruit huskers and seed eaters: Family Pitheciidae

The pitheciids include Saki Monkeys, Bearded Sakis, Uacaris, and the Titi and Owl Monkeys (fig. 2.6). The taxonomic definition of the pitheciids presented here differs from some researchers who base theirs exclusively on molecular studies. I include the Owl Monkey genus in this group; others allocate owl monkeys to the cebids. Many of these monkeys are difficult to study because they are rare, often live in low densities, may travel large distances, and occupy heavily flooded terrains. The Owl Monkey, *Aotus*, is nocturnal, making it particularly challenging to observe in the dark of night. It was not until the 1980s and 1990s that the first field reports based on long-term field studies in the wild described groups belonging to the genera *Chiropotes* and *Cacajao*. Few colonies of sakis, bearded sakis, and uacaris have ever been maintained in zoos or labs, and so they have been little studied in captivity.

Pitheciids, of all the extant New World monkeys, have the most prominent, fleshy noses. The sakis, bearded sakis, and uacaris have a very broad flap of skin between the nostrils, supported by very wide nasal bones. The reason for this ultrawide fleshy nose is not clear. It may simply reflect the cranial anatomy upon which it rests, rather than having an adaptive explanation of its own. However, the explanation for the prominent bulbous nose of the Owl Monkey is clearer. Owl monkey noses are sensitive tactile and olfactory organs involved in their specialized system of communication that is geared to nocturnality. A variety of their nocturnal adaptations are discussed in chapter 8, on sociality. The fifth pitheciid genus, the Titi Monkey, *Callicebus*, does not have

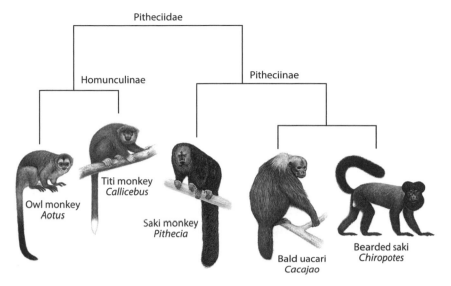

FIG. 2.6. Cladistic relationships and classification of living pitheciid genera, not to scale. Images courtesy of Stephen Nash.

a very wide external nose, which appears to be a consequence of a divergent modification of the cranium. Big noses are a family-level trait, so the loss of prominence in the titi is probably tied to the evolutionary reduction of their snout and a reorganization of their front teeth for dietary and social reasons, also discussed later.

As an ecological unit, pitheciids are nestled in the small- to medium-sized monkey niche in the New World, weighing roughly 0.9–3.6 kg, 2–8 lbs. In other respects, pitheciids are difficult to define by a consistent set of anatomical characters because of the disparate and discontinuous morphologies exhibited by the five living genera. This is not unusual in evolutionary biology. It is more common when examining the phylogenetic links among groups that are classified in different families or other higher taxonomic categories, and when very old, primitive fossils are compared with modern forms. However, molecular studies have firmly corroborated that one of these anatomically obscured linkages exists, linking Titi Monkeys with the Sakis, Bearded Sakis, and Uacaris. Yet the molecules and morphology do not align in a simple, entirely satisfactory way with regard to the position of the Owl Monkey, neither as a pitheciid nor as a cebid, the two families with which this genus has most often been associated.

On the basis of morphology, pitheciids are grouped phylogenetically by a set of three interconnected taxonomic linkages, as follows: *Pithecia, Chiropotes,*

Cacajao + Callicebus + Aotus. The Sakis, Bearded Sakis, and Uacaris are connected to *Callicebus,* and Titi Monkeys, in turn, are connected to *Aotus.* The
three saki-uacari genera are a cohesive subgroup, with well-defined, unusual
cranial and dental features. The other two genera, the Titi Monkey and the
Owl Monkey, are quite different and they also contrast with one another significantly in craniodental form.

On one level this taxonomic pattern is explained by remarkable divergent
adaptations that have left their marks on the morphology of the cranium and
dentition, superficially summarized as: dietary adaptation in the sakis-uacari
groups; social and feeding adaptation in titi monkeys; and, nocturnality in
owl monkeys. Given such dramatically distinct sources of selective pressure,
trenchant anatomical differences might be expected. But deeper analyses place
the titis closer to sakis, bearded sakis, and uacaris in some features and to owl
monkeys in others, signs that all five are closely related, with a few subtle
mandibular and dental traits binding them all together as well. The details are
further considered in connection with the dietary specializations of pitheciids
in chapter 5. They revolve around eating hard-husked fruit and tough seeds,
a diet that has selected for specializations in the front teeth that facilitate the
harvesting aspect of food-getting, that is, biting through tough skins or woody
protective casings to access seeds or fruit pulp.

A rethinking of the nature of extant pitheciids had profound consequences
for interpreting New World monkey evolution. Until then the three saki-
uacari genera were considered platyrrhine outliers, an anatomically odd,
minor twig on the phylogenetic tree, with no apparent ties to others and
little bearing on platyrrhine history. In revising that view to accommodate
the Titi and Owl Monkey genera, the twig became a sturdy branch, bearing important information that integrated sakis, bearded sakis, and uacaris
ecologically.

Under this broader concept of the family-level clade, the fossil record of
pitheciids also grew in size. It now includes more than a dozen genera and goes
back 20 million years. This indicates that pitheciids have long been a major
factor in the evolution and ecological framework of the platyrrhine adaptive
radiation. The fossils reveal two phases of this family's radiation. The two oldest genera from Patagonia show that specialized anterior teeth, as well as an
unusual wedge-shaped premolar set, had already been modified in the direction evident in extant pitheciines, though situated in a lower jaw that was
somewhat primitive. The younger forms, about 12–14 million years old, show
further advances in evolving very much the same types of canines, cheek teeth,
and jaw shapes found in the extant sakis, bearded sakis and uacaris, craniodentally the most modified members of the clade.

Subfamily Pitheciinae

Pithecia, Saki Monkey; *Chiropotes*, Bearded Saki; *Cacajao*, Uacari

The middle-sized pitheciines, weighing roughly 1.8–4 kg, 4–9 lb, compose the larger-sized subfamily of pitheciids (fig. 2.2). The Saki Monkeys, Bearded Sakis, and Uacaris are quadrupeds with relatively large heads, relatively short, bushy tails (table 2.2), and moderate or long forelimbs and hindlimbs in relation to the trunk, making them quadrupeds that are adept at leaping. But their most distinctive anatomical traits are concentrated in the skull and dentition, the foundation of the masticatory apparatus. Pitheciines have often been thought of as seed predators, a term that may exaggerate the impact of their diet on the trees they use for food. They are a clade more committed to eating seeds than any other primate group, yet their functional morphology and behavior paints a more complex and interesting picture.

Exhibiting an anatomy without parallel among the primates, these monkeys have large, tusklike lower canine teeth and a scooplike wedge formed by four slender lower incisors, an arrangement that supports their specialized diet consisting of the seeds and pulp of fruits that are protected by tough husks and woody shells. This feature set is exhibited in the most extreme fashion in the genera *Chiropotes* and *Cacajao*, and less so in *Pithecia*. The impressive lower canines may be useful in some food-piercing actions, but to get at the nutritious interior of a hard-shelled fruit the pitheciines often breach woody exteriors first in a viselike system made up of the canines and the adjacent premolar teeth. The fruit is placed directly behind the lower canine so it is stabilized as the jaws close and crack the casing under occlusal pressure, which is maximized by the wedge-shaped crown of the most anterior premolar. To extract the seeds from inside the shell, pitheciines use their lower incisors to pick them out. They also use these teeth to shave off slivers of fruit pulp. The crowns of the molar teeth are essentially flat, and virtually cuspless, but the enamel cap is crinkled by a random array of shallow furrows and ridges. The role of these crenulations is to assist in crushing and grinding ingested material. In lieu of a mortar-and-pestle-like organization of cusps and basins, the textured enamel surfaces of the cheek teeth trap and help stabilize food particles that get progressively smaller during repetitive chewing cycles. The design of their cheek teeth may also be an advantage as their feeding habits expose the animals to hard exogenous materials, which may include fragments of woody shells that could promote tooth wear and even cusp breakage.

The craniodental morphology of pitheciines varies taxonomically as a morphocline, meaning a series of anatomical gradations among taxonomic groups. The Saki Monkey is the genus at one end of the series and reflects the primitive

Morphocline

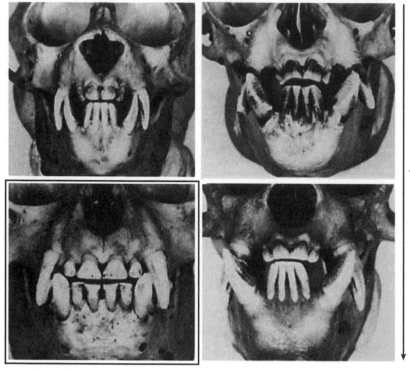

Morphocline

FIG. 2.7. Incisors and canines of a saki, bearded saki, and uacari arranged as a morphocline (clockwise from top left) and compared with a spider monkey. Adapted from Rosenberger and Tejedor (2013).

form, while the Bearded Saki and Uacari are the two genera that jointly occupy the more derived, opposite end of the cline (fig. 2.7). Brain size follows a similar pattern: relatively smaller in the Saki Monkey and larger in the Bearded Saki and Uacari. The morphocline concept is an important tool in reconstructing the primitive and derived states of characteristics, which is the first step in inferring the phylogenetic interrelationships of the animals in question and framing questions about *how* and *why* the anatomy was transformed over time.

Another intriguing resemblance that exclusively links *Chiropotes* and *Cacajao* is that male offspring delay sexual maturity, as indicated by the fact that their testes do not descend from the abdomen into the scrotum, and become visible, until about four years of age. The reason for this is obscure but it may be associated with a prolongation of adolescence, which keeps younger males from exhibiting the full range of behaviors and interactions experienced by adult males. That may be advantageous to young males as a safety device in species that are sexually dimorphic, where males have overly large and visually prominent canines that may be used aggressively.

Pithecia—The Saki Monkeys are the smallest members of the subfamily, weighing 1.3–1.8 kg, 3–4 lb, though their fluffy fur makes for a rather stocky appearance (plate 8). As noted, *Pithecia* tend to have the most primitive craniodental traits of the pitheciine group, which corresponds with their being the least specialized dietetically. For instance, like many other platyrrhines, sakis supplement their diet with leaves and insects when their preferred fruits are not available, but bearded sakis and uacaris manage to remain almost entirely frugivorous.

Most of what we know about these animals in the wild comes from studying the White-faced Saki, *Pithecia pithecia*, a bushy-tailed, long-haired, sexually dimorphic species in which males and females present markedly distinct coats, in color and pattern, a phenomenon called sexual dichromatism. The adult male is essentially black, with black fur and a black-skinned snout. White hair surrounds the eyes and muzzle, reaching back to cover the ears, forehead and chin. The adult female has salt-and-pepper fur covering most of her body, including a flat, helmet-like coif with pixie-like bangs. The black face is outlined by a pair of white stripes arching from the medial corners of each eye to a spot below the corners of the mouth. Males are about 25% heavier than females in weight, making them sexually dimorphic in size as well. These are shy monkeys that live in small social groups, commonly composed of two to five adults, but groups of 12 have also been documented. They appear to be monogamous and pair-bonded, but that is still unclear, and it also presents a puzzle. Typically, there is a strong association between pair-bonding, monogamy, and sexual monomorphism, meaning males and females look much the same. But the sexually dichromatic and physically dimorphic White-faced Saki pattern is not consistent with the monogamy model, which leaves several questions yet to be answered.

Saki monkeys live in a wide variety of forested habitats, from tropical rainforests to dry tropical forests, and others. They are widely distributed in the north and south of the Amazon River. Their taxonomy is a matter of debate. There are five or more species; up to sixteen according to some. With a facility for vertical clinging postures and leaping locomotion, sakis tend to prefer the middle and lower levels of the tree canopy where this style of positional behavior is an advantage in crossing gaps between slim tree trunks. In their 2013 review of saki monkey ecology and behavior, Marilyn Norconk, the foremost authority on the behavior and ecology of these and other pitheciines and their foremost scientific advocate, and her colleague Eleonore Setz noted that local names for the genus sometimes reflect this style of movement, by referring to them as *volador* in Colombia, which translated from the Spanish means flying.

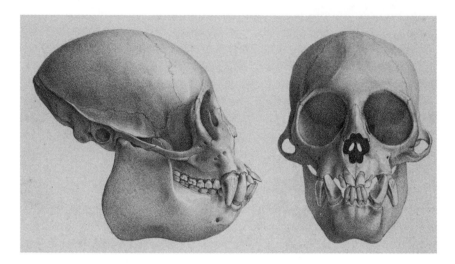

FIG. 2.8. Skull of *Cacajao*, the Uacari. From Blainville (1839).

Chiropotes—The genera *Chiropotes* and *Cacajao* are similar craniodentally and skeletally (fig. 2.8). They exhibit the most biomechanically specialized anterior teeth and jaws of the pitheciine subfamily, which are designed to pry open large, hard-shelled fruits to get at the seeds. Up to 75% of the diet of bearded sakis and uacaris can consist of such hard-to-access seeds. Both genera share a propensity to hang by their feet during feeding, freeing the hands to manipulate and stabilize an armored fruit behind a canine tooth as they crack it, or to secure an opened fruit while they pluck out the seeds using the lower incisors. The geographic distributions of these genera around the Amazon basin are similar. Their combined range closely matches the distribution of the large, woody fruits they prefer. Their tastes are rather idiosyncratic. In contrast to most primate frugivores that eat from a large number of tree species, bearded sakis and uacaris rely on a relatively small number from a tree family that is not taxonomically diverse, the Lecythidaceae, which include Brazil nuts, whose shells are hard as rocks. These monkeys are also distinctive in being able to rely on immature fruits, which is one reason why in some cases they can subsist on a diet comprising about 90% fruits and seeds.

Weighing about 2.7 kg, 6 lb, the Bearded Saki genus comprises large, conspicuous pitheciines (plate 9). Some species have sleek coats that are all black and others are mostly brown-backed with black extremities. Their most striking features are a long fluffy tail that is often carried in a wide curve arching above the back with the tip facing forward, and a head surrounded by dense fur. The pattern is exaggerated in males and it is especially trenchant in *Chi-*

ropotes satanus, now called the Black Bearded Saki but originally named after the devil, which has the most dramatic beard of any New World monkey. It is plush and bushy, beginning at the eyes and extending downward, ending at the chest. Above the eyes, a pair of poufy balls of fur part in the middle and reach back covering the ears.

There are about five *Chiropotes* species. Their preferred habitat is primary forest, including both non-inundated and flooded terrain. Unlike the relatively small troops encountered among saki monkeys, groups of bearded sakis may include as many as 65 individuals. They are organized in a fission-fusion social system, in which the aggregate of individuals forming a group breaks up into temporary subgroups of varying sizes and proportions of males and females, and then coalesces again. It is a social pattern that we take up in chapter 8. These large groups may have huge home ranges in continuous mature rainforest, some larger than 2,500 acres, more than 2 square miles. But smaller troops are known to live in forest fragments and on small islands with standing forests less than one-tenth that size.

A segment of the long lineage leading to the sister genera of Bearded Sakis and Uacaris is documented in the fossil record by the 12–14-million-year-old *Cebupithecia*, from Colombia. It is represented by one of the few partial skeletons of a fossil New World monkey, which also has dental remains preserving upper and lower teeth and a partial mandible, all very modern in appearance. The massive upper and lower canines are indications that this fossil genus is more closely related to the *Chiropotes-Cacajao* group than to the Saki Monkey genus, *Pithecia*.

Cacajao—The Uacaris have been perhaps the most daunting field subjects of any New World monkey. The swampy, seasonally flooded habitats in which they often live make it difficult to follow them. They tend to exploit higher levels in the canopy, making it difficult to see and track them; have enormous home ranges; live in low densities at the population level; maintain large groups that subdivide up into smaller, mobile foraging parties; and individuals are not easily distinguishable. Consequently, little was known about their wild ecology and behavior until the 1980s, and most of what we do know relates to the most basic matters that field-workers are able to observe, such as habitat preferences, ranging patterns, and estimated group sizes. Much less is known about how the animals budget their time, how groups are composed, that is, the numbers of adult males, females, and subadults of all ages, social behaviors, mating strategies, and modes of communication.

Cacajao is the only New World monkey with an extremely short tail (plate 10). This is a variation of the short-tailed pattern exhibited by the other pitheciines. It is about 40% the length of the animal's head and body measurement, and 65% the length of the trunk (table 2.2). In all other platyrrhines, the tail is longer than head and body length. An explanation for the short tail remains elusive. One of the uacari species has a very red face and a bald head, reflected in its formal species name, *Cacajao calvus*; *calvus* is Latin for hairless. Because they have no fur on their head, it is possible to see a pair of bilateral mounds bulging at the top of the cranium beneath the skin. These are the much enlarged right and left temporal muscles, an important part of the heavily built masticatory system which is also seen in the bearded sakis.

Like its sister genus, *Chiropotes*, *Cacajao* is sexually dimorphic in size, largely quadrupedal, and often hangs by the feet during feeding. However, while a uacari is suspended, its short tail typically cannot be draped over a tree limb to gain extra support, a maneuver often seen in bearded sakis. There are four species, living mostly in forests alongside the Amazon River and its tributaries, which become flooded during the rainy season. *Cacajao* have the most limited geographic distribution of any pitheciine genus, perhaps because they have a strong preference for inundated forest. Groups of uacaris can comprise 200 individuals and very large home ranges, up to 1,200 hectares, nearly 5 square miles. Like bearded sakis, they live within a fission-fusion social system. In a rare behavior among primates, species of *Cacajao* appear to migrate seasonally from flooded forests to drier, better-drained, botanically diverse forests, called terra firma ("firm earth"), where they may exploit specific crops of fruit that are more plentiful there, including highly desirable Brazil nuts.

Subfamily Homunculinae

Callicebus, Titi Monkey; *Aotus*, Owl Monkey

The subfamily name for this group, Homunculinae, is based on a fossil genus, *Homunculus*, unlike the etymology for the other New World monkey families and subfamilies, which are named after extant genera. The reason is that in addition to Titi and Owl Monkeys, the fossil *Homunculus*, considered further in chapter 9, belongs to this group, too. This grouping is based on a phylogenetic hypothesis. All taxonomic decisions reflect hypotheses and are subject to change, but the actual taxonomic terms devised for purposes of classification are determined by explicit instructions, written to resist potentially confusing, destabilizing change. Following nomenclatural rules for zoology, the first term properly given to a taxon of family-level rank, which must always be based on a genus contained within that group for reference, has seniority over all subsequent terms.

The expression "family level" is a technical umbrella that refers to the three Linnaean titles above the genus rank, Superfamily, Family, and Subfamily. The

Family Homunculidae was published in 1894 shortly after the genus *Homunculus* was named, and before the application of any family-level labels based on the other genera contained in the same group, *Aotus* in 1908 and *Callicebus* in 1925. This does not mean names like Family Aotidae or Subfamily Aotinae, as seen in some current platyrrhine classifications, are incorrect when applied to a family-level group attributed exclusively to *Aotus*; both are available taxonomic terms. Yet, since it has become a priority to infuse classification with phylogenetic meaning, it seems inconsistent with this philosophy to place the Owl Monkey in a family of its own because there is controversy over its systematic status, as some have argued. There is another category designed explicitly for that purpose, called *incertae sedis*, meaning of uncertain position. On the other hand, if classifications are built exclusively on the basis of molecular evidence without being informed by fossils—now a common trend—or vice versa, a significant amount of valuable evolutionary information is lost. As to the other family-level names used in this book—Cebidae, given in 1831, Pitheciidae in 1865, and Atelidae in 1825—all were proposed before any of the fossils now classified in those groups were discovered, given formal genus names, or used as a root-word for family-level taxa.

The morphology of the two living homunculines is different from other pitheciids. In being medium-sized genera, weighing roughly 900 g, 2 lb, they are smaller (fig. 2.2). And they lack the core, hyperspecialized dental anatomy exhibited by living pitheciines, the shell-cracking architectures of the canine and premolars, and the flat-crowned molars. However, titis and owl monkeys do present a few dental features reflecting the importance of a tough-fruit dietary niche, which logically would have preceded the shell-cracking niche.

As will be discussed at length below, titis and owl monkeys share a range of behavioral novelties, including a pair-bonded, monogamous mating system that involves intensive paternal care of offspring, rare features among other primates. They have small canines and are monomorphic in body size, which are also rarities (table 2.1). This is a tightly organized, functionally interconnected set of traits, and it includes more that will be discussed. The other New World monkey group that is monomorphic, the callitrichines, does not exhibit the same combination of characteristics. For example, they have enlarged canines in both sexes and are not monogamously pair-bonded. These and other uniquely shared features are indications of the close phylogenetic relationship between *Callicebus* and *Aotus*. In fact, they provide one of the outstanding examples of a primate clade that is in some ways better defined by behavior than by morphology. Postcranially, titi and owl monkeys are rather nondescript. They are quadrupeds that also engage in leaping, a typical locomotor profile for monkeys in their size class.

Long before knowledge of their behavior was available and the association of pair-bonding and small canines was understood, their unremarkable dental morphology led anatomists to identify *Callicebus* and *Aotus* as very primitive New World monkeys, but that is not the case at all. The most fundamental fact of Owl Monkey life belies that supposition. They are nocturnal. This is undoubtedly a derived adaptation in anthropoids, which we can establish by employing the character analysis method used in investigating the evolution of nose shape described in chapter 1, or the morphocline analysis pertaining to the pitheciine dentition, and the evidence of other anatomical and behavioral systems. The alternative to nocturnality, the diurnal activity rhythm, occurs universally among the other platyrrhines, and in living Old World monkeys and apes, and it is the standard pattern among the early anthropoids from Africa. Occam's Razor thus requires that a diurnal ancestral condition, the rule rather than the exception, is the correct working hypothesis for platyrrhines. The morphological details of the *Aotus* eye corroborate this hypothesis because, apart from its large size, it is fundamentally designed for daytime activity.

Several adaptations of the Titi Monkey point in the same direction, as non-primitive patterns. Functional morphology leads to this conclusion and so does the application of the morphocline concept. The rare, specialized, hard-fruit husking adaptations seen in the lower incisor teeth of Saki Monkeys, Bearded Sakis, and Uacaris extend to include the titis, in detail, and the system as a whole differs from the more common pattern of frugivores seen among other platyrrhines (fig. 2.7). Most frugivores have low-crowned lower incisors designed to nip at or pluck fruits rather than scraping their rind, or teasing out small seeds using high-crowned lowers shaped like scoops or probes. *Callicebus* is a frugivore with close-packed, tall, slender lower incisors that closely match the derived pitheciine pattern.

In addition to hardness, fruit size can influence the natural selection of traits in frugivores of a given body size. That insight offers an explanation relating to the unique jaw morphology shared by titi and owl monkeys. The mandible is hinged to the cranium at the temporomandibular joint (TMJ). One of the properties of TMJ height is that it determines how much clearance space occurs at the front of the mandible, between the upper and lower incisors or canines, when the mouth is open. In *Aotus* and *Callicebus* the TMJ is significantly elevated in comparison to those of other platyrrhines of comparable size. This enables them to produce a relatively large gape at a comparatively small body size. Other novel details include an arrangement of muscle attachments in a biomechanically complementary way above and below the toothrow to power the system. In other words, the novel jaw configuration makes it possible for titis and owl monkeys to effectively use the anterior teeth in dealing with fruits that are, for them, large in size, which is an advantage in peeling tough husks and scraping fruit pulp that adheres to large seeds or pits, like a mango's. As to the morphology of the third genus of this clade, the fossil *Homunculus*, the information we have on jaw anatomy conforms closely to *Aotus*, the cranial anatomy matches favorably with *Callicebus*, and the more fragmentary dental anatomy is comparable to both.

Callicebus—The monogamous Titi Monkey may have the strongest pair bond among primates. The adult male and female are very attentive to one another and they coordinate certain behaviors, most obviously the early-morning, long-distance calls they emit as a duet. At roughly 900 g, 2 lb, these smallish monkeys are not the loudest platyrrhines in the forest, an honor reserved for the large, roaring Howler Monkey, but their voices produce a piercing, unmistakably rhythmic, staccato song (plate 11).

Callicebus noses are not as dramatic as the flaring nasal shapes seen among several other pitheciids. The reason for this relates to the bony structure underlying the external nose, which is conditioned by several elements. The whole picture offers a lesson in how anatomical parts are subtly put together and interconnected, sometimes without natural selection driving conspicuous distinguishing traits.

The morphology of the snout is unusually abbreviated in titi monkeys, modified in connection with another atypical trait, small, non-projecting, monomorphic canines (fig. 2.9). Small canine crowns mean small canine roots, so the midface is not enlarged as in pitheciines, where massive, barrel-shaped canine roots dramatically dominate the contour and broaden the face. Titis also lack the widely flaring nasal bones that roof the bony opening of the nose in the saki-uacari group, where the nasals tend to extend like an awning beyond its margins. In addition, the small bone (premaxillary) at the tip of the rostrum, in which the upper incisors are embedded, stops growing at a young age, contributing to facial foreshortening. All this makes for a small nasal aperture, a short snout, and a trim face, the foundation upon which the external nose is fashioned. Another effect, relating most of all to the small canines and small premaxilla, is osteological and geometric. The compactness at the front end of the face influences the shape of the palate and the whole dental arcade. Rather than the toothrow conforming to a U-shaped outline, squared-off at the front to accommodate a pair of large canines adjacent to a wide battery of incisors, as in pitheciines and to a lesser extent owl monkeys, where the front of the face is broad because the upper middle incisors are very wide, the palate and dental arcade of titis takes on a parabolic shape, which is also highly unusual in primates—*Homo* is the only other example.

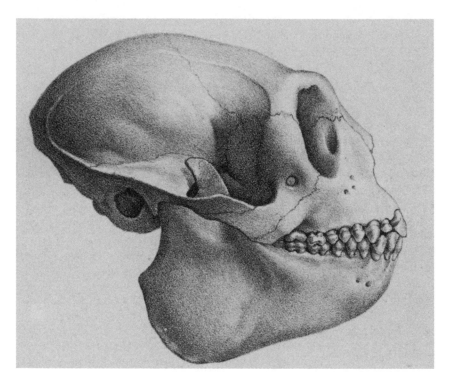

FIG. 2.9. Skull of *Callicebus*, the Titi Monkey. From Blainville (1839).

The titis' monomorphism extends to body measurements and pelage color, and they live in enduring, potentially lifelong, pair-bonded groups consisting of an adult male and female and immature offspring. The adults maintain social harmony by intensely grooming one another while resting or before sleep, by playing with infants and juveniles, and by providing frequent physical contact, sitting against each other in a cluster while at rest for long periods of time, and while sleeping. They normally engage in a unique form of interpersonal contact, tail-twining, as two or more individuals dangle their tails below the branch on which they are perched and coil them together into a single braid. This is a common, daily practice in titi monkeys. It is also seen in the other genus belonging to this subfamily, the Owl Monkey, though more rarely.

Titi monkeys are relatively quiet during their daily activities except for mornings, when they are one of the most easily heard platyrrhines. When they wake, titis utter booming dawn calls, alerting other groups who are within earshot of their presence. Their dawn calls can be heard half a mile away, through dense forest vegetation. The male and female adults often vocalize together, as a duet lasting several minutes. This collaboration exemplifies the cooperative nature of the mated pair. In all activities, neither one is dominant over the other and they are equally attentive to one another.

The species-and genus-level taxonomy of titis is a matter of some debate and it is a primary example of taxonomic inflation, as mentioned above. The Convention on International Trade in Endangered Species of Wild Fauna and Flora (CITES) recognizes 30 species of *Callicebus*. Other studies conducted over the course of 50 years during the latter decades of the 1900s and early 2000s identified three or six or twenty, with the number increasing as researchers adopted alternative definitions of the species concept and favored molecular methods to the exclusion of morphology and behavior. Ultimately, three genera have been recognized by some, but without any evidence that the groups identified are in any significant way distinguished in lifestyle. Geographically, *Callicebus* are widely distributed in a great variety of forested habitats, including the Atlantic Forest and vast regions of the Amazon. They are predominantly fruit eaters that consume significant amounts of seed in addition to fruit pulp. The degree to which they eat insects or young leaves as supplements depends on the species and the particular habitats in which they live.

One fossil, *Miocallicebus*, from Colombia, though still very poorly known, has been identified to fall securely within the *Callicebus* lineage, demonstrating that titis have existed for at least 12–14 million years. The second platyrrhine fossil ever discovered, Argentina's 16-million-year-old *Homunuclus*, found in the 1890s, may be a very close relative, or an actual member, of the Titi Monkey line, which would make the lineage even older.

Aotus—Aotus, monkeys of medium-sized build with a dark, leathery muzzle, are the only nocturnal platyrrhine and the only night-adapted anthropoid primate. It is no surprise that they have the largest eyes of any New World monkey or that their large orbits have a profound influence on the shape of the cranium (fig. 2.10). Nocturnality has even influenced the evolution of the Owl Monkey's pelage in the form of a heart-shaped pattern of white facial hair that highlights the large eyes (plate 12). In spite of this, anatomically, ecologically, and behaviorally, *Aotus* strongly resemble the closely related Titi Monkey, as the previous discussions of the homunculine subfamily and *Callicebus* have stressed. Some researchers think of the pair as diurnal and nocturnal versions of one another, and hypothesize that the Owl and Titi monkey lineages shared a unique common ancestor that was much like the titi, with the *Aotus* branch departing only subtly from ancestral homunculine patterns as it evolved nocturnal adaptations.

This perspective was a long time coming. *Callicebus* had been a subject of serious study in the lab since the 1960s and in the field since the early 1970s.

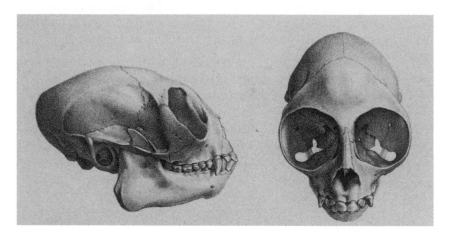

FIG. 2.10. Skull of *Aotus*, the Owl Monkey. From Blainville (1839).

But no one attempted to doggedly follow owl monkeys around at night in the forest for extended periods to collect data on their ecology and behavior until the primate biologist and conservationist Patricia Wright, equipped with only a flashlight, compass, and notebook, did it for several months even before she was a graduate student. In the 1980s, as the core of her thesis dissertation research, she developed a long-term project that simultaneously compared *Callicebus* and *Aotus* where they lived together, sympatrically, in the remote Manu National Park in Peru, to ask: What makes them different?

The nocturnal lifestyle of *Aotus* is an evolutionary strategy of great significance because it is the opposite activity rhythm from the anthropoid norm, the diurnal cycle, which was central to the adaptive breakthrough of anthropoids 50 million years ago or more. The anthropoids, among the most visually adapted and visually dependent mammals, have eyes and a neural processing system that are super-sensitive to imaging in daylight. Vision is a rich information input modality, and daylight provides a far more revealing environment than nightlight. The large size of owl monkey eyes is compensation for an eyeball that is structurally unsuited for sharp night vision, as with all other anthropoids and even the nocturnal tarsiers in some ways. All of them lack the histological features that make night vision a primary adaptation of the wet-nosed, strepsirhine primates and many other nocturnal mammals. As discussed further in chapter 8, in addition to having enhanced night vision, owl monkeys have other sensory systems that are augmented appropriately to life with little light; hearing, touch, and smell are used effectively as modes of communication. Being nocturnally adapted does not mean the Owl Monkeys' modus operandi is to go about life in complete darkness, ignorant of the light. To the contrary. They are active at dawn and dusk and actually rely on moonlight at night, and it influences their quadrupedal locomotor behavior and the extent to which they travel from a nest tree, for example.

Why did the *Aotus* lineage become nocturnal? We presume there must have been ecological benefits. Owl monkeys weigh about 900 g, 2 lb, which puts them in the same size class as the diurnal Titi Monkey and the Squirrel Monkey, genera with which *Aotus* is often sympatric. By being active at night, owl monkeys have less competition from these other monkeys that would potentially exploit the same resources, including food, space to operate, and shelter. By being nocturnal, *Aotus* also avoid the relatively large array of diurnal raptors that habitually prey on platyrrhines, including eagles, hawks, and falcons. Nocturnal predators, such as owls and snakes, are still a threat, and *Aotus* are vulnerable during the day at their sleeping sites. But to their advantage, *Aotus* are well camouflaged by a dull outward appearance, with coloration that blends into the green and brownish hues of the canopy and understory, and by being vocally quiet monkeys. Their loudest calls, given expressly during territorial encounters by both adults together, are muted, carrying for only about 50 yards, far less than the dawn calls of *Callicebus* and no match for Howler Monkey roars.

The evolution of nocturnality of Owl Monkeys has been the root of their success, which is evident in several ways. They are found coast to coast in forests along the Amazon in South America, and north-south from Panama to northern Argentina. Owl Monkeys are able to range so widely because they are versatile, mostly fruit eaters that balance their diets with either leaves or insects, depending on locality and habitat type. The southern limits of their range also illustrate how adaptable *Aotus* are in adjusting the day-night activity cycle. Not all the species are strictly nocturnal. In northern Argentina, in the most seasonal and harsh environments in which they live, the local species, *Aotus azarae*, limits its strictly nocturnal activity to warmer months. At other, cooler times, its most active periods occur during the day and at night. How many species exist is debatable. CITES lists nine species, others now identify eleven, while earlier studies estimated there is only one.

Another measure of the Owl Monkey's success is the longevity of the *Aotus* lineage in geological time. It stretches back about 20 million years, as shown by the fossil cranium belonging to the genus *Tremacebus*, further discussed below. The fossil has relatively large orbits, and the size of its olfactory bulb, which can be reconstructed from the impressions left by the bulb on the inside of the braincase, indicates an enhanced sense of smell, as is found in extant *Aotus*. From younger deposits in Colombia, dating 12–14 million years, there

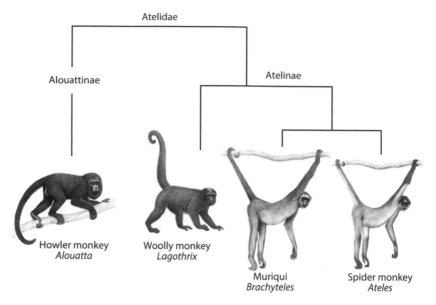

FIG. 2.11. Cladistic relationships and classification of living atelid genera, not to scale. Images courtesy of Stephen Nash.

are fossils that are actually classified in the same genus as the living animals but as a distinct species, *Aotus dindensis*.

Prehensile-tailed frugivore-folivores: Family Atelidae

More than a dozen atelid species occupy the large New World monkey niche, with a considerable range in body mass (fig. 2.2). A number of them are sexually dimorphic, in such features as body size, pelage, and development of the canine teeth (table 2.1). Consequently, the body size range among the living atelids, in which the biggest is twice the mass of the smallest, results from two phenomena: there are size differences among species and between conspecific males and females. Body weight can be less than 5 kg, 11 lb, in some small female howler monkeys, and about 11 kg, 24 lb, in large male Muriqui and the Spider Monkeys.

The fully prehensile tail, which is connected to foraging and feeding behavior and even social interactions, is the most striking adaptation shared by all atelids (fig. 2.11). The form-function complex of the prehensile tail involves a wide range of novel features not seen in other primates that lack one, made possible by evolutionary modification of bones, muscles, skin, nerves, and a renovation of brain parts. It involves the skin in the form of a long hairless strip on the underside of the tail, beginning two-thirds of the way down and ending at the tip, which exposes a soft, textured belt-like pad. With a fingertip-like frictional surface, this specialized skin enhances the tail's tactile sensitivity and grasping ability. Seeing a large atelid in the field against the leafy backdrop of the canopy, it is hard to mistake it for any other platyrrhine, especially when it moves. Atelids have relatively large, barrel-shaped chests, rotund bellies, and very long tails (table 2.2). Two of the atelines, the Spider Monkey genus and Muriqui, are especially conspicuous, wheeling through the canopy as if they have five long limbs: legs and forearms plus a tail.

The four genera of this family range in locomotor style from a slow, sauntering quadrupedal gait atop the branches to a fast-moving, two-armed style of gliding beneath them, to swinging, herky-jerky ricochetal moves through the myriad obstacles presented by the canopy's branches, boughs, and flexible twigs The differences in motor style are a manifestation of contrasting ecological strategies that have evolved in the two atelid subfamilies. One is an energy-minimizer, the almost slothlike alouattine Howler Monkeys, while the three ateline genera are dynamic energy-maximizers, especially the Muriquis and Spider Monkeys. The atelid tail always takes part in posture and locomotion. Even the least energetic genus, the Howler Monkey, will wrap its tail around a branch or tree trunk to support its body weight while walking quadrupedally or while hanging head-down to free both hands for feeding purposes.

Subfamily Atelinae

Ateles, Spider Monkey; *Brachyteles*, Muriqui;
Lagothrix, Woolly Monkey

One of the three ateline genera, the Woolly Monkey, *Lagothrix*, is a model of ancestral atelid anatomy and behavior. The other two, the Spider Monkey, *Ateles*, and the Muriqui, *Brachyteles*, are more divergent. This is especially evident when considering the postcranial skeleton. In this way their structural variations resemble the morphocline phenomenon, the graded series of form that we encountered in discussing the crania of the Saki Monkey, Bearded Saki and Uacari, for example.

A morphocline distribution of traits among taxa that make up a clade is a common evolutionary motif. Identifying these patterns helps unravel the evolutionary trajectories taken by related animals as their ecological niches became sorted out over evolutionary time. The atelines' striking, specialized skeletal attributes reflect the evolution of acrobatic, tail-assisted suspensory locomotion in its utmost expression among New World monkeys, and are unlike all other primates. The biological roles of their hypermobile, hypermaneuverable prehensile tails are linked to foraging, feeding, and social behaviors, and provide important clues to the evolution of the group.

There is one anatomical curiosity shared by the Spider Monkey and Muriqui, immediately noticed by early naturalists, that seems like a skeletal anomaly that has never been satisfactorily explained. Both are four-fingered. The thumb is either absent altogether or it appears as a stunted vestige in the palm of the hand. That is why the Spider Monkey genus is called *Ateles*. Named in 1806, it means "imperfect" in Greek. The same idea was used again in 1823, for the Muriqui, which used to be called the Woolly Spider Monkey, by combining the term with another Greek word, *brachy*, meaning "short." Thus both the formal and common names of these animals reflect anatomical features: long limbs and tails and thumbless hands.

Why the thumb became excessively small and/or functionless in *Ateles*, *Brachyteles*, and Old World colobine monkeys via parallel evolution, is not well understood. To "get it out of the way" so it does not interfere with hand-holds, especially in a rapid arm-swinging primate such as the Spider Monkey or Muriqui, has been suggested but it does not seem like an adequate reason. Gibbons and siamangs, the Old World lesser apes, for example, the most advanced primate brachiators, have hands with a deep cleft that actually offsets the thumb in a unique way that augments its clasping potential. That does not mean the nuisance explanation is irrelevant; both modifications hint at the idea that "doing something" evolutionarily with the first digit may benefit hand use among some arboreal primates of larger size. Of course, there may be no uniform adaptive explanation. It may boil down to a chance genetic effect in the monkeys.

The fact that the full anatomical and behavioral expression of the advanced ateline locomotor complex occurs only in the most acrobatic and thumbless *Ateles* and *Brachyteles* indicates that these genera shared an exclusive common ancestor. The extensive, detailed resemblances in which they depart from the patterns in other atelids and platyrrhines involve body proportions, hands, forelimbs, upper arms, elbows, shoulder blades, clavicle, rib cage, vertebral column, pelvis, ankles, and hyperbendable tails. This phylogenetic interpretation differs from molecular analyses, which suggest the Muriqui is more closely related to the Woolly Monkey, *Lagothrix*, and implies that the resemblances shared by *Ateles* and *Brachyteles* evolved through a process that involved an extensive amount of parallel evolution across all these subsystems of the skeleton.

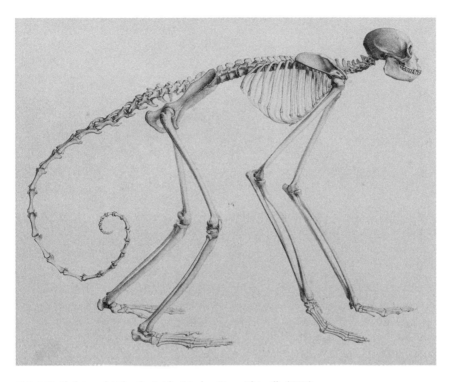

FIG. 2.12. Skeleton of *Ateles*, the Spider Monkey. From Blainville (1839).

Cranially and dentally *Ateles* and *Brachyteles* are quite distinct, as is *Lago-thrix*. Spider Monkeys have evolved a dentition suited to ripe fruit-eating while the Muriqui have convergently evolved a Howler Monkey-style of leaf-eating molars and incisors. The Woolly Monkeys, in contrast, have evolved a frugivo-rous dentition well suited to harder fleshy fruits, and seed eating.

Ateles—No other platyrrhine could be mistaken for the Spider Monkey when it is in motion (plate 13). Large in size, with many weighing more than 9 kg, 20 lb, and with a relatively small, round head, rotund belly, long, thin arms, and long, sinewy tail, they speedily power their way through the forest canopy in a brachiating style (fig. 2.12). With two arms and a tail grabbing holds on overhead branches and twigs repeatedly, they appear to fly through the treetops, undaunted by the challenge of woody and leafy barriers, twigs that bend and break carrying their weight, gaps that separate the crowns of trees, while they swoop along paths from bough to bough. Their skeletons are designed to dynamically suspend and hurl their body weight from a fixed point above, getting a boost from the pull of gravity like a pendulum. They are climbers and clamberers, with a long reach, supple forelimbs and hips, long and strong gripping feet, and mobile ankles. Some of the routes they take to navigate their home ranges are habitual, including stout branches appropriate for their heavy weight, but improvised itineraries are inevitable and often occur.

Ateles are built to roam widely and quickly through the day in search of patches of trees that provide quick nutrition in the form of energy-rich ripe fruit, the main component of their diet. Ripe fruit can be masticated with comparatively small cheek teeth, jaws, and chewing muscles, which is why their faces are delicately built and proportionately small in comparison with the braincase. Adaptations to diet and locomotion extend to social behavior. Their large social groups, for example, may consist of 40 or more individuals that split up into small foraging parties. These subgroups may act as a unit for several days as the animals roam about their home range in search of trees with fruit ready to eat. Large foraging groups are a disadvantage here since ripe fruits tend to be scattered about in the forest in small clusters that may mature at different times.

Spider Monkeys are essentially monomorphic in body size and the sexes are not differentially colored. This makes it difficult for observers to distinguish males and females at a distance. There is, however, one anatomical feature that separates them. Females have a greatly elongated, pendulous clitoris, and it cannot be mistaken for a male's flaccid penis, which is retracted in its neutral state. Evidence from the field suggests the specialized *Ateles* clitoris is important in two ways. It is used as a visual aid that enables males to identify females at a distance, and it is used in olfactory communication related to mating, as males appear to sniff and handle the clitoris to assess a female's reproductive status.

There are four species of *Ateles*, three living in South America and one in Central America. The latter form is the most northerly distributed Neotropical primate, ranging into southern Mexico. Habitats vary, from wet, evergreen rainforest to dry, deciduous forest, where trees lose their leaves seasonally. Spider monkeys prefer undisturbed forest, where most of their time is spent in the middle story or in emergent trees that overtop the continuous canopy.

The Brazilian fossil *Caipora* may be an extinct representative of the Spider Monkey and/or the Muriqui lineage, but its age cannot be determined. It was collected in a dry cave without associated, datable rocks or mineral deposits. It is assumed to be of Pleistocene age, less than 2.5 million years old, perhaps much less. The locality is situated well outside the geographic range of extant Spider Monkey species but close to the Atlantic Forest where muriquis now live, a reminder that the environments of South America are constantly in flux, and that the current distribution of platyrrhines is a product of today's climatic regime.

Brachyteles—The Spider Monkey's sister genus, the Muriqui, is built precisely along the same lines (plate 14; fig. 2.12). Both are large, long-limbed, and thumbless with powerful, sensitive, ropelike prehensile tails they control to coil, twist, and spiral. The only skeletal differences of note are that spiders are slender and muriquis are stocky, with more robust bones. In the wild, both genera exhibit the same lithe, fluid acrobatic style, with the tail acting as a grasping, fifth limb. They even use their bodies similarly during interactive social performances. Without taking color and geography into account, it would be easy to mistake a group of four or five adult muriquis for a similar, though perhaps smaller, group of spiders engaged in a multi-individual embrace, all hanging by their tails, the arms of each one wrapped around another, emitting high-pitched, squealing noises that sound like laughter.

The third ateline genus, *Lagothrix*, the Woolly Monkey, behaves this way, too, though with less intensity, according to field biologist Marilyn Norconk. She also noted that, behaviorally, *Lagothrix* and *Brachyteles* are "a matter of extremes with the Muriqui at the end of the highly social continuum of atelids." This description is comparable to the morphoclines of craniodental and skeletal morphologies. This complex set of social behaviors is an example of a fixed-action pattern, an instinctive routine that occurs under specific conditions. Since it is manifest in all three ateline genera, it is likely to have been part of the genetic and behavioral makeup of their ancestral stock. In other words, these gestures seen today in modern atelines are at least as old as the ateline clade, which is probably more than 10 million years old, according to the molecular analyses.

There is probably only one species of the blond Muriqui, though some researchers identify two forms as either species or subspecies based on differentiating traits such as canine length in males, pigmentation of the face, coloration of offspring pelage, genetic and behavioral differences. As mentioned above, until the native name Muriqui became widely adopted in the 1980s, *Brachyteles* was aptly known in the English-language literature as the Woolly Spider Monkey because its dense fur compares favorably with the Woolly Monkey's pelage, both of which are unlike the silken, smooth coat of the Spider Monkey.

Brachyteles and *Ateles* are a divergent duality, two lines each with its own evolutionary trajectory but founded on what appears to be an identical body plan. They are distinguished by their food preferences, which reflect the

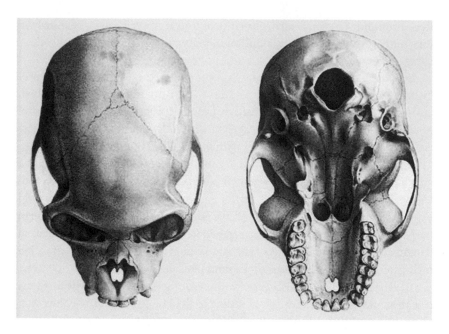

FIG. 2.13. Cranium of *Brachyteles*, the Muriqui. From Blainville (1839).

ecology of the habitats they occupy and the communities in which they live. Anatomically, this plays out most clearly in their heads, which are quite different. *Brachyteles* have a relatively larger head, with larger jaws, cheek teeth, and chewing muscles, consistent with a masticatory system made for leaf eating (fig. 2.13). Muriqui have cheek teeth designed to shred and tear rather than crush and grind, the spider's métier, and their incisors are quite small, another pattern typical of leaf eaters. The easily chewed, ripe fruits favored by *Ateles* can be eaten with a less robust system, reflecting the soft, three-dimensional foods that they chew, and broad, shovel-like incisors for stripping away pliable fruit rinds.

The Muriqui diet evolved in the Atlantic Forest of eastern Brazil, in a habitat where ripe fruits are seasonally scarce and an alternative staple appropriate for a large platyrrhine is required. Spider Monkeys live in wetter, botanically richer habitats, where ripe fruits are more available but not overly utilized by other primates. The fascinating undercurrent here is that the *Brachyteles* masticatory system is clearly built to process a low-quality, energy-poor food, while their body and brain are built to take advantage of, or require, high-quality, energy-rich foods. As with the explanation for their resemblances, predicated on a close phylogenetic relationship, the best way to make sense of this discordance is that the Muriqui shifted away from an ancestral, unspecialized fruit-based diet as it evolved in more limiting habitats outside Amazonia, in the environs of the formative Atlantic Forest.

In addition to the locomotor and behavioral features retained by *Brachyteles* and *Ateles* as descendants from a common ancestor, there are other highly interesting traits they share, with ecological and social consequences, such as body size monomorphism, females with a large clitoris, and a fission-fusion group structure that subdivides into smaller units or foraging parties, sometimes for days at a time. There are also novel Muriqui behaviors, like turn-taking mating practices, when sexually receptive females mate with multiple males in a single session; the males line up to wait for their opportunity to copulate.

Lagothrix—Anatomically, the Woolly Monkey is the least modified of the three ateline genera. A variety of their features are models of hypothetical ancestral patterns that changed adaptively in the line leading to *Ateles* and *Brachyteles*. These include characteristics of both the cranium and the postcranial skeleton. At a glance, Woolly Monkeys are easily distinguished from the other two genera by their dense, dark brown fur, the large head, and the more deliberate way they move about the trees quadrupedally, even when using the tail for overhead support (plate 15).

The locomotor behavior of the Woolly Monkey is less acrobatic when compared with the dynamic Spider Monkey and the Muriqui, and their skeletons lack the derived specializations that make possible the latter pair's rapid, stylized, arm-and-tail swinging locomotion. Woollys are more robustly built, the forelimbs are shorter proportionately, the shoulders are not as mobile to facilitate climbing, clambering, and arm-swinging movements, and they have a fully developed thumb. They are sturdy, climbing quadrupeds. Though the tail is fully prehensile and equipped with the same friction skin on the underside, it is often used simply as a stabilizing aid during quadrupedalism. The *Lagothrix* cranium is also more robustly built than an *Ateles* and has a longer face than *Brachyteles*. Its brain is relatively smaller, another indication that this genus is a good model for an ancestral ateline. Woolly Monkeys have very large incisors and cheek teeth crowned with puffy cusps, unlike Spider Monkeys and Muriqui. Such rounded cusps are typical when there are selective advantages to resisting heavy tooth wear during crushing and grinding actions, as when eating harder fruits or seeds. *Lagothrix* are sexually dimorphic, unlike the sexually monomorphic *Ateles* and *Brachyteles*. Males are larger than females, more robustly built in skull and skeleton, and they have larger canine teeth.

There are two species of *Lagothrix*, both living in the western Amazon region. Neither is widely distributed. The critically endangered Yellow-tailed

Woolly Monkey, *Lagothrix flavicauda*, sometimes called genus *Oreonax*, an old genus name that was briefly resurrected in the early 2000s, lives in a few parcels of land along the eastern slope of the Andes. In 2012 it was estimated by IUCN that only 1,000 individuals survive in the wild. More recent surveys have expanded the known geographic range of the Yellow-tailed Woolly Monkey, and they suggest several thousand individuals exist, but the long-term prospects for the species are no less dire.

Subfamily Alouattinae

Alouatta, Howler Monkey

There is only one living genus of this subfamily, which means the attributes of *Alouatta* characterize the clade. Divergent adaptations for communication, diet, and locomotion make for the distinct morphological oddity of the genus—one of the most divergent among primates—and illustrate how crania are shaped to balance complex anatomical needs. The highly unusual, very large skull of the Howler Monkey comprises an adaptive package that incorporates an enormous sound-production device situated in the neck, while also supporting the masticatory apparatus and serving as the head end of the skeleton, which is naturally integrated with the rest of the locomotor system. Driven largely by the evolution of a greatly enlarged hyoid bone and thyroid cartilage in the throat, the anatomy of the cranium and mandible was reshaped to accommodate these varied needs (fig. 2.14).

The large dentition and chewing muscles evolved in response to the mechanical difficulties of masticating leaves. This evolution, in turn, selected for a reduction in the size of the brain and its braincase, because nutritionally poor leaves cannot sustain a large, energy-hungry brain. The massive, heavy head is dominated by a very large protruding face, with consequences for head carriage, how it is attached to and supported by the vertebral column, and how it works with the lungs that power sound production. Behind the head, the postcranial skeleton is also designed unusually to support a kind of crouched, quadrupedal style of locomotion, with elbows and knees bent, which also requires adjustments in the cranium to resolve head carriage. Howler locomotion incorporates slow clambering and climbing, but not the arm-plus-tail-swinging pattern seen in other atelids. Nor is the fully prehensile tail used as dynamically or as often in locomotion, though it is recruited in postural modes, especially during feeding. These attributes are part of the overall Howler Monkey ecological strategy of minimizing the expenditure of energy.

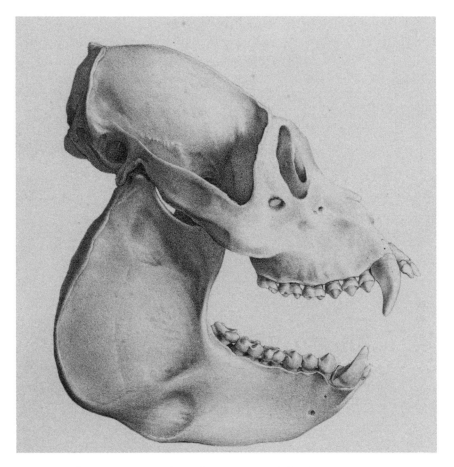

FIG. 2.14. Skull of *Alouatta*, the Howler Monkey. From Blainville (1839).

Alouatta—Howler Monkeys are the great growling beasts of the Neotropical rainforest, the loudest and, not coincidentally, the most lethargically mannered platyrrhine, an association of behaviors connected with food above all else (plate 16). Their roars are deeper than a lion's and, in spite of the colossal body size difference, *Alouatta* calls are nearly as deep as an elephant's. At about 150 decibels their calls are comparable to the ear-shattering noise of a jet engine, according to Purdue University's College of Engineering. The sound can be heard miles away.

Together with other structures of the larynx, the system supports lengthened vocal folds (cords) for generating low-pitched sound waves as they vibrate. This provides for novel acoustic properties. Shaped like a large, hollow pouch, the howler hyoid is permanently inflated and open at the top. Hardened as sound-reflecting bone rather than the more absorbent cartilage that typically forms the mammalian hyoid, the empty space is a resonating chamber, working much like the hollow body of an acoustic guitar to amplify the intensity of sound waves that come as a gush of air rushes across the opening. The animal's entire body is involved in sound production. When howlers make loud calls they adopt a posture that helps them generate a powerful, unimpeded airstream through the windpipe driven by the diaphragm, the pumping action of which can often be seen as contractions of the abdomen. While in a quadrupedal stance, they drop the shoulders and extend the head upward while howling, taking advantage of the natural set of the skull on the vertebral column which is also quite distinctive, joined to the spine in a cantilevered fashion, at the rear, like a dog's skull, rather than toward the underside as in other atelids.

For the loud-calling howlers there is significant variation in the size of the hyoid between males and females of the same species, and among the several species. This complements variations in body size, as *Alouatta* is also sexually dimorphic in body mass and robusticity. Males can be 30%–75% heavier than females, and the sexes can be colored quite differently in some species. Why do they roar? One hypothesis is that roaring is a form of male swagger, a common predilection in sexually dimorphic species. It is thought to be a form of sociosexual advertisement involved in making breeding choices, giving females the opportunity to assess the vigor of males and their potential to help maintain territory and protect offspring. But its role within a group may be more complicated than that because howling typically turns into a group chorus once it is initiated by a male. These social groups may consist of 5 to 15 individuals, with one or more adult males, several adult females, and their offspring. The calls are usually heard in the early morning, but in places where the density of howler troops is high the calls may be heard all day long, and even during the night.

Other mammals are also prone to evolving sexually dimorphic traits that are exaggerated in males, such as antlers in moose and manes in lions. Living in an arboreal environment where visual signals cannot be effective over long distances, *Alouatta* have evolved an acoustic alternative, using very low frequency sound that travels well in an environment riddled with physical obstructions, so males can broadcast how fit they are and keep non-resident males at bay. Since all howler groups have their loud-calling males as well as group choruses, and it is advantageous to maintain small territories as part of their energy-minimizing strategy, every howler troop within earshot of another knows the locations of its potential competitors. Loud calling thus becomes an effective, safe way to publicize one's presence and safeguard food and other resources without conflict. The males safeguard females and females benefit by being protected from the possibility that marauding males from other groups will infiltrate and kill their offspring. Howlers are known to be infanticidal, a topic further discussed in chapter 8.

All Howler Monkey species are committed leaf eaters but they will eat fruits opportunistically, if not preferentially, when their favorites, such as figs, are available. Leaf eating gives them advantages because all forests produce leaves in abundance and there is little need to search widely for feeding sites, so home ranges can be relatively small and group sizes modest. This means *Alouatta* have environmental flexibility which allows them to live in a very wide range of habitats, including areas that are degraded, small forest fragments, and others where few large, frugivorous platyrrhines can exist.

Having the ability to survive on ubiquitous, nutritionally poor food, howlers are accomplished pioneers in landscapes outside of Amazonia, even expanding through regions where only narrow bands of forest exist alongside watercourses. As a result, they are one of the most widespread platyrrhine genera, with a geographic range that reaches from southern Mexico to northern Argentina. Over this large expanse, the taxonomy of the genus is uncertain. Six to ten or more species have been recognized, including two that live only in Central America.

How several of the unique features of *Alouatta* and this subfamily evolved over time is revealed by several very important fossils. There are four fossil genera assignable to this subfamily, two of which point to the long history experienced by this group. The fossil *Stirtonia* demonstrates unequivocally that a form of the Howler Monkey lineage existed in the evolving Amazon basin 12–14 million years ago, living within a community of primates that resembled, and has been continuous with the modern platyrrhine community. This point is reinforced by the contemporary presence at the same site of a fossil Owl Monkey, *Aotus dindensis*. In fact, it has been suggested that *Stirtonia* should be reclassified as *Alouatta* because its known parts, which are limited to the dentition, are a match with the Howler Monkey in the minutest anatomical details. A much younger fossil from a cave in Brazil that cannot be well dated but is presumed Pleistocene in age (less than 2.6 million years old) is an almost complete skeleton that has a variety of the special features of howler-like locomotion and the design of the head, including the small brain. But the teeth of this fossil monkey, *Cartelles*, are definitely not those of a leaf eater. They suggest that a fruit-eating diet which emphasized seed eating may have preceded folivory, preadapting the ancestors of howlers to adopt a leafy diet when that became advantageous. Documenting shifts like this, from one dietary category to another, can be achieved only when closely related fossils show an intermediate condition, which appears to be the case with *Cartelles*.

Another alouattine fossil from Cuba, *Paralouatta*, is known by the skull, dentition, and parts of the skeleton. One specimen of this genus may be about 16 million years old, but that age estimate must be considered provisional. It is based on only an ankle bone, which is difficult to assign taxonomically, and we also have no particular reason to assign it to the genus *Paralouatta*, first diagnosed by specimens that are many millions of years younger—other than the fact that it is currently the only primate known to have existed on Cuba. Another fossil is *Solimoea*, represented by a single, remarkably *Alouatta*-like lower molar that is 7–11 million years old, found in Brazil.

CHAPTER 3

WHAT'S IN A NAME?

The word for Howler Monkey differs from place to place: Barbado (southeast Brazil), Bugio (the vast Brazilian interior corridor of shrubland, tropical savanna, and wetland), Guariba (Brazilian Amazon), Mono aullador (the Amazon, Argentina, Bolivia, Mesoamerica), Araguato (Venezuela), Baboon (the Guianas), and the pervasive Spanish terms Mono and Macaco. This is not surprising, as humans need to label things. Yet it means that the potential for misunderstanding would be acute if scientists were to communicate using the common, folk names traditionally given to animals and plants by local people, and even by scientists writing in their own native languages. More examples: Singe hurleurs (French), Scimmiae reggenti (Italian), and Brüllaffen (German). Taxonomy evolved as the solution to this problem by providing a universal language to name and classify. Its intent is to promote accurate fluency, preserve quality, and improve the transmission of technical information about animals, plants, planets, rocks, stars, chemicals, and all material objects within the purview of science. Because taxonomic terms, and classifications in particular, are constructs developed explicitly for the purpose of doing science, they have evolved to become more than identifying labels. They also embody scientific hypotheses, and this makes taxonomy subject to change as knowledge advances, making it imperfect and imprecise at times.

Avoiding a taxonomic Tower of Babel became fundamentally important as the world opened up to European explorers during the Age of Discovery, which began in the 15th century. The bounty brought back from far-flung ocean voyages included new foods, cultural artifacts, and remains of exotic flora and fauna, often embellished with the inaccurate descriptions of travelers' tales and alien names given in a mangled foreign dialect. The mounting collections obtained by expeditions, intentional surveys, and random discoveries encouraged a lasting interest in natural history and classification, and a competitive attitude among naturalists eager to publish their findings. That led to a proliferation of biological names. Many of these were valid, but countless others were duplicative, or even based on imaginary sightings or myths.

Carolus Linnaeus, the 18th-century Swedish botanist, is credited with standardizing the rudiments of a biological taxonomic language, the system called binomial (or binominal) nomenclature, to scientifically organize the rapidly growing numbers of newly found plant and animal species being identified and appearing in the literature. His system became the basis of the International Code of Zoological Nomenclature, a set of rules governing how animals should be named and classified that continues to be followed voluntarily by the scientific community.

The most obvious need for this set of rules in biology involves genus and species names, the critical reference terms used to record, describe, and measure the biodiversity of life on earth. The first rule is that each species must be signified linguistically by a two-name, binomial pairing, preferably fashioned from Latin words and written in italics, such as the zoological name for humans, *Homo sapiens*. The first part of this set is the genus name, *Homo*. The second, *sapiens*, is the species term but, technically, species names nominally do not exist outside the combination. So we, as a species, are formally known as *Homo sapiens*, meaning wise or knowing man, but without an official common name. There are no rules for the usage of common names in the International Code.

The binomial method is analogous to the traditions and laws we use in many parts of the world to form our personal names, which are combinations of family names, or surnames, and given names. In this analogy, surnames are equivalent to genus names. They are the umbrella terms. Your given name has only limited standing as an identifier without being attached to your family name. And, just as your siblings can share the same family name, a group of species can share the same genus name to signify that they are closely related. To extend the analogy further, while there are many stylistic variations in how different cultures form personal names—in number of terms, with surnames following matrilines or patrilines, etc.—the genus-species sequence is comparable to the last-name-first formality followed in many countries.

The early evolution of the binomial system no doubt drew upon the new norms that were becoming widespread in European society at the time, including the formalization of people's names for recordkeeping purposes. Before that, there had been no consistent use of official family names in a number of countries, as the traditions and laws were influenced by shifting cultures. Individuals often came to be known informally by modified given names that included terms that provided context, to evoke a place of origin, an avocation, a physical characteristic, or parentage. Leon of the Lake; Sam the Smithy; Steven the Small One; James, Son of Jack—the added descriptors were adapted to become now-familiar last names: Leon Lake, Sam Smith, Steven Small,

James Jackson. Similarly, genus and species names are also often devised in reference to an animal's place of origin, specific characteristics, or associations with another biological entity.

Linnaeus, working a century before Darwin, was not an evolutionist. He did not have the theory of evolution in mind as he classified, yet his classifications have endured because he operated with a modern ethos, adhering to method and working objectively. Linnaeus classified humans using anatomy, the same criterion he applied to nonhuman primates and all other animals and plants. He arranged his most influential book, the 10th (1758) edition of *Systema Naturae*, around our own zoological-evolutionary group by introducing the title Order Primates, from the Latin root *prima*, meaning the first, on the opening page of the section on mammals. It started off with the classification of *Homo* as the first heading. Immediately following on the same line was the declaration "*nosce Te ipsum* (*)," meaning know thyself, with the asterisk leading to more than three pages describing the genus *Homo*. Latin was at that time the language of science and scholarship, so it is no surprise that a directive to Latinize names became one of the fundamental rules devised by Linnaeus and his followers to manage the construction of a taxonomic directory, and its dissemination through publication.

Another core idea introduced later is that a tangible voucher, a type specimen, must be attached to each species name and preserved to permit inspection. This type specimen guards against misidentifications. The type specimen serves as the exemplar of the species itself, physical proof of its existence and, ideally, a demonstration of attributes that justify recognizing its uniqueness in nature. It became customary for the type specimens, and others, to be housed in museums. By Linnaeus's time, natural history museums were already well-established research centers in several major European cities and many became cultural institutions of national pride, presenting exhibitions that were open to the public. The museums were growing repositories of preserved, exotic samples of life often collected on long ocean voyages that had naturalists on board to gather and identify animals that were interesting, and plants that were possibly valuable as commodities. That was Darwin's job—what today we might call a baccalaureate internship—during his five-year voyage around the world, 1831–1836, on board a British surveying ship, the HMS *Beagle*. Many of his materials went to the Natural History Museum, in London. Linnaeus's personal collection of more than 40,000 specimens is now cared for by the Linnaean Society of London.

As to *Homo sapiens*, Linnaeus did not designate a type specimen because it was not a requirement at the time he worked. It was only in 1959 that the British botanist W. T. Stearn formally proposed one in the literature. In this case, the specimen is not likely to ever be inspected, for it is Carolus Linnaeus himself, buried in Uppsala Cathedral, Uppsala, Sweden.

A new fossil gets a title

It was rumored in 2005 that a fossil primate specimen had been found in one of the rocks that fell from the beachfront palisades of the Rio Gallegos in southern Argentina and I went to examine it, together with my Argentine colleagues, paleontologist Marcelo Tejedor and Adan Tauber, a geologist. The site where the material originated, which had been known for its fossil troves for more than a century and continues to be productive, is stunning, with cliff faces exposed for many kilometers, rising dozens of meters above the beach on the north side of the narrow river which flows directly into the Atlantic Ocean (fig. 3.1). Tidal rhythm determines when fossil collectors can work the site. Only at low tide can the strip of beach be walked and the countless tumbled-down boulders inspected after they have crashed onto the beach and broken apart, revealing the fossils of land vertebrates cemented inside them. These petrified bones, sometimes complete skeletons, had been buried in layers of sediment that had solidified over the ages.

What I saw when I was handed one small piece of rock was a flattened surface where several dark grey knobs of pearly enamel, only a few millimeters in size, were exposed. The bumps were arranged in a pattern that could mean only one thing: they were cusps of molar teeth. As platyrrhine specialists, Marcelo and I could immediately confirm that these cusps once belonged to a long-extinct New World monkey; teeth are one of the most valuable assets for mammal paleontology given their diagnostic and adaptive importance. The rock had been given to John and Monica Blake, the owners of the property where the fossil-rich cliffs were located. They had kept it in a curio cabinet along with a collection of other artifacts found on their land.

An appointment to have the rock X-rayed was arranged for the next day by Maria Palacios, who accompanied us. Maria was a paleontology curator at the local museum, Museo Regional Provincial "Padre Manuel Jesús Molina," in the city of Rio Gallegos, several kilometers away across the river. It was Maria who originally facilitated our meeting. The Blakes graciously agreed to donate the rock to the museum, and they handed it to us. We left after tea that evening with a rock-fossil wrapped in newspaper and bubble wrap, nestled inside a brown paper shopping bag.

We hoped that the X-ray on the following day would show all that was encased in the rock, which was the size of a fist, large enough to contain a whole cranium of a medium-sized New World monkey. But Marcelo, Adan, and I were too anxious to wait until the radiology lab opened in the morning. That

FIG. 3.1. Fossiliferous bluffs at Killik Aike Norte, Argentina (top), the palate of the fossil *Killikaike blakei* encased in the rock and fully cleaned (bottom right).

night, we took out the tools we had at hand in our hotel room: toothbrush, toothpick, a small magnifying optical loop, an ashtray full of water, and some paper towels. We took turns, wetting, picking at and brushing the rock. It dissolved easily around the teeth to reveal essentially a whole, undamaged palate framed by an almost complete dental arcade of pristine teeth from the canines to the third molars (fig. 3.1). We knew immediately that this was the best-preserved fossil primate dentition ever found in Argentina. But how much more of the cranium was buried inside the rock, and to what monkey did it belong? Our best guess was that it could be *Homunculus patagonicus*, an iconic primate previously known from the region but only by very poor fossil material, nothing like what we already knew our specimen to be.

Early the next morning the three of us arrived at the storefront radiology establishment and took a seat on a bench in the waiting room alongside the patients. The receptionist called out for us to sign in. We walked up to her desk, paper bag in hand. Adan explained that Maria had set up our appointment. Yes, she said, but registration is necessary: "Patient's name?" she asked. There was a long pause as we looked at one another, sleep-deprived and giddy from what we had discovered the night before. I, the American who could afford to look foolish in a foreign-language conversation, piped up: *"Homunculus patagonicus."* She began to fill in the form as we held out the bag. It was not about us, we tried to explain, it was about the contents of the bag. She made a note and asked, "Age?" Another pause. We spoke amongst ourselves in English and Spanish, and ventured a guess: "About 14 million years." She caught on, smiled, and went to call the radiologist. We were shown into the lab and began to tell him the backstory, who we were, why Maria sent us there, what our objective was, what we needed to know. The radiologist was glad to participate and eager to test the capabilities of the new, recently installed X-ray machine to see how well it penetrated material that was far denser than human tissue.

We were soon able to see the X-ray images on a light box. The first few views were disappointing, nothing but tiny mineral flecks that seemed to float in the stone. Then we began to make out what looked like segments of bone surrounded by rock. Gradually moving through the images one by one revealed more anatomy, in cross-sectional views, beginning with a dark arc that was unmistakably a piece of skull. Then another and another. It was a bit confusing at first. We were accustomed to holding a fossil, looking at it as a whole, as a three-dimensional object with surface contour and distinctive anatomical landmarks, not as a series of two-dimensional slices seen on end and out of standard anatomical orientation.

Knowing there was a palate in the rock enabled us to get our bearings, and we began to mentally match the curves we saw through the light box to our stored knowledge of what a primate skull looks like. Above the palate we could identify the tip of the snout below the opening of the nose and both sides of the lower face building up in cross section behind it, framing the nasal aperture. Nothing was warped or crushed. As the images moved farther back, through the rock, we saw the two circles that are the rims of the orbits, intact and almost perfectly symmetrical. We had a whole face. At that point we knew this was more than a great dentition. It was the finest fossil primate specimen ever found from the middle Miocene time period, the geological age of the beachfront deposit along the Rio Gallegos. Then we saw a rounded forehead that arced smoothly above the nasal bridge between the eye sockets. The level of excitement in the lab revved up. Could the whole cranium be there?

Unfortunately, we soon came upon slices that were empty of bone and the images turned again to speckled flecks on an opaque canvas, the rock. We were disappointed and yet we knew how important this fine specimen was (fig. 3.1). We immediately began to plan our next steps: to expose, prepare and preserve the whole fossil; to produce a microCT scan of the specimen so it could be studied inside and out in great detail, in 3-D, and to eventually share it with the scientific community as a virtual fossil. We agreed to publish an article jointly and we planned for the figures, tables, and text. In order to date the fossil, Adan, the geologist, went back to the cliffs and selected a rock from the same geological layer that yielded the fossil. I took it back to the United States to have it dated by Carl Swisher III, who specialized in a method called potassium-argon dating. It turned out to be 16.5 million years old.

Later, with the actual, cleaned specimen in hand, we determined that we were wrong in our guess that the fossil was *Homunculus patagonicus* because it was different from the only specimen of *Homunculus* with which it could be compared, a broken, partial face. The type specimen of *Homunculus*, a lower jaw, had unfortunately been lost, sometime during the 1930s, which meant we had to use a stand-in for the authentic voucher. We interpreted the Blake specimen as a new genus and species, which warranted a new taxonomic name. The genus name we chose, *Killikaike*, is from the fossil locality's place name, Killik Aike Norte, which comes from the native language of the Tehelche people who once lived in the region. The species name *blakei* is a tribute to John and Monica Blake. To name it formally, following the rules of zoological nomenclature, we designated the Blakes' fossil as the type specimen of *Killikaike blakei*, and it is now curated in the museum in Rio Gallegos. However, because of the loss of the *Homunculus* type specimen, fossil crania discovered later at the Killik Aike Norte locality have engendered the same challenge of identification we encountered, eliminating the possibility that they actually belong to the original *Homunculus* taxon. This continues to be a source of controversy among some primate paleontologists and, as of this writing, the taxonomy of *Homunculus* and *Killikaike* is a matter of debate. The excellent fossil face and a second specimen we found at Killik Aike Norte are now known in the literature by two different generic names, *Killikaike* and *Homunculus*.

While the type specimen rule is rigorously enforced and scientists are prohibited, with the help of peer review, from naming species without designating a voucher specimen, there is a contingent of taxonomists from many zoological fields who support the use of photographs as alternatives to physical specimens in cases where one cannot be collected, for example, if an individual animal was seen for a brief moment in the field and then escaped capture. In the era of Photoshop, the dangers of mistakes, misunderstandings, and misbehavior are self-evident, yet the debate over what is acceptable as a type continues.

In November 2016, I joined 490 colleagues from all over the world who signed on as authors, making the effort more of a petition than an article, to argue those shortcomings in a piece titled "Photography-based taxonomy is inadequate, unnecessary, and potentially harmful for biological sciences." It was published in the journal *Zootaxa*. The intensity of interest in the issue has been tracked electronically, and it is acute. After six months, the article had been read online by about 39,000 people, and I continue to receive notices almost daily indicating it has been viewed yet again.

It is interesting to note that paleontologists have been well aware of this area, Killik Aike Norte, since it was explored in the 1890s by a famed American fossil hunter, John Bell Hatcher, a curator at Princeton University's museum. This museum housed one of the world's great collections of early fossil mammals until it closed in 1985, and the collections were sent to Yale's Peabody Museum of Natural History. Artifacts of Hatcher's expeditions were still evident at Killik Aike Norte more than 115 years later, gradually weathering: a steel-wheeled, horse-drawn covered wagon and a row of craggy pilings standing upright in the river. Hatcher built a pier so that he could load a steamship to haul his fossils, tons of material trucked from various places in Patagonia, for transport to the coast, the first leg of a 6,000-mile ocean voyage to Princeton, New Jersey, in the United States.

Names can reflect evolutionary hypotheses

The same naming principles we applied to the fossil *Killikaike blakei* have been applied to all New World monkeys, extinct and extant, but arriving at a relatively stable set of names for the living platyrrhines has taken time. As an illustration, Linnaeus in 1758 and 1766 named 11 New World monkey species; however, none of them remain classified in the genus he employed then, *Simia*, which he applied to 30 different monkeys and one ape. Scientists' ideas of the genus have changed radically since Linnaeus. By 1840, scholars had transferred his 11 platyrrhine species to 8 different genera.

Species names also have their own evolutionary histories. The Buffy-eared Marmoset, *Callithrix flaviceps*, provides an interesting example. It is one of the earliest instances of a platyrrhine monkey brought back to Europe shortly after Columbus landed in the Americas, and it appears in artwork produced in the early 1500s. Colin Groves, a leading primatologist and primate historian, identified *Callithrix flaviceps* in the foreground of a woodcut by the artist Albrecht Dürer, depicting a courtlike scene in 1517–18. The monkey has a rope tied around its waist like a leash, so the drawing was probably made while the animal was alive. In a painting done between 1532 and 1536 by portrait artist Hans Holbein, Groves also recognized a *Callithrix flaviceps* held in the crook of the arm of a young, unnamed boy, the subject of the portrait, who is believed by some to be a son of King Henry VIII of England.

It took four more centuries for the scientific name of this eastern Brazilian species to become standardized. How many other names, formal and informal, *Callithrix flaviceps* carried before that is impossible to know. The common names appearing in the current scientific literature are likely only a fraction of those used over the past 500 years. They include Yellow-headed Marmoset and Buffy-headed Marmoset in English; Blonde-headed Monkey in Spanish (*titi de cabeza rubia*); Mountain Marmoset (*sagüi-da-serra*), Yellow-faced Marmoset (*sagui-da-cara-amarela*), and Bamboo Marmoset (*sagui-taquara*) in Portuguese.

Developing the Linnaean binomial *Callithrix flaviceps* that is used today involved three major taxonomic steps taken across multiple generations. The genus name, which covers a variety of marmoset species, was fixed in 1777 by a German naturalist, Johann Christian Polycarp Erxleben. *Callithrix* is a compound term based on two Greek words, *kallos* and *thrix* (or, *thric*), meaning "beautiful hair." Though Latin is predominant in taxonomic names, Greek words are also traditionally used. The Greek word for hair, *thrix*, has been applied in assorted combinations by many authors who coined the names of primates with plush fur (e.g., *Lagothrix*, *Pygathrix*). Over the years it has become almost synonymous with the word monkey, especially for platyrrhines, sometimes for extinct species (e.g., *Xenothrix*, *Antillothrix*) whose hair we will never see. Even after 1777, the monkey species we now attribute to *Callithrix* were called by at least 21 different genus names, including misidentifications and misspellings.

The species name, *flaviceps*, of more recent vintage, is another compound term derived from Latin, meaning yellow-headed. It was first applied to the Buffy-headed Marmoset in 1903 by the British zoologist Oldfield Thomas. In an effort to fix the identity of the animal and the species name *flaviceps* so material of this species would not be misidentified in the future, Thomas designated a very informative type specimen, which had not been done before: the pelt and cranium of an adult female collected in the wild. It is curated in the Natural History Museum in London.

The final step of the naming saga occurred 10 years later, when the American zoologist Daniel Giraud Elliot first applied the genus and species names together to form the combination *Callithrix flaviceps*. Since then the proper name has vacillated in a different way, but not one that confuses the identity of the animal. Its taxonomic status as a species, therefore its name, has been called into question. Is it distinct from its geographic neighbor *Callithrix aurita*, the Buffy Tufted-ear Marmoset, or are they each a subspecies of the same species? If they are deemed subspecies, *aurita*, coined in 1812 would supplant *flaviceps*, of 1903, as the species designation.

The nomenclatural history of *Callithrix flaviceps* follows the trajectory of evolutionary thought and practice over centuries. Erxleben's crafting of the genus name *Callithrix* fulfilled one logical requirement, that the marmoset he was writing about, which was actually the Common Marmoset, *Callithrix jacchus*, be given a distinctive name to separate it from other New World monkeys. In fact, Erxleben was also moving Linnaeus's *jacchus* out from the genus *Simia*. That kind of decision took on evolutionary importance 101 years after Linnaeus published his foundational work on classification with the advent of Darwinian thinking, the revolution inspired by his book *On the Origin of Species*, as scholars settled on a scientific rationale for deploying the taxon of genus as a concept. The names were no longer just labels. They became the equivalent of scientific ideas, hypotheses about genetic relationships.

Darwin's principles of evolution introduced two concepts as taxonomic factors to be considered in naming. Unity of descent, or monophyly, the sharing of a unique common ancestor, became the principal basis, as a hypothesis, for assembling groups of species into genera. Adaptation, the evolutionary adjustment of an organism to its ecological situation, which was already a working idea among naturalists, was used to characterize taxa and set the boundaries on how much species diversity, in anatomical terms, should be accommodated in a particular genus. When Oldfield Thomas was preparing to list *flaviceps* as a new species in the early 20th century, he knew that a number of adaptively similar eastern and central Brazilian marmoset species were being aggregated into a natural group. The Buffy-headed Marmoset was a new addition to the genus *Callithrix* and was not to be confused phylogenetically with other small platyrrhines that belonged to other genera, such as *Cebuella*, the Pygmy Marmoset, *Saguinus*, the Tamarin, or *Leontopithecus*, the Lion Marmoset.

Changing ideas can result in name changes

These criteria for classification, monophyly and adaptive similarity, are only guidelines, and their implementation can stir debate when scholars rely on the idea of adaptation as a taxonomic frame around the genus. Why? Because the imperative is to stick with objectivity and minimize subjectivity in developing scientific arguments. Phylogeny is an intrinsic, fixed property of each species. Whether or not we can accurately reconstruct the kinship-history of a group of species, their genetic relationships are an objective fact of life and we can measure—in theory at least—how closely related one species is to another. Adaptation, however, is a more subjective scientific inference. Scientists can evaluate the lifestyle and ecological and behavioral similarities and differences among species using quantifiable criteria such as ranging patterns, use of space, sleeping habits, food choices, locomotor abilities, vocalizations, reproductive output, and more. Still, it remains difficult in many cases to objectively compare these factors, one-to-one or in combination, across putative species and genera. There is no agreed upon yardstick for measuring one genus and distinguishing it from another.

The taxonomy of the Capuchin Monkey is a case in point. Today, the number of capuchin genera has become a point of contention. Some primatologists divide capuchins into two genera, *Cebus* and *Sapajus*, and as many as 11 species, whereas the traditional count is one genus and four or five species. Considering the reasons for the two-genus arrangement is instructive in illustrating the subjective nature of the genus category and taxonomic decisions pertaining to it. It has long been known that on craniodental and skeletal grounds capuchin species can be sorted into a more heavily built, robust form and a more lightly built, gracile form, with considerable overlap in the morphology. The essence of the two-genus argument is that such differences, highlighted in figure 3.2, warrant generic separation because they reflect trenchant adaptive disparities. In addition, they confirm genetic evidence that the species cluster into two groups.

That interpretation seems to be a stretch, however. All Capuchin Monkeys, no matter how they are classified, form a unified, coherent evolutionary and ecological unit distinguished from all other New World monkeys. The new knowledge brought to bear as the rationale for dividing them into two distinct genera is largely a refinement of observations and discoveries made generations ago. Fine-tuning them has now become possible by using new tools and analytical methods to better understand the morphology, and new information on the genetic relatedness of capuchin populations, a welcome addition to knowledge. Nevertheless, information about the behavior and ecology of

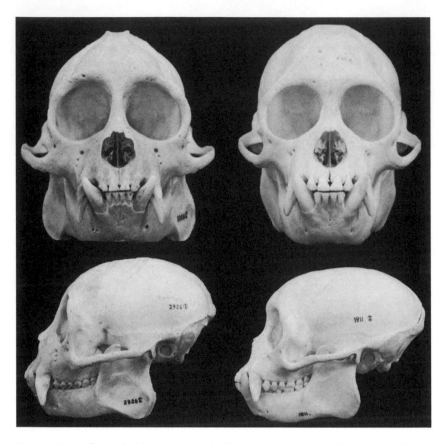

FIG. 3.2. Crania of two male *Cebus* species considered by some to represent two distinct genera. Adapted from the Mammalian Crania Photographic Archive, Dokkyo Medical University, http://1kai.dokkyomed.ac.jp/mammal/.

wild capuchins—what they actually do in nature—is perhaps the most effective demonstration of their adaptive continuity and the collective singularity of a Capuchin Monkey, that is, the *Cebus* lifestyle. This lifestyle varies between species seasonally, not fundamentally, not annually or inferentially over longer periods of evolutionary time, which is demonstrated in their relatively uniform morphology. The differences between gracile and robust capuchins are matters of degree rather than kind (fig. 3.2). They relate to selection for routine shifts in the strategies used by the gracile and robust forms to acquire foods periodically, during the lean seasons, as discussed in chapter 5, not to the evolution of different manners of existence, to different *lifestyles*, which ought to be the defining characteristic of a platyrrhine genus.

CHAPTER 4

EVOLUTIONARY MODELS

Because classifications embody evolutionary hypotheses, it is only natural to expect that the taxonomy of New World monkeys would undergo change as we learn more. In fact, there was a dramatic scientific shift that began in the late 1970s and 1980s that translated into a major revision in platyrrhine classification. It accelerated in the sense that the new arrangement was soon adopted by a broad range of researchers working in many fields of study concerning platyrrhines, like systematic biology, paleontology, ecology, behavior, physiology, and conservation biology. The power of this new classification was that it provided a sound, phylogenetically based perspective that guided researchers to generate and test explicit evolutionary hypotheses. By the mid-1990s, the system that had prevailed for much of the 20th century was replaced (table 4.1).

The reasons for the shift involved a global scientific movement, an intense new interest in phylogeny reconstruction among morphologists and paleontologists, driven by conceptual refinements associated with cladistic analysis, the desktop computer revolution, and the capacity to tap DNA. Other technical advances included sophisticated new ways to work out phylogenetic relationships mathematically using computer-based algorithms able to manage very large volumes of morphological and molecular data. New information was also pouring in from first-ever field studies of platyrrhines, genus after genus, with crossover interdisciplinary appeal that invigorated the phylogeny research program and looped back to become significant in behavioral and ecological studies. This was an infusion of knowledge with an ecological orientation, both conceptual and factual, that did not previously exist. Merging phylogeny with ecology and behavior stimulated a new way of looking at New World monkeys. It had the potential to make sense of it all by engendering a coherent narrative, a story of what happened. New ideas suggested not only how the platyrrhines were interrelated but also how their history unfolded adaptively, as a pattern. I call this the Ecophylogenetics Hypothesis.

TABLE 4.1. Revised classification of New World monkeys

Old classification[a]				Revised classification	
Families	Subfamilies	Genus	Common Name	Families	Subfamilies
Callitrichidae	N/A	*Cebuella*	Pygmy Marmoset	Cebidae	Callitrichinae
Callitrichidae	N/A	*Callithrix*	Marmoset	Cebidae	Callitrichinae
Callitrichidae	N/A	*Leontopithecus*	Lion Marmoset	Cebidae	Callitrichinae
Callitrichidae	N/A	*Saguinus*	Tamarin	Cebidae	Callitrichinae
Callimiconidae	N/A	*Callimico*	Goeldi's Monkey	Cebidae	Callitrichinae
Cebidae	Saimiriinae	*Saimiri*	Squirrel Monkey	Cebidae	Cebinae
Cebidae	Cebinae	*Cebus*	Capuchin Monkey	Cebidae	Cebinae
Cebidae	Aotinae	*Aotus*	Owl Monkey	Pitheciidae	Homunculinae
Cebidae	Callicebinae	*Callicebus*	Titi Monkey	Pitheciidae	Homunculinae
Cebidae	Pitheciinae	*Pithecia*	Saki Monkey	Pitheciidae	Pitheciinae
Cebidae	Pitheciinae	*Chiropotes*	Bearded Saki	Pitheciidae	Pitheciinae
Cebidae	Pitheciinae	*Cacajao*	Uacari	Pitheciidae	Pitheciinae
Cebidae	Alouattinae	*Alouatta*	Howler Monkey	Atelidae	Alouattinae
Cebidae	Atelinae	*Lagothrix*	Woolly Monkey	Atelidae	Atelinae
Cebidae	Atelinae	*Ateles*	Spider Monkey	Atelidae	Atelinae
Cebidae	Atelinae	*Brachyteles*	Muriqui	Atelidae	Atelinae

[a] From Hershkovitz (1977)

The paradigm shift involved a rejection of the gradistic thinking that was formerly the organizational basis of platyrrhine classification. The gradistic ideas concerning platyrrhines had remained much the same for about 150 years, going as far back as the first attempts to organize primate genera into a true Linnaean hierarchy of various ranks in the early 1800s. Gradistics was an outgrowth of a pre-Darwinian viewpoint that had aimed to place taxonomic groups at points along the Great Chain of Being, the *Scala Naturae* or Scale of Nature, a metaphysical, nonscientific notion inspired by antiquated western thinking about the natural world. Its medieval ideas that were carried over into the Darwinian era essentially followed the same tendency by designating species and other groups, in their entirely, as being either less or more advanced—lower or higher—along a continuum leading toward an arbitrary standard of evolutionary achievement.

Marmosets, for example, were considered a primitive group because they were assumed to have begun small and remained small, while other platyrrhines like the Howler Monkey genus, *Alouatta*, were thought to have risen to an advanced evolutionary status as evidenced by their large body size. This kind of

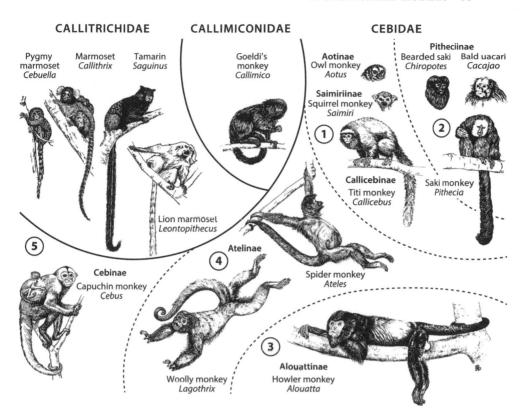

FIG. 4.1. Gradistic model of platyrrhine evolution and classification with five grades of cebids numbered (1–5). Note that *Brachyteles*, the Muriqui, was not represented. Adapted from Hershkovitz (1977). Originally printed in Philip Hershkovitz (1969), The Recent Mammals of the Neotropical Region: A Zoogeographic and Ecological Review, in *Quarterly Review of Biology* 44(1), pp. 1–70, later adapted in Hershkovitz (1977), *Living New World Monkeys (Platyrrhini)*. Illustrations by John Pfiffner and Field Museum staff artist Marion Pahl. ©1977 by The University of Chicago. Reproduced by permission from the Field Museum and University of Chicago Press.

one-way, typological thinking was used to justify a linear, steplike arrangement of platyrrhines into taxonomic groups according to their body sizes, the sizes of their brains, the complexity of their teeth, and more. This is a perspective that is a-phylogenetic and it marginalizes, or ignores, the impact of natural selection on the evolutionary process. In the case of the Marmoset, *Callithrix*, we now know that the gradistic assessment was wrong—their small stature, for example, is a derived trait, not a primitive one. Their small size is therefore a characteristic of cladistic value that helps place them in the platyrrhine Tree of Life and factors into the classification of callitrichines as a distinct group.

Modern classifications begin by establishing the cladistic relationships of the taxa. The aim is then to organize the animals into a hierarchy that reflects those relationships. Groups such as the family and subfamily are constructed by filling them with lower-level taxa: families contain subfamilies and subfamilies contain genera. Each unit is formed as a distinct monophyletic set, each one hypothesized to contain the descendants of a common ancestor unique to that set. Modern classifications are good guides in the compare-and-contrast motif of evolutionary studies, where to assess its importance a new observation often needs to be checked for a likeness in a related animal and squared against a difference found in another, more distantly related form.

The notion of staging and classifying taxa according to their wholesale primitiveness or derivedness was turned on its head by the fundamental insight of cladistics. The notion of primitiveness as a criterion for associating one animal with another was invalidated by the cladistic approach. This generally meant that the old gradistic classifications could not be seen as embodying reliable phylogenetic hypotheses.

The most basic tenet of cladistics is that only unique, non-primitive features can be considered credible traits indicative of a close phylogenetic relationship. The derived features show where in the Tree of Life a group belongs. Derived characteristics indicate that the taxa exhibiting them are the descendants of a common ancestor that existed at the root of the branch that represents them. Primitive characteristics cannot show those relationships because they do not signify anything unique about genetic ancestry. Distantly related taxa situated on disparate branches of the tree can share primitive traits as inheritance from the same remote ancestor that later gave rise to other distinct branches. The presence of a typical platyrrhine nose in Spider Monkeys and Howler Monkeys, therefore, is not a sign that these monkeys belong to a separate clade. All New World monkeys have platyrrhine noses. It is a primitive trait inherited by all the lines of New World monkeys from the last common ancestor they all shared. As we have seen in chapter 1, going further down toward the base of the primate Tree of Life, it becomes evident from character analysis that the first anthropoids must have had a platyrrhine nose as well, inherited from an even more remote ancestor that was shared with the wide-nosed tarsiers (fig. 1.1).

Noses can tell us nothing about the affinities of howlers and spiders. To establish that they do share a unique ancestry, we need to discover in them something that other platyrrhines never evolved. The skin-patch on the underside of their tail is a good example. That is a derived condition that evolved from a primitive one that is widely shared among the other platyrrhines, a fully furred tail that was mobile but not prehensile. As we shall see in chapter 6, the uniqueness of the atelid prehensile tail is affirmed by establishing that the tail of the Capuchin Monkey evolved its semiprehensile abilities independently by parallel evolution, a fact that is, in turn, corroborated by independent knowledge of the phylogenetics of the Capuchin and Spider Monkeys.

The common ancestor in the cladistic paradigm inferred by this method, as in the expression "that would have existed at the root branch," is hypothetical. The cladistic approach, which was developed by the entomologist Willi Hennig and first published in 1950, invokes the existence of a common ancestor as a way of explaining how a set of taxa came to inherit derived traits and its association with an important aspect of phylogeny, monophyly: two or more taxa related to one another by virtue of their uniquely shared ancestry. In practice, we do not anticipate discovering actual ancestors as fossils because chances are so small that any given species will ever become fossilized, no less be discovered by a paleontologist—authentic ancestors even more rarely. Indeed, one of the beauties of the cladistic approach is that it applies to reconstructing the histories of both living and fossil forms, and even a combination of the two, and to information that rarely, if ever, fossilizes, namely DNA. Fossils of the genuine ancestors are not absolutely necessary to discover the cladistic outlines of phylogenetic history. Yet they can help in specific ways, as we shall see below, providing information that cannot be gained from studying only living forms, either morphologically or genetically. It is worth noting that the morphology-based, cladistic modernization of platyrrhine history and classification that began in the 1970s was based almost entirely on the living forms because the fossil record was so meager at that time.

How do diverse genera coexist in one patch of forest?

The Ecophylogenetic Hypothesis

The objective of ecophylogenetics is to develop a model of the evolutionary history of a local community, or an adaptive radiation, by synthesizing information gleaned from ecology and phylogeny. These models entail discussions of the central adaptations that define species' lifestyles and offer hypotheses that explain how their ecological interactions evolved over time. At a more fundamental level, ecophylogenetics recognizes the inescapable role of phylogeny in shaping the evolving adaptations of species, how they partition resources, and how this genetic history governs their interactions over time. The term ecophylogenetics echoes the expression "ecomorphology," also relatively new to the evolutionary biology lexicon, which denotes an approach to anatomy that relates form, function, and performance to the ecological setting of a species. In 2009, two independently written articles introduced the word "ecophylogenetics." The ecologist and evolutionary biologist Marc Gadotte discussed the idea in the online Encyclopedia of the Earth. My colleagues Marcelo Tejedor, Siobhán Cooke, Stephen Pekkar, and I used the term in the title of our article, "Platyrrhine Ecophylogenetics in Space and Time," which was prepared for an edited volume on New World monkeys. It was a concept whose time had come. A leading scientific journal, *Biological Reviews*, highlighted the idea in 2012, in a piece titled "Ecophylogenetics: Advances and Perspectives."

Primatologists had for a long time segregated phylogeny and adaptation, as well as the method of study underlying the latter, functional morphology, as separate research domains. That stance ignores a fundamental fact of evolution, that phylogenetic relatedness literally breeds resemblance in form, ecology, and behavior. To arrive at a full picture of evolution's products, it is obvious that a synthesis of biological features and adaptations is required. Darwin's iconic words describing evolution as "the theory of descent with modification" define the interplay between phylogeny and adaptation.

Phylogeny and adaptation are the obverse sides of the evolutionary coin, meaning they are mutually inextricable. Clades become successful because they evolve something new, usually a favorable adaptive syndrome. That means lines of descent can be identified and traced historically by the unique adaptations they alone have evolved, which are derived traits and complexes. This is why, in technical terms, the first-order explanation for shared similarity is the *null hypothesis*, that is, the explanation for why there is no significant difference between groups: genetic relatedness. Discovering the converse, that a set of similarities arose independently, such as a grasping tail in the Capuchin Monkey and the Spider Monkey, then becomes an important contribution to other facets of evolutionary knowledge, often pointing to the selective contexts and advantages that promoted their evolution by natural selection in more than one instance.

The ecophylogenetic model of New World monkeys was a new idea in 1980 when I first proposed it, though I did not use the term at that time. It used the cladistic method, then still seen as an intellectually challenging method among many primatologists, and it was developed within a taxonomic framework that had not been applied to platyrrhines before. The state of research into platyrrhine phylogeny had not yet caught up to the growing wave that was asserting itself in primate systematics, a deep concern for phylogeny. There was not a single article devoted to New World monkeys in a state-of-the-art volume published in 1975, *Phylogeny of the Primates*, edited by W. P. Luckett and F. S. Szalay. Nevertheless, that compilation of studies by a diverse set of authors emphasizing morphology, behavior, genetics, and living and extinct species across the order was transformational. It was immensely important in fostering the concept of cladistics.

At that time, the prevailing ideas about platyrrhine evolution were championed by Philip Hershkovitz, who was a leading figure in South American mammalogy for decades and Curator of Mammals at the Field Museum of Natural History in Chicago. His view of platyrrhine classification (fig. 4.1; table 4.1) was not very different from that of George Gaylord Simpson, for years a luminary at the American Museum of Natural History in New York, and one of the most respected mammal paleontologists and evolutionists of the 20th century. Simpson's classic 1945 monograph, *The Principles of Classification and a Classification of Mammals*, was paramount in its influence. Nor were the ideas of Hershkovitz and Simpson substantively different from those of influential European zoologists, such as Reginald Innes Pocock at the London Zoo, who wrote several seminal papers on primates from 1917 onward, or Sir W. E. Le Gros Clark, whose book-length syntheses of primate evolution—living and fossil—in 1934 and 1959 provided the foundation of all that was to come.

The Ecophylogenetic Hypothesis proposes that platyrrhine evolution has unfolded along two major adaptive dimensions involving, most substantively, the interplay of diet and locomotion to produce the 16 genera we see today in South and Central America. It is important to stress that the present array represents survivors of clades that can be traced back to a unique origin that goes back more than 20 million years. As discussed in later chapters, the platyrrhine radiation probably began more than 45 million years ago, but we have little information pertaining to the first 25 million years of its existence. The Ecophylogenetic Hypothesis pertains only to the available evidence of the more recent animals, that is, approximately the last 20-million-years-plus of living New World monkey history.

The essential point of the hypothesis is that the interaction between phylogeny and ecology has shaped the contours of the modern platyrrhine fauna. Each of the major taxonomic groups that we define phylogenetically is also an ecological unit—opposite reciprocals of the same evolutionary coin. The ecological adaptations that determine the lifestyles of New World monkey genera, predicated and constrained by the adaptations of the clade to which each one belongs, are played out everywhere in the Neotropics, whether the animals live in primate communities composed of a handful of genera or more than a dozen. For example, the Atlantic Forest community includes only six genera, but each of the three families and five of the six subfamilies occur there, living lifestyles closely comparable if not indistinguishable from the phylogenetic and ecological mosaic that composes other communities on the continent where 12 or more species are found sympatrically. This demonstrates a balance that the platyrrhine adaptive radiation maintains because it has an intrinsic structure, held together by the fabric of phylogeny.

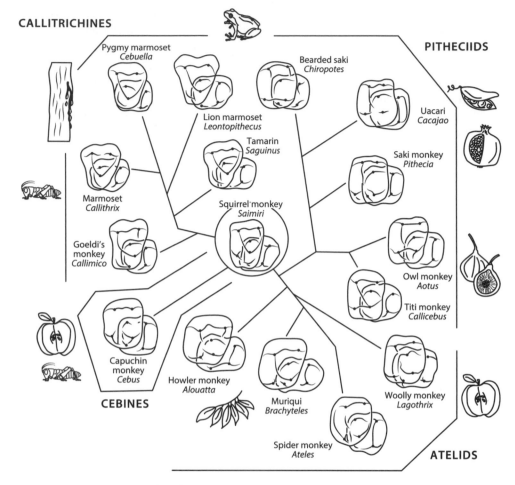

CALLITRICHINES

Pygmy marmoset
Cebuella

Lion marmoset
Leontopithecus

Tamarin
Saguinus

Marmoset
Callithrix

Squirrel monkey
Saimiri

Goeldi's
monkey
Callimico

Capuchin
monkey
Cebus

CEBINES

Howler monkey
Alouatta

Muriqui
Brachyteles

Spider monkey
Ateles

PITHECIIDS

Bearded saki
Chiropotes

Uacari
Cacajao

Saki monkey
Pithecia

Owl monkey
Aotus

Titi monkey
Callicebus

Woolly monkey
Lagothrix

ATELIDS

FIG. 4.2. Schematic overlays of upper and lower first molar sets of the 16 living genera aligned at maximal inter-cuspation and arranged according to their cladistic relationships. Squirrel monkeys are in the center as a model of the ancestral pattern. Adapted from Rosenberger (1992).

The Ecophylogenetic Hypothesis suggests that living platyrrhines initially diverged, ecologically and phylogenetically, to inhabit two sectors of the environment delimited primarily by diet. Types of food and the pursuit of it are most important. The two divisions are broadly defined as the frugivore-folivore and frugivore-insectivore Adaptive Zones. The 16-genus array of molar morphologies, for example, is structurally organized around phylogenetic groups, as seen in fig. 4.2. The pitheciids and atelids are frugivore-folivores and the cebids are frugivore-insectivores—again, using general terms to signify the combination of dietary specializations as well as tendencies. The clades occupying each zone became further differentiated along dietary lines by evolving different food preferences and ways of obtaining food within each zone. These are the Adaptive Modes, lifestyles made possible by the evolution of additional contrasts in body size combined with innovative locomotor and foraging strategies. The same type of dimensionality and layering is found in other primate communities living in Eurasia and Africa. The various ways in which New and Old World monkey genera have evolved adjustments to the interactions of diet, size, and locomotion allow for a partitioning of the local ecology, which enables them to coexist.

The fossil record, while still limited in scope, suggests that this organization of platyrrhine communities has been relatively stable since about 20 million years ago. This pattern of relative stasis is a fundamental tenet of the Long-Lineage Hypothesis, the temporal complement to the Ecophylogenetic Hypothesis, and will be discussed further below. Although the Old World evidence suggests that the ecological components of the Ecophylogenetic Hypothesis are repetitive phenomena and may be due to natural process rules that guide the evolution of primate adaptive radiations, it is still reasonable to ask whether the link to phylogenetics is a random outcome. Did the shape of the platyrrhine radiation come about by chance as a result of ecological opportunity and circumstance? This question was explored mathematically and tested against the fossil record in a project with my Argentine colleagues Leandro Aristede, Ivan Perez, and Marcelo Tejedor. We found that this is not a case of chance. The genetic and adaptive structure of the radiation coevolved, and an early shift in body size signaled the moves into distinct Adaptive Zones. Using molecular data as a check, we also found evidence for a slowdown in clade formation over time, as niches driven by diet were filling up and the radiation attained an enduring ecological balance.

What explains the stability and persistence of platyrrhine communities? There are several ideas that merit further exploration. One clarifies why clades are ecologically divided and constrained, with limited opportunity to make wholesale shifts from one Adaptive Zone into another. Body size may be a crucial factor here, natural size thresholds that determine dietary-locomotor niches. Small callitrichines, for example, cannot possibly survive on a leafy diet for physiological reasons, nor can their bodies climb, clamber, swing, and hang like the Spider Monkey. By the same token, a large spider monkey cannot survive nutritionally by perpetually scouring the trees for tiny insects, or globules of gum. As we shall see, the monkeys occupying the middle size range have more dietary flexibility but they, too, are constrained by their position in the community, based on their ecophylogenetic heritage.

The modern radiation may have remained stable in an ecophylogenetic sense because the platyrrhine community as a whole had not been radically disturbed by outside taxonomic forces. There have certainly been major environmental changes on the continent, but South America was largely isolated as a landmass throughout the millions of years of platyrrhines' experience. No competing groups of primates ever attained a foothold, so there were no zero-sum competitive conditions, wholesale extinctions, or faunal turnovers. Evolutionary changes occurred from within. Nor were the majority of the continent's other resident arboreal mammals, mostly marsupials, direct dietary competitors.

DNA and anatomy

Molecules and morphology

By the 1990s it had been corroborated that there are three major clades of platyrrhines, not two groups as per the old gradistic model. The three clades are now classified as the three families, the cebids, pitheciids, and atelids. Horacio Schneider, the leading primate molecular phylogeneticist in Brazil, assembled a team to methodically work out platyrrhine interrelationships using DNA. Schneider and his group converged on many of the same interpretations arrived at by using morphology and behavior. "Molecules, Morphology and Platyrrhine Systematics" is the title of a paper he and I jointly wrote in 1996, summarizing our independent phylogenetic analyses and emphasizing their correspondence. This article accelerated the acceptance of the three-family organization of platyrrhine phylogeny and many of the linkages among the genera contained in each group. Morphology and molecules had both proven that the traditional platyrrhine classifications needed to be overhauled in a major way.

Our intentions went beyond redrawing the platyrrhine Tree of Life. At that time when it came to ideas of phylogenetic history, primatology had already shifted its focus from the fossil record, thus morphology, to the study of molecules. We showed that morphology and DNA are both credible, robust sources of phylogenetic information, though one should not expect the results of these studies to always be identical. They are different sources of information, each with its own strengths and weaknesses. As alternative investigative approaches, when applied to the same problem, they provide us with a system of checks and balances that may shed light on matters needing special attention, concerning the problem at hand as well as the methods used to tackle it.

In the Schneider and Rosenberger study, it is important to note that there was complete agreement between the morphological and molecular methods on ten phylogenetic linkages at the genus level and above in the large and diverse radiation of platyrrhines. Only three substantive differences between the approaches emerged, pertaining to the cladistic positions of the Owl Monkey,

Aotus, Goeldi's Monkey, *Callimico*, and the Muriqui, *Brachyteles*. The morphological view placed *Aotus* among pitheciids; *Callimico* as the most basal, or first-branching callitrichine; and *Brachyteles* as the sister genus to *Ateles*. The molecular view was that Owl Monkeys are cebids; Goeldi's are most closely related to the Marmoset and Pygmy Marmoset; and the Muriqui is the sister genus to the Woolly Monkey, *Lagothrix*. On the whole, the extent to which the phylogenies and splitting times generated by molecular and morphological data, including fossil evidence, do conform outweighs these three points of disagreement. Both approaches showed strong mutual support in the shape and timing of platyrrhine evolution.

The discrepancy involving the Muriqui—fascinating in its own right—is not highly significant to the big picture of platyrrhine evolution because the controversy is narrowly confined to one subfamily-level clade, a trio of three living genera. It influences mainly how we view the likelihood that the link to the Woolly Monkey signaled by DNA is accurate, or that monophyletic descent is responsible for the resemblances shared by the Spider Monkey and the Muriqui. More significant are the different views about the Goeldi's Monkey because they influence how we reconstruct the origins of callitrichines, one of the most intriguing, adaptively unique radiations produced by the primates.

The incompatible views regarding the Owl Monkey are very important because the morphological and molecular approaches place *Aotus* in entirely different family-level clades, an unexpected outcome given how compatible results of morphology and molecules are with respect to all of the other platyrrhines. Sorting this out remains a challenge. Which of the two proposals is currently the strongest working hypothesis? As with all scientific hypotheses, one way to assess their quality is to consider their heuristic value. What other testable hypotheses do these ideas stimulate, and what other phenomena do they help explain?

As discussed in many parts of this book, the hypothesis that *Aotus* is a pitheciid most closely related to *Callicebus* has the power to explain a suite of detailed, apparently derived morphological, behavioral, and ecological resemblances that are uniquely shared by them. If Owl Monkeys are cebids, we would expect that such a pattern of resemblances would tie *Aotus* to cebines, callitrichines, or both. This is not the case. There are no studies that demonstrate continuity or exclusivity between the characteristics or lifestyles of owl monkeys and any of the cebids, other than those that are general platyrrhine-wide or size-related patterns. For the molecular studies, there are also well-known, persistent discrepancies in the precise cladistic placement of the Owl Monkey genus among the cebids themselves, essentially three different alternatives, which is a level of uncertainty that does not affect the molecular results for any other platyrrhine genus. Investigating why that is the case may be a good first step toward resolving the "*Aotus* problem."

Cebines and callitrichines share a unique common ancestor

As the morphology of Goeldi's Monkey came to be known in the 1910s and 1920s, it was clear there would be no easy solutions to the question of its classification given the mix of traits *Callimico* exhibited, which defied a simple typology. In traditional taxonomic terms, did this monkey belong to the same stage or grade as the small-sized, clawed marmosets and tamarins, in which case it could be classified as a "callitrichid?" Or was its natural position aligned with the 12 other platyrrhines that had the same dental formula, justifying placement in the "cebids"? A third proposal saw it as neither "callitrichid" nor "cebid," but representing another gradistic unit of family rank, the Callimiconidae.

Little evolutionary insight could be gained by consigning Goeldi's Monkey to an isolated family of its own kind. Elemental questions pertinent to *Callimico* and to the evolution of the callitrichine morphology in general were matters needing resolution; for example, are small body size and claws primitive or derived among platyrrhines? A related question that was far more consequential to unraveling the evolution of the platyrrhine radiation remained unarticulated: To which platyrrhines are these clawed New World monkeys most closely related?

That was one of the riddles driving my early morphological research, and the work of my colleague Susan Ford, who specialized in the evolution of the postcranium. We both concluded that *Callimico* and the other callitrichines were evolutionary dwarfs and their claws were derived, not primitive. That set the stage for ending the taxonomic isolation of callitrichines, including Goeldi's Monkey, as a phyletically disconnected group. It was preparation for posing a more intriguing follow-up question. If clawed digits were specializations among the platyrrhines, from which taxon, or which common ancestor, might they have derived? Put another way, which of the nailed New World monkey genera could serve as a good model of the morphological antecedent?

We researched the non-clawed platyrrhines, looking for anatomical and behavioral models that might forecast the evolutionary history of the Callitrichinae, and their claws. We found it would join clawed and non-clawed platyrrhines in a new clade with an ecological theme of its own. Also, it would mean that the traditional assemblage of the non-clawed monkeys as "cebids" did not constitute a monophyletic group. Since so many genera, essentially all the major lineages of platyrrhines, were involved in such an analysis, this discovery would totally rewrite the taxonomic baseline, the background, for investigating platyrrhine evolution.

The cladistic bridge to callitrichines is embodied in the Subfamily Cebinae. Like Goeldi's Monkey, the morphology of which serves to close the gap between the old "callitrichids" and "cebids," the Squirrel Monkey genus, *Saimiri*, is also a morphological intermediate (table 4.1). In size, it establishes a degree of anatomical continuity between the dwarfed callitrichines and the largest cebine, the Capuchin Monkey, *Cebus*. But other connecting features are more trenchant. *Saimiri* fingertips, apart from the flatly nailed thumb and large toe, are fitted with nails that are narrow, convex, and end in an apex, though they do not have a prolonged, downwardly curved, pointy tip. Dentally, squirrel monkeys sit at the primitive pole of a morphocline. Their third molars are very small. The crown morphology of the first two upper molars is four-cusped, but the fourth cusp is also quite small, especially on the second molar. *Saimiri* molars, in morphology and proportions, approach the pattern of *Callimico* and both foreshadow the simple, three-cusped molars, and two-molar dental formula of the clawed callitrichins (fig. 4.3). Robert Martin, a leading expert of the study of allometry in primates—how changes in body size relate to changes in proportions and morphological shapes—argued convincingly that the unusual reduction of callitrichine molars is correlated with dwarfism, which would be another indication of directional change within the molar morphocline bracketed by *Saimiri* and the two-molared callitrichines.

A molar series that is reduced in size toward the rear, with reduced prominence of the fourth upper molar cusp, the hypocone, is a derived pattern shared by cebines and callitrichines. The other living cebine, the Capuchin Monkey, is widely accepted to be an exceptional primate with many highly derived features of its own. Thus its markedly contrasting anatomical adaptations, such as its enlarged hypocone cusps and other features of the molar crowns, cannot be construed as evidence contradicting a cebine-callitrichine connection. As we shall see time and again in chapters to follow, the whole body of *Cebus* is that of a large, specialized, heavily predaceous, frugivore-insectivore, a variation on the same adaptive theme shared by callitrichines and squirrel monkeys. Rather than being a challenge to the callitrichine-cebine phylogenetic hypothesis, it is a powerful sign of extended adaptive continuity, which has taken on a new face in capuchins in response to an additional set of selective pressures and ecological opportunities present in the same Adaptive Zone. Functional adaptations to food and foraging clarify the cladistic affinities of callitrichines. Callitrichines are more closely related to cebines than to any other platyrrhines.

By the early 2000s, a sufficient number of molecular phylogenetic studies had been produced that soundly corroborated the major points of the thesis

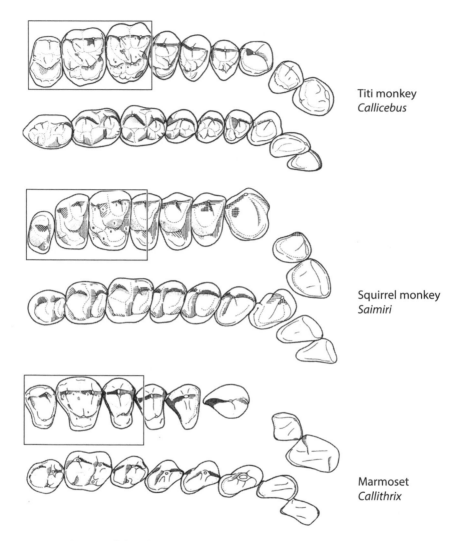

Titi monkey
Callicebus

Squirrel monkey
Saimiri

Marmoset
Callithrix

FIG. 4.3. Three sets of platyrrhine dentitions (upper teeth above, lowers below) showing progressive reduction in size and shape of the molars as a morphocline. Rectangles represent the combined area of all three upper molars of the titi. Adapted from Maier (1984).

Schneider and I proposed a decade earlier, that Capuchin Monkeys and Squirrel Monkeys comprise sister genera—another new result generated using cladistic analysis—more closely affiliated with the clawed callitrichines than with the other platyrrhines. Removing these genera from their alignment with the other three-molared platyrrhines validated the taxonomic shift away from the classic "callitrichid" vs. "cebid" classification of platyrrhines, and its replacement by the three-family arrangement that is followed here.

Chasing monkeys

Synthesizing behavior, ecology, and morphology

This new view of platyrrhine interrelationships could probably not have come about, or been seen as a robust interpretation, without fieldwork, because what we see in the living animals today, observing them in their natural habitat, is a key to understanding their morphology and evolutionary history. One of the most important early projects that drove this home was summarized in a 1980 paper by John G. Fleagle and Russell A. Mittermeier, two innovative scientists—leaders in the study of platyrrhines—who, respectively, made major contributions to the understanding of platyrrhine morphology and evolution, and conservation. They described and quantified the locomotor behavior of seven species of Suriname monkeys living in the same forest, and representing all the major clades. They showed how variety in body size and locomotor behavior enabled the animals to use different parts of the forest and how monkeys with similar diets tended to diverge in behavior and habitat preferences. With each genus presenting its own set of behaviors, tied to anatomical patterns customized by selection, it is also logical to conclude that what they do now to make their way in the world is comparable to what they have done in the past. That means the living New World monkeys are also living fossils, carrying forward for us to see the various systems of inherited behaviors that evolved long ago, and continue to evolve.

Observing animals in the zoo, examining fossils in museums, measuring skulls and teeth and bones of modern species, and studying living primates in the field, is what led to the Ecophylogenetic and Long-Lineage Hypotheses. The experience has a time machine–like quality, a way of seeing the past while seated in the present, seeing the beauty of evolution in action, how it situates a species in nature and how different genera and species have found ecological balance within a primate radiation.

Scientists are interested in discovery, and there may be no other places where discoveries in primatology are more likely to happen than in the museum and in the field. They are both highly complicated environments that are impossible for humans to take in fully at any given moment. In the field, sometimes you know that you are the only person who has ever walked that piece of forest, possibly observing monkey behaviors not seen elsewhere. With specimens in the museum, sometimes you feel that you are seeing an anatomical pattern that has never been described that way, and you begin to make sense of it on your own. Doing research in both worlds, I found that I did not see fossils as lifeless objects, or living animals as lacking a history.

The first time I went to the field in Peru in the summer of 1974, I was a graduate student who had grown up in an urban environment, who had never even gone camping. After a daylong trip by wooden boat from Iquitos, I entered the forest at the head of a trail leading into the Peruvian Amazon, at night, with only a small flashlight in hand. The field site was an hour's hike through the bush from the tiny village of Mishana. I was there to study titi monkeys with my college mentor, Warren Kinzey, and a small team. *Callicebus* are pair bonded, and by using these monkeys as a case study Kinzey aimed to build a behavioral and ecological profile to examine the conditions under which human monogamy might have evolved. We found a family group of *mico leon dorado*, the Red-bellied or Collared Titi Monkey, that appeared not to be shy of humans so we focused on them.

Our campsite was a platform raised a few feet off the ground. It had a thatched roof of palm fronds, a pair of benches, and two tables, one for eating and one for working, all made from trees that had been cut down to open a clearing in the forest. That was our base camp for the summer. There was room on the platform to pitch a tent in which several of us slept, and from the pillars supporting the roof Kinzey and I hung our hammocks for sleeping outdoors. There was a stream nearby where we got our drinking water, bathed, washed dishes and clothes. We dug a few latrines; we broke a few shovels doing that. For light we had a single lantern, and for cooking we had a propane stove. When we ran out of the local drip coffee filters made of cloth suspended from a ring of wire, all we had for an alternative was a clean, white sock.

Periodically, one of us would make the two-hour round-trip hike to the village and back, or the two-day riverboat trip to Iquitos to get supplies. Canned food, batteries, and weeks' old *Time* magazines were always high on our shopping lists. I still recall hearing one of our team members scrambling through the forest looking more disheveled than usual in his rush to get back from the city, waving soggy folds of paper and shouting out the news headline, "Nixon resigned!" August 1974 was a momentous month.

Being isolated in the remote jungle was uncomfortable at times, but I liked being with the monkeys, following them around, observing them in real time, collecting and analyzing data, and beginning to understand patterns in their behavior. The first task in camp was to open up trails where we thought the monkeys would spend most of their time. This was the era before the technologies were introduced for GPS mapping, radio collars for tracking monkeys, or digital devices for recording data.

We cut trails with machetes and plotted them with a compass to get straight lines through the forest. We marked the trail system by tying plastic ribbons to wooden stakes or trees and labeling them, Trail A, B, C, each marker set 50 meters apart, then at right angles, Trail 1, 2, 3, 4, 5, so we had a labeled grid system, A1, B3, etc., to follow monkeys and note their exact location. When off the trail, we measured distance by pacing—every pace was a meter. We were able to locate monkeys' sleeping trees, and the trees on which they fed. We collected observations every five minutes, writing them down in a little spiral-bound notebook. There was a lot to keep track of and we had prepared a mental list of "all things monkey." After dinner and cleanup, we tallied our "time samples" and figured out how much monkeytime was spent feeding, moving, resting, sleeping, interacting socially, etc. Our last chore before going to bed each night was to transfer critical information from scribbled notes onto a large pane of waterproof graph paper we made by taping several 9 × 12 sheets together. The compass came out again and we reconstructed their exact travel paths, scaling down the meters measured in the forest to the millimeters of our ruler, tied to the grid and tied to the clock, based on the location of each feeding tree and the sleeping trees where they began and ended the day.

Putting in 12-hour days on the trails, Warren and I worked as a team. There was a lot to do and our time was limited. He was the recorder and I was the spotter. Every five minutes Warren would call out to me: "time check." He would collect data from what he could see and wrote down what I could see at that moment. *Feeding: on what? Locomotion: what kind? Location: how many paces/meters from the previous scan? At what compass angle? Where in the canopy are they: branch, bough, or terminal twig? Estimated height above the ground? How many individuals are in sight: what are they doing? Who is doing what? Can you identify them? How far apart are they? Is the feeding tree part of our database? Is it numbered? How many paces out to the nearest trail marker? What angle? What's the coordinate of that marker? Are there any databased trees in sight where they fed or slept before? How far away? What time is it? Watch out for the snake! Thirty seconds to go until the next recording bout* . . . Warren then paced out the distance to the nearest marked tree where a feeding or resting bout had occurred. We spent a lot of time retracing our paces and searching for ground-truthing coordinates to fill in the gaps in our notes and verify our course, which was the monkeys' course. Whenever we had the time, we would collect samples of leaves, fruit, and bark from feeding trees and save them in labeled plastic bags so the trees could be identified later by our botanist, looking especially for partly eaten, dropped fruit. Whenever possible, we collected and bagged feces to identify the fruits the animals ate and the seeds they appeared to swallow.

When you locate the monkeys' sleeping trees in the evening, you know where they will be the next morning and you can make sure to get there before they wake up. Titi monkeys choose a different sleeping tree every night and the group sleeps together. At dusk they prepare to bed down on a sturdy bough, with male and female twining their tails together, with their progeny's tail braiding in between. When we saw that, we knew the monkeys were settling in for the night. There was always a rush getting back to camp, to transcribe the mapping data from our notebooks, plot it on the grid, and set a course for the next day. We worked into the night to do this. We had to be able to find that sleeping tree in the dark at five o'clock the next morning. The sleeping trees could be a 45-minute walk from camp through the forest, sometimes off the grid system where there were no trails and few marked trees, if any, to guide us. We would set out at 4 a.m., hoping the troop had not moved during the night, which they did only rarely, and hoping to hear their loud, exciting "dawn call duet" up close. Whenever we lost track of the animals, despite our best efforts, the surest way to find them again was by hearing their long calls. Sometimes it took several days of looking and listening. Warren and I would then post ourselves several hundred meters apart from one another with compasses in hand, long before dawn, waiting in the darkness for the monkey calls to come, hoping we could triangulate to the source of the animals before they took off to find food.

The Titi Monkey project was our priority, but while doing this fieldwork we heard from the Mishana villagers that there were *leoncito*, pygmy marmosets, a little downstream on the Nanay River. We wanted to see them. So, on one of our supply trips we borrowed a handmade dugout canoe and a paddle and got into the boat. Leaving the river for a small tributary, we had to push and hack our way through low-lying trees at the water's edge to enter a larger stream, all the while trying to be very quiet, so as not to alert the monkeys to our presence, and so we could hear them. A pygmy marmoset is tiny, about the size of a small banana, including the length of its tail. They are well camouflaged by mottled gray fur, and are not easy to see. You tend to hear them first, which was our experience with titi monkeys, but not by the rustling of leaves as they move because they are so light in weight; rather, by the very high-pitched, whistling-type sound they make to keep in touch with one another.

We heard them, got out of the canoe and spent a few hours watching them on a miniscule island which had one tree on it. It was amazing to see the animals, a troop of nine, leaping and jumping a lot, and spread-eagled while clinging to large-diameter trees with their clawed hands and feet, gouging into the bark and eating licks and globules of tree gum. To our surprise, we saw that their postures and locomotor style frequently employed vertical clinging and leaping. This unique locomotor pattern had never before been documented for an anthropoid primate, though it is common among some lemurs and tarsiers.

Scientists had not known how these animals' claws were used, and why callitrichine claws evolved. The clinging behavior, which Warren and I could plainly see, demonstrated that the claws worked like little grappling hooks, allowing the monkeys to hang on to the tree while they were scraping holes in it to obtain gum for food. There was no other way these tiny monkeys could perch themselves on large, vertical tree trunks. Claw-clinging was, therefore, a feeding adaptation built into the locomotor system. In contrast, in clawless lemurs and tarsiers vertical clinging and leaping is an obvious locomotor adaptation used for traveling between trees.

This was an important discovery leading to the understanding that callitrichines are not the very small, primitive primates they were often made out to be. To the contrary, they are ecological specialists. Three articles were subsequently published based on that one field trip; I was coauthor on two. It was my introduction to scientific publishing. More importantly, I began to see that diet and locomotion were intimately linked in the biology of a monkey's lifestyle. There is the food, and there must be a way to get at it. Feeding and foraging coevolve.

Kinzey and company continued to work at Mishana for a few more field seasons, and he attracted interest from botanists who were drawn to the ecology of the region because it contained starkly different soil types supporting different plant and animal communities. Today, the area is part of a large protected territory, the Allpahuayo-Mishana National Reserve, which hosts many endemic plants and animals, including endangered populations of the Titi Monkey and the Saki Monkey.

The platyrrhine Tree of Life

Once it was realized that the historical linchpin of the classical Family Cebidae, the genus *Cebus*, did not belong with the other platyrrhine genera it used to be grouped with—the pitheciids and atelids—the next question was: What are the cladistic interrelationships among the remaining nine? What does the platyrrhine Tree of Life look like? The answers were developed with the following hypotheses: (1) Atelids, the four, larger, prehensile-tailed platyrrhines, were a single monophyletic group, separated from *Cebus* in spite of its semiprehensile tail, and despite the many differences that set one of them, the Howler Monkey, *Alouatta*, apart from the others, namely the Spider Monkey, Muriqui, and Woolly Monkey, *Ateles, Brachyteles*, and *Lagothrix*, respectively. The gradistic viewpoint had been so powerful that howlers were hardly ever thought to be linked with the other atelids in classification or in any evolutionary sense. (2) Saki Monkeys, Bearded Sakis, and Uacaris were part of another group consisting of five genera, the pitheciids. Three were easily grouped together: *Pithecia, Chiropotes*, and *Cacajao*. A fourth, *Callicebus*, the Titi Monkey, was part of that group—a connection that was not previously known—and it formed a link to the Owl Monkey, *Aotus*.

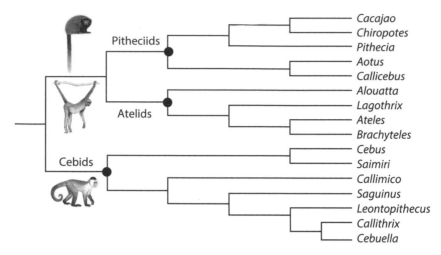

FIG. 4.4. Morphology-based cladogram of the living platyrrhine genera and families. Images courtesy of Stephen Nash.

This model of platyrrhine affinities (fig. 4.4) is based on cladistic analysis, which is a method that seeks to find monophyletic relationships, clusters of taxa that share a unique ancestry unto themselves based on their derived features. The diagram that illustrates these relationships, the cladogram, is sometimes referred to as a phylogenetic tree but the two are not the same thing. True phylogenetic trees are like genealogies. They stipulate ancestry and descent, where an animal came from evolutionarily, as one species evolved into another, or as a lineage continued its evolution over time. This kind of diagram requires fossil evidence. Phylogenetic trees also depict collateral, or sister group relationships, how sets of taxa that do not represent an ancestral-descendant sequence link up with one another in the tree. Cladograms are quite different in that they focus exclusively on sister group relationships. They can be constructed using living species alone, without fossils. This is important because, as has been shown, adaptive radiations can be understood even if all the evidence comes from living species, which was pretty much the case for New World monkeys until interest in platyrrhine fossils began to grow during the 1980s and 1990s.

The model of cladistic interrelationships among atelids clarifies how this is done. It was developed in 1999 by me and Karen Strier, the leading authority on Muriqui behavior and ecology. It began by employing the character analysis approach to sort through features such as body mass, sexual dimorphism, dental morphology, locomotion, and aspects of social organization, trying to establish the primitive and derived versions of those characteristics or complexes. Groups exhibiting the same derived features were interpreted as having a unique common ancestor. For example, the body weight of ancestral atelids was estimated to have been about 5.5 kg, 12 lb, which we took as the primitive condition because all other extant platyrrhines are significantly smaller than an atelid. In this regard, the actual weight value has little significance. The smaller end of the scale would resemble pitheciids and the largest cebids, and the heavier weights seen in all atelids, we inferred, would represent the derived conditions.

In another example, quadrupedalism, because it is the most common form of locomotion in platyrrhines and is the basic style of *Alouatta*, without the swinging flourishes of the atelines, is the ancestral atelid condition, and other behaviors, like the tail-assisted climbing-clambering and acrobatic locomotion, were derived from it. This is why it is understood that *Lagothrix*, *Ateles* and *Brachyteles* form one separate phylogenetic group and *Ateles* and *Brachyteles* form another, a derivative subgroup, characterized by additional advances in the locomotor skeleton and locomotor behavior. This character-by-character, iterative analytical process can be carried out with software as well, which allows for a very large number of characters to be evaluated simultaneously and played off one another in order to find inconsistencies. However, in general, even when databases are large, it often takes a surprisingly small number of traits to confidently identify clades.

Strier and I thought of morphology and behavior as being integrated in every way, and so the resulting cladogram was effectively an ecophylogenetic model, which can be summarized this way: The Muriqui and Spider Monkey are sister genera that are more closely related to the Woolly Monkey than any of these three genera are related to the Howler Monkey (see fig. 2.11). In other words, the atelids form a monophyletic group, with the Woolly Monkey, Spider Monkey, and Muriqui making up a monophyletic subgroup in which the Spider Monkey and Muriqui are sister taxa. In reconstructing the characteristics projected to have been present in the last common ancestor of atelids, we were able to infer that the evolution of the clade proceeded within a frugivorous-folivorous Adaptive Zone. Ecological distinctions among the genera were attained as they evolved new ways of inhabiting the frugivorous-folivorous zone, eating different types of foods and using different styles of locomotion to forage. These are new Adaptive Modes. The anatomical and behavioral correlates relating to them are discussed in later chapters dealing with diet, locomotion, and social organization.

The pitheciids are linked phylogenetically by their dentitions and a variety of cranial features, as described above and further explained below. Their jaws and anterior teeth are designed to ingest relatively large and tough fruits, and to masticate seeds, in an anatomical configuration that is unique. This is clearly shown by resemblances shared among *Callicebus*, *Pithecia*, *Chiropotes*, and *Cacajao*, titis, saki monkeys, bearded sakis, and uacaris.

Aotus is different. Though Owl Monkeys share the general outlines of the pitheciid pattern, they are more primitive, especially in the cheek teeth. The main ways in which *Aotus* links with pitheciids cladistically is through a set of resemblances shared exclusively with *Callicebus*, especially in cranial anatomy, jaw shape, communication, social organization, and mating strategies, as discussed in later chapters. That hypothesis can be further extended, because the Owl Monkey dentition also emphasizes harvesting in having anatomically specialized incisor teeth, although their form is distinctly different from the pitheciines and *Callicebus*. The central upper incisors of *Aotus* are exceptionally wide and thick, forming a large, solid platform to stabilize a fruit while the lower incisors are raised against the item.

Molecular studies have presented a different phylogenetic interpretation of *Aotus*. They regard Owl Monkeys as belonging to the cebid clade, although there is no broad agreement about the specifics of the linkage, namely, which genus or set of genera are the animal's closest relatives. It is rare for phylogenetic studies based on morphology and molecules to produce such radically different results, incompatibly placing one genus on a very different branch of a robust tree. The dispute remains unresolved.

How do the three major clades of platyrrhines, cebids, pitheciids and atelids relate to one another cladistically? According to the morphological interpretation presented here, the atelids and pitheciids are a unified group that links up with the cebids (fig. 4.4). Some of the most recent molecular studies, however, suggest that atelids and cebids are a unified group separate from the pitheciids. Be that as it may, the important thing is how well the Ecophylogenetic Model presented here explains what we see in the living animals today and how it syncs with Darwin's description of evolution: "descent with modification." How do these hypotheses of descent square with the animals' ecological adaptations, morphologically and behaviorally? How do they align with diet, locomotion, brain morphology and function, and sociality? The answers are in the following chapters.

CHAPTER 5

HOW TO EAT LIKE A MONKEY

One might think the rainforest is a monkey buffet—all you can eat, anything you can eat, an all-for-the-taking movable feast of fruits, leaves, seeds, gums, nuts, nectar, pith and flowers, insects, frogs, lizards, bird's eggs—on which the monkeys can gorge themselves. In fact, that is not the case. Because of the structure of their teeth and guts, body size, metabolic needs, and locomotor systems, and depending on what season it is, each species employs a different feeding strategy that has been engineered over geological time via natural selection. For each of these species, there are hidden costs and difficulties encountered when dealing with any potential food, resulting from their own biology as well as the biology of the life forms they eat, their physical structure, chemical composition, and evolutionary strategies, not to mention the presence of other monkeys that may be potential competitors because they have similar needs.

With all the nutrition contained in the forest, primates eat a wide variety of food types from a very large number of plant species, testimony to the extraordinary richness of the rainforest flora and to the core primate feature of versatility. Vegetation is the predominant resource, but fruits are at the dietary epicenter of primate life. The biological relationship between primates and fruit trees is complicated and mutually beneficial. Because primates have been coevolving with flowering land plants, angiosperms, since their inception, each one benefiting from the other's goods and activities, primates are partly responsible for the enormous diversity and success of fruit-bearing trees, shrubs, and lianas, and vice versa. The angiosperms produce edible fruit surrounding their seeds, and many primates, like birds and bats, excrete the seeds when defecating, which assists in seed dispersal and propagation. The fossil record suggests this feedback loop started with the beginnings of the modern primate radiation at least 55 million years ago, and probably before that, as forest life rebounded after the Chicxulub asteroid tore into Earth 66 million years ago and decimated biodiversity globally, contributing to the extinction of dinosaurs in the process. Afterward, paleontology shows that the transformative power of the coevolutionary, ecological interplay between primates and angiosperms contributed to building the rainforest as we know it. The appearance of early primates coincided with the emergence and rapid diversification of their primary habitat, the first closed-canopy tropical rainforests.

Charles Janson, one of the most insightful platyrrhine behavioral ecologists, explored the hypothesis of coevolution by studying 258 species of fruit in a tropical rainforest in Peru, which constituted more than half the diversity of fleshy fruits in the area. He demonstrated the coordination between fruit morphology and the abilities of two major dispersers living in the forest, New World monkeys and birds. He showed that each group tended to select from different fruit categories based on size, toughness, and color. Birds preferred fruits that were smaller, less than 14 mm, or about half an inch in diameter, were thinly protected by a skin (like a cherry), and were colored red, black, white, purple, or mixed in color. Monkeys preferred larger fruits, larger than 14 mm diameter, that were covered in a thicker husk (like an orange), and were orange, yellow, brown, and green. As Janson pointed out, monkeys may see these color hues better than others and in comparison to the beaked birds, the primates are better able to manipulate fruits with their hands, teeth, and mouth. In a comprehensive literature review accounting for more than 70 years of research, he and his colleagues also demonstrated that the vast majority of platyrrhines that ingest seeds also defecate them. The only exceptions appear to be the pitheciines, monkeys that specialize on eating seeds as opposed to the fleshy parts of fruits.

In the 1980s, I visited the small forest preserve at Fazenda Montes Claros, a research and conservation site in a remnant patch of Atlantic Forest, near the Brazilian city of Belo Horizonte in the State of Minas Gerais. It is now home to the Muriqui Project of Caratinga (Projeto Muriqui de Caratinga), one of two conservation success stories discussed in chapter 11. Four species of monkeys live there: the Buffy-headed Marmoset, *Callithrix flaviceps*, that eats primarily fruit, insects, and some gum; the Black or Black-horned Capuchin, *Cebus nigritus*, that tends to eat fruit, insects, and pith; the Brown Howler Monkey, *Alouatta fusca*, and the Muriqui, *Brachyteles arachnoides*, that eat mostly fruit and leaves. The kind of fruit each monkey selects depends on body size, tooth morphology, and locomotion. The different combinations of species-specific anatomies and behaviors limit how their needs overlap with respect to the resources each species requires to survive.

At Montes Claros, and in every other community of platyrrhines, evolution has struck a balance that allows the bounty of nature to be spread among the resident species, each occupying its own ecological niche defined by the anatomical and behavioral parameters of the genus to which it belongs. Each genus and species has evolved feeding preferences to sustain life in its own way, which limits competition with their monkey neighbors. It's a balance between what each species is best suited to eat and what they are unable to eat efficiently.

Different teeth for different foods

Teeth evolve in response to the most challenging foods that are eaten, even though those foods may make up a small portion of a primate's diet. That portion may be vital to an animal's well-being throughout the year, or during seasonal shortages. In 1976, Warren Kinzey and I called this the Critical Function Hypothesis in our study of the functional morphology of platyrrhine molar teeth. An ecological derivative of this concept later became the Fallback Food Hypothesis. This hypothesis emphasizes dietary shifts that occur in connection with seasonal variation in food availability that can contribute to dietary stress. Seasonal shortages are known to impact tropical animals, and for some primates it is a documented cause for weight loss. Some species shift their feeding foci to foods with lower nutritional value, that are harder to access, harder to masticate, require more travel and handling time, or contain a higher proportions of toxins that limit how much can be consumed and may require a detoxification system in the gut. Such shifts may be decisive in the life of an individual, so they are at a selective premium. The Critical Function and Fallback Food models predict that dental adaptations should not be correlated with easy-to-access or easy-to-process foods that may be preferred by the animal and widely available. Instead, it predicts that dental adaptations are selectively determined in relation to hard-to-access or hard-to-process foods that are obtainable whenever required.

A new approach to the study of mammalian and primate molar teeth was developed in the lab during the late 1960s and 1970s by the groundbreaking functional morphologists Alfred Crompton and Karen Hiiemae. Their studies integrated the macroscopic shape-geometry of upper and lower molars with analyses of microscopic tooth-scratch effects produced as the teeth occlude, and the jaw-and-tooth movements that occur during chewing, the latter revealed by X-ray cinematography taken during experimental feeding trials. Working closely with Hiiemae was Richard Kay, then a graduate student. Kay went on to apply the concepts and methods they developed during this project and reset the basis for studying dental functional morphology in living and fossil primates, beginning with a classic paper published in 1975, "The functional adaptations of primate molar teeth." Also during that time period, the evolutionary biologist and ornithologist Walter Bock, and Gerd von Wahlert, an ichthyologist, analyzed the conceptual relationships between functional morphology and adaptive inference, and developed the Form-Function-Biological Role model of adaptation.

A broad agreement resulted: the physical properties of foods, such as their mechanical strength and deformability, determine the shapes of primate molar teeth via natural selection. This led to several general formulations linking the mechanical potential of tooth anatomy with the structure of food, and explained why there are folivorous molar shapes, frugivorous molar shapes, and insectivorous molar shapes. Molars designed for processing pliable, fibrous food—leaves—have long, accentuated edges and crests that shred and shear. Molars built to masticate more dense three-dimensional material—fruit—have rounded cusp surfaces and basins that trap and mash. Molar architecture geared to treat brittle food—insects—exhibit pointy cusps and sharp crests to puncture and mince exoskeletons (fig. 5.1). Even with these insights there was a catch.

While these are useful correlative relationships they do not provide a causal model, either proximally, for how a tooth actually performs, or, ultimately, for what food, specifically, the tooth was adapted. The catch was in how we use shorthand to link teeth and food type. Associating molar shape with a broadly defined dietary category, such as leaves, insects, and fruit, skirts around what we acknowledge as the principal source of selection—the specific mechanical and chemical compositions of the items eaten. Fruits may be soft or hard; insects may be inelastic or squishy; the leaves chosen may be young or mature, tender or strong. If frugivory is the adaptive explanation behind Spider Monkey and Saki Monkey dentitions, for example, why do their incisors look so radically different (fig. 2.7)? Why are other basic parts of the masticatory apparatus—the canines, molars, mandible and chewing musculature—so different as well?

The lessons drawn from the Form-Function-Biological Role model helped move functional dental morphology to another level of subtlety that bound together field and lab work as indispensible for analysis. In this model a sharp crest is simply designed to cut. What it cuts can be determined only by field observations. A crested molar may be used to shear leaves in some species and to dice insects in others. The food type to which the crested molar is primarily adapted may not be determinable without additional information, and it may not be optimized for any one kind. In some situations, molars cannot even be predictive of diet no matter how distinctive the form is. The features critical to the animal's feeding behavior may reside elsewhere, as is the case with the gum-eating Marmoset and Pygmy Marmoset: incisors and canines are the key (fig. 5.2). A more sensitive morphological analysis ought to also integrate premolars, jaw shape, muscles of mastication, etc.

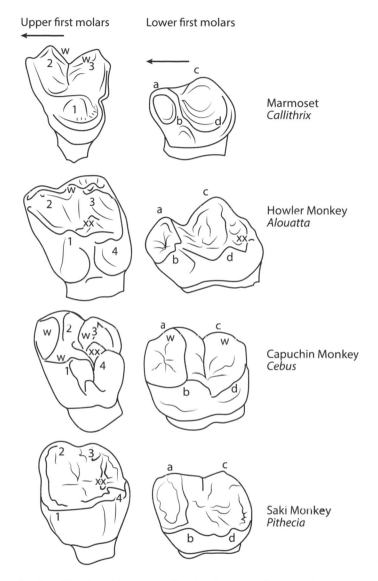

FIG. 5.1. Sketches of four platyrrhine upper and lower molar sets emphasizing differences in cusp relief (1–4; a–d) and development of ridges relating to varied dietary adaptations. Worn surfaces indicated by w and xx. Adapted from Rosenberger and Kinzey (1976).

It also became obvious that feeding behavior—thus morphological/biomechanical analyses—had to be studied along with other natural life parameters in order to develop a full appreciation of dental and dietary adaptation. Ingestion is one thing; mastication is another. Getting at the food with incisors and canines is just as important as processing the food with the premolars

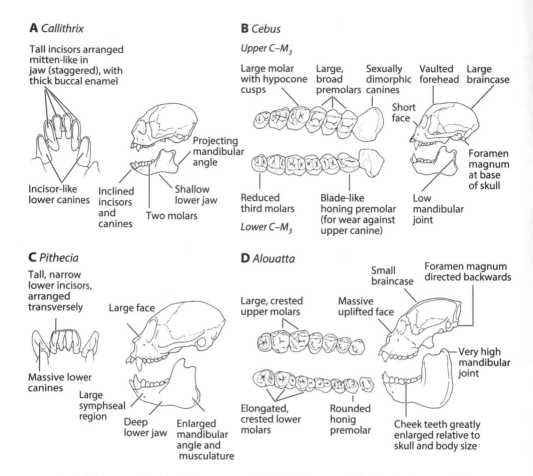

A *Callithrix*

Tall incisors arranged mitten-like in jaw (staggered), with thick buccal enamel

Incisor-like lower canines

Inclined incisors and canines

Projecting mandibular angle

Shallow lower jaw

Two molars

B *Cebus*

Upper C–M₃

Large molar with hypocone cusps

Large, broad premolars

Sexually dimorphic canines

Vaulted forehead

Large braincase

Short face

Foramen magnum at base of skull

Reduced third molars

Blade-like honing premolar (for wear against upper canine)

Low mandibular joint

Lower C–M₃

C *Pithecia*

Tall, narrow lower incisors, arranged transversely

Large face

Massive lower canines

Large symphseal region

Deep lower jaw

Enlarged mandibular angle and musculature

D *Alouatta*

Small braincase

Foramen magnum directed backwards

Large, crested upper molars

Massive uplifted face

Very high mandibular joint

Elongated, crested lower molars

Rounded honig premolar

Cheek teeth greatly enlarged relative to skull and body size

FIG. 5.2. Key cranial and dental features associated with dietary specializations: (A) gum eating in a marmoset; (B) eating hard fruit and other vegetation in a capuchin; (C) husking shells in a saki monkey; (D) leaf eating in a howler monkey. From Rosenberger (1992), copyright Cambridge University Press, reprinted with permission.

and molars. As the most frugivorous order of mammals, primates display the greatest variety of ingestive tools, incisors, and incisor-canine combinations, as well as biting premolars, and platyrrhines are more varied in this respect than any other primate group. As a consequence, the forms of their rather simple front teeth as well as the intricacies of their postcanines are important keys to understanding New World monkey ecophylogenetics (figs. 4.2, 5.2). At another level, posture and locomotion, and dexterity, come into play: how the environment shapes selection of the methods by which foods are approached and acquired. From the animal's point of view—which should be ours as well, scientifically—dietary adaptation is more than molar tooth design. Often, it is a lifestyle.

By the 1990s and 2000s, field-workers had studied many species of primates and discovered that seasonality, in particular the leaner periods when foods are scarcer, leads to changes in diet that often involve feeding on more challenging foods than the preferred ones that are out of season. The structural adaptations of the dentition of various species appeared to be correlated with the mechanical properties of the specific types of foods eaten at those critical times. New World monkeys led the way toward this realization because of their exceptional degree of morphological and dietary diversity, which sometimes made for striking contrasts even among species living together at the same site, and also when comparing platyrrhines with other groups of primates. The vital data collected to amplify the models of dental adaptation included measurements of the physical properties of foods. Warren Kinzey and Marilyn Norconk pioneered the efforts to quantify the properties of foods consumed by their subjects in situ, using simple instruments to measure the toughness of fruits eaten by spider monkeys and saki monkeys in the wild before the fruits' structure changed after falling or removal from the trees. They found significant differences that related to the biomechanical requirements for breaching and consuming fruit. Fieldwork was at that time proving the principles of the Critical Function model and validating its ecological complement, the Fallback Food model.

What do they eat?

Rainforest primates are literally immersed in potential edibles because so much of the forest is made up of materials that are nutritious. Carbohydrates, sugars, and starches are efficient, rich sources of food energy, and they are superabundant in living plants. Fifty to 80% of the dry matter in leaves, fruit, and seeds is composed of carbohydrates, which contribute 40% of the metabolizable energy of all primates. While the amount of protein is meager in most plants, primates actually do not need much protein. In fact, the nutritional ecologist Olav T. Oftedal has argued that primates are generally adapted to subsist on foods with only moderate nutritional density. For example, they need to produce only small amounts of milk for suckling their young because the infants develop slowly. In this way they are able to avoid the critical energy bottlenecks encountered by other mammal mothers during growth and lactation.

One of the first and most influential studies of primate diets and communities, published in 1969, was conducted by the pioneering French ecologists and nutritional botanists Claude Marcel Hladik and Annette Hladik. They studied and quantified the feeding behavior of five platyrrhine species living on Barro Colorado Island in the Panama Canal, at a hilltop research station of the Smithsonian Institution, and on the mainland. The composition of primates in their study was very similar to those in the Fazenda Montes Claros group at Caratinga. These results were the baseline from which it became evident that rainforest primate communities are constructed in much the same ways ecologically and phylogenetically.

Species of the Howler Monkey, *Alouatta*, and the Capuchin Monkey, *Cebus*, are present at both places. The Brazilian Muriqui, *Brachyteles*, is a close relative of the off-island Spider Monkey, *Ateles*, in Panama, similar in body size, body build, and lifestyles, though differing in diet. The Marmoset living at Montes Claros, *Callithrix*, is a small callitrichine like the Tamarin at Barro Colorado, *Saguinus*. Only the Owl Monkey, *Aotus*, lacks an analog at Montes Claros, although the Titi Monkey, *Callicebus*, which in many ways is a diurnal version of *Aotus*, is relatively common in many areas of the Atlantic Forest. Both belong to the same subfamily and have similar dietary needs.

Hladik and Hladik organized the diets of their subjects into three main categories—leaves, fruits, and prey—and 36 subcategories. Several refinements of the Hladiks' system have evolved, though these core classes are still widely used half a century later. One difference is substituting the term insects for prey, as primates tend not to feed on a wide variety of animals. Primate prey usually consists of terrestrial invertebrates, arthropods, insects, including beetles, moths, and crickets, spiders, centipedes, and the like. For cebids that do eat a wider variety of other animals, including small vertebrates, the term faunivore is useful. We also use some dietary terms in combinations that expand the categories, such as frugivore-folivore and frugivore-insectivore, for reasons that will become clear. Colloquial expressions like gum eating, seed eating, hard- and soft-fruit feeding, and hard- and soft-bodied insect feeding have become increasingly important—as have formidable terms in the realm of professional jargon. Examples are phytophagous (plant eating), durophagous (hard-fruit eating), sclerocarpic (hard seed-coat cracking), and mycophagous (fungus eating), terms that have largely been created by morphologists concerned with the physical tasks of ingesting and masticating particular foods, that is, applying the Form-Function model with rigor.

The Hladiks showed that the diverse group of primates they studied, drawn from major lineages of New World monkeys, favor fruit by a wide margin. No matter whether the species is an atelid, such as a howler or spider monkey, a cebine like a capuchin, a pitheciid like an owl monkey, or a callitrichine like a tamarin—it will favor some kinds of fruit. Few of these fruit varieties are found in the urban supermarkets to which many of us are accustomed. Some are noxious to humans.

In wild fruits eaten by platyrrhines the edible parts may be a thin coating of soft, slimy aril that surrounds a seed and can be easily removed, or a thick lining that adheres to an outer shell and needs to be scraped away. Primates contend with all that, as well as a phenology schedule, meaning the times during the year when fruits and leaves appear, and the natural distribution of fruit-tree species in the forest, which may be clumped together or spread like a patchwork at varying distances. Primates must also deal with longer-term climatic influences that determine the yields of fruit crops over a lifetime. In some cases, when territories are small and one or two food trees are the primary staple, as with the gum-eating Pygmy Marmoset, they may have to contend with the death of a feeding tree, a consequence of the monkeys' own activities. Primates have to be smart and versatile to survive.

The most avid fruit eater, *Ateles*, a ripe-fruit specialist, is a broad-spectrum feeder. Its diet typically consists of 80% or more fruit at any field site. The number of fruit species eaten by spider monkeys is astonishing. At Yasuni National Park in Ecuador, an exceptionally rich forest where they have been well studied, they eat fruit from 238 species. Their principal fruit sources are more limited, however, the top five species accounting for 23% of their diet. These facts explain why spider monkeys are also locomotor specialists, traveling far and wide, at considerable speed, to survey destinations where a group of trees they may know is producing ripe fruit, which might be available for only a few days. Spider monkeys are not alone in being able to manage such a large volume of food-related, temporospatial information. Paul Garber has shown that the Saddle-backed Tamarin, *Saguinus fuscicollis*, and the Mustached Tamarin, *Saguinus mystax*, are able to remember the locations of more than 300 feeding trees visited during the course of a year.

Secondary food preferences

According to the studies at Barro Colorado Island, the second-most important food item in each primate species' diet is also a major factor in their lives, and with that it becomes clear how the monkeys manage to stay out of each other's way in an ecological sense. Comparable studies at many sites have shown that the large howlers monkeys go for foliage at very high rates as a supplement to fruit. Nevertheless, though leaves comprise as much as 80% of the *Alouatta* diet, howlers gorge themselves when a favored fruit is ripe and ready to be eaten. In contrast, the equally large, frugivorous spider monkeys go for foliage at a third or half that rate. The middle-sized capuchin monkeys complement their fruit intake with prey and vegetation. The smaller-bodied, nocturnal owl monkeys consume foliage as a supplementary food, like the howlers. The still smaller tamarins focus on prey.

The reason primates have primary and secondary preferences is that no single food resource can fulfill all the nutritional needs crucial to the production of energy, year-round. That is why primates eat from a large variety of fruit trees. Each fruit type has its own constituent chemical properties. The mixtures of fruit, leaves, and insects provide the animals with their basic dietary needs, mainly the carbohydrates, proteins, fats and lipids, oils, fatty acids and vitamins. Within primate communities, each genus consumes a different combination of foods and has different ways of accessing them, effectively dividing the resources. Natural selection enables the monkeys to develop their various niches to limit competition for the same foods.

This highly mixed and seasonal diet helps explain the morphological diversity of platyrrhine, and primate, teeth, which are generally thought by mammalogists as being rather unspecialized by comparison with most other orders of mammals. Primates have retained the distinctive mammalian feature of four tooth groups—incisors, canines, premolars, and molars—each with distinctive shapes and different biomechanical potential. In contrast, most other orders of mammals have narrowed their foraging tendencies or feeding biomechanics, and accordingly, via selection, this ancient pattern was modified in many different ways. Teeth have been added, subtracted, or lost altogether. Premolar and molar postcanine teeth have become homogenized or radically differentiated morphologically. Cusps and ridges have been transformed into corrugated bands of enamel and dentine with sharp edges that shred material like a food grater. Crowns have become open-rooted to produce ever-growing teeth. Primates have evolved more modestly, by slightly shifting the positions of postcanine cusps into new alignments; occasionally adding or subtracting a cusp; pairing cusps and merging them at their bases; creasing the inner sidewalls of rounded cusps to form blunt edges; sharpening the leading edges of crests; elevating or lowering cusp relief; flattening or deepening the basins that cusps slot into during occlusion. Shape changes in the simple designs of the incisor teeth are even more minor. They evolved to become taller or shorter, wider or narrower, rounded in cross section or relatively compressed from front to back, or reinforced by thickening the enamel on the outer surface where incisors are most likely to be heavily worn. The results make for a versatile dentition that balances easy-to-process fruits and also operates efficiently to acquire or process challenging vegetable foods as well as small prey, depending on the foraging habits and lifestyle of a species.

The staple fruits consumed by primates tend to be protein-poor but they provide lots of carbohydrates. Monkeys such as *Ateles* that overwhelmingly rely on ripe fruit, may thus require some protein to balance their nutritional needs. This suggests that their long-distance roaming to forage may not only be geared to finding fruit. It may also reflect their need to find protein in the form of leaves they can manage to eat with a fruit-eating masticatory and digestive anatomy, meaning leaves that are young, thus protein-rich, and more easily chewed and metabolized than mature leaves. The spider's fruit-focused diet is high in carbohydrate content, and easily and rapidly convertible to energy, which they burn at a high rate given their freewheeling locomotor style and long daily travels. The organ that benefits most directly from carbohydrates is the brain, which is fueled almost exclusively by the glucose derived from that compound.

To obtain the proper balance of carbohydrates and protein, larger primates tend to augment fruit with leaves, while smaller primates tend to eat insects. Hence the broad descriptive combinations: frugivore-folivore and frugivore-insectivore. These alternative feeding strategies result from the fact that the energy requirements of animals are related to body size. Small animals expend energy at higher rates than larger animals for sheer maintenance needs, so their diets need to be proportionately richer in nutritional value and more easily metabolizable. Animal prey is a good source of digestible protein and fats, so small primates tend to favor insect-based diets. Large primates would have difficulty doing that.

For large genera such as the Howler Monkey, even leaves present dietary challenges, for a variety of reasons. Leaves are not very nutritious, so they have to be consumed in bulk to get a worthwhile net reward in energy, which can be measured as calories. Additionally, unless leaves are very young, they may be hard to masticate and difficult to digest, partly because they may contain toxic substances. Another reason is structural; leaves contain cellulose, a woody material, and lignin, another biomolecular compound that strengthens plant cell walls. No primates naturally produce the chemical enzymes needed to break down these materials in the digestive tract. To get around this, folivorous primates host microorganisms in their gut biomes to help release the energy from leaves by the process of fermentation. That is why folivorous primates tend to have large, bloated pot bellies, where digestion slowly ferments and detoxifies large quantities of leaves.

In addition to tooth crown shape, another functionally important feature of molar teeth is the enamel cap that becomes the crown. It is thin in some species and thick in others. Primates with thicker enamel tend to have blunter cusps and ridges, a pattern that spreads the pressure applied to food when upper and lower molars occlude. It is a crushing and grinding adaptation. The thick enamel also makes the surface of the tooth more resistant to normal tooth wear. Capuchin monkeys, for example, have exceptionally thick enamel on their rather blunt molar teeth, though they may not need such reinforcement throughout the year. That observation, through a combination of lab-based morphology and research in the wild, led to an important insight that has been applied to many primates, the Fallback Food model. The definitive study that revealed the fallback pattern was conducted by tropical ecologists John Terborgh and Charles Janson, then his undergraduate student. They showed the importance of critical functions associated with backup foods during times of scarcity as a likely source of selection, as opposed to measures of food intake experienced on an annual basis or when food is plentiful.

Surviving preferred-food scarcity

At Cocha Cashu, Peru, in the 1970s, Terborgh and his team studied the community of platyrrhines living in a pristine Amazonian rainforest in a remote national park, one of the largest and most biodiverse in South America. It has since been designated the Manu Biosphere Reserve, a UNESCO World Heritage Center. In *Five New World Primates*, a classic in primate socioecology published in 1983, Terborgh described what happened to the monkeys during the dry season, and how they changed their ranging and feeding behavior. Each species had its own strategy, but the two capuchin species and the squirrel monkeys were most illuminating.

When overall fruit productivity went down, 77% of feeding time in *Saimiri* became devoted to figs, up from the average of 50% across the seasons. When not gorging on figs, the squirrel monkeys resorted to "pure insectivory." This means that during the leanest season tough insect exoskeletons need to be masticated as well as soft fruit. The Critical Function Hypothesis predicts that under these circumstances the dominant features of *Saimiri* molar crowns would be pointy cusps and slicing crests for puncturing and mincing exoskeletons, and they are; they are not the puffy cusps and broad basins that are ideally suited for mashing soft material like figs. This is an effective integration of molar mechanics, because the soft fruit can also be easily chewed with pointy, crested molars and premolars.

The two *Cebus* species present a more trenchant example because they are very closely related and share the same habitat at Manu, so their morphologies and lifestyles are naturally expected to be very similar. They have dull, rounded molars without crests, not well designed for eating leaves. The capuchins tend to ignore leaves even though they would seem to be the right body size to take advantage of them. The reason they don't eat leaves is because of their ecophylogenetic heritage: the cebid family is intrinsically faunivorous. They have evolved molars with thick, wear-resistant enamel because at the onset of the five-month dry season, *Cebus albifrons*, the White-fronted Capuchin, and *Cebus apella*, the Tufted or Brown Capuchin, begin to eat the nutritious endosperm from inside the hard-shelled *Astrocaryum* palm fruits, when they start to ripen. Endosperm is the softer tissue inside the nut that surrounds the seed and is rich in carbohydrates, protein, and lipids, supplying nutrients for the seed's growth.

But the ways in which these monkeys get to the endosperm differ because of their distinctly different body types and masticatory systems (fig. 3.2). *Cebus apella* is the more robust animal, more muscular and thick-boned. Consistent with its stocky build, it has powerful chewing muscles. *Cebus albifrons* is a more delicately built animal, cranially and in its physique. When the *Astrocaryum* nut is very green, before the shell is hardened and woody, *Cebus apella* is able to crush it with a single bite, whereas *Cebus albifrons* must wait until the fruits fall to the ground and rot on the forest floor, which makes them softer, and easier to crack open.

For the first three months of the dry season both species rely heavily on *Astrocaryum*. It makes up 56% of the diet of *Cebus albifrons* and 64% of the diet of *Cebus apella*. What access techniques do they employ? *Cebus albifrons*, which is otherwise fully arboreal, spends many hours on the ground under palms where large numbers of fruits have fallen and begun to rot. On the forest floor the fruits are invaded by bruchid beetles that feed on *Astrocaryum* seed after boring into them. This hollows out and weakens the nut, while also removing a portion of the nutritious endosperm that surrounds the seed. The White-fronted Capuchins sniff their way through a pile of nuts to identify the infested ones and pick out those that are accessible. After carrying a nut into a nearby tree, they smash it against a branch with their hands until it can be further bitten open. They may take 5–10 minutes to find and open each nut, and to eat the endosperm left over by the beetles.

Tufted Capuchins are more efficient. Juveniles, already larger and stronger than fully grown White-fronted Capuchins, can breach hard-shelled *Astrocaryum* in the trees and thus get to eat the full content of endosperm. With its stronger bite, *Cebus apella* also makes use of other parts of the *Astrocaryum* palm that *Cebus albifrons* physically cannot, such as the pith in the bases of hard fibrous leafstalks. This ability makes the Tufted Capuchin's dry-season reliance on the tree even more effective. This extraordinary capacity to eat the very hardest parts of palms may have led *Cebus apella* to evolve into a palm specialist. They even select palm trees as sleeping sites.

To get an idea of how tough an *Astrocaryum* nut is, primatologists have tested what it takes to produce an initial crack that can lead to the nut finally splitting open. In one experiment, a mechanical pressure-tester was used to measure indentation hardness in a standard way. Cracking the *Astrocaryum* nut required 197 units of force, more than the force required to dent the shell that encloses a Macadamia nut (which is actually a seed), which registered 162 units. For reference, the experiment showed indentation hardness of carrots at about 1, the value for almond shells at 2.5, and for Brazil nut shells it was 7. We have, as yet, no measure of the toughness of the *Astrocaryum* used by the Cocha Cashu monkeys, which probably select less woody fruits, not yet mature and somewhat softer.

The energy costs to both capuchin species highlight the advantages gained by *Cebus apella* through its novel feeding adaptations. Without having to find rotten, infested nuts, they can harvest more *Astrocaryum* while operating in much smaller home ranges, thus minimizing travel time and effort. Their efficient nut-cracking skill also favors a large energy return per time invested. As calculated by Terborgh, *Cebus apella* devote 45 minutes of their monkey-day feeding on *Astrocaryum* while *Cebus albifrons* spend 130 minutes for a smaller return. *Cebus apella* survive the dry season by concentrating on a few patches of *Astrocaryum* palms while *Cebus albifrons* must travel widely in search of figs as well, to supplement their diet. While both species mitigate seasonal shortages by switching to fallback foods, the enhanced critical function adaptations of *Cebus apella* stand out to the extent in which its anatomy and behavior have been modified by natural selection for an express purpose during a specific time of the year. As to the other Cocha Cashu primates, when their foods were scarce they also altered their diets to make do with what was available. Spider monkeys, howler monkeys, and titi monkeys all ate more leaves.

Gouging tree bark to eat the tree gum

For the Marmoset and Pygmy Marmoset, the biomechanical challenge of feeding on gum is experienced year-round. Gum is an exudate, a substance that seeps from trees, in addition to other exudates such as sap, latex, and resin, each with distinctive biochemical characteristics. Primate ecologist A. C. Smith demonstrated how specialized the Pygmy and Common Marmosets, *Cebuella pygmaea* and *Callithrix jacchus*, are in their reliance on gums as their principal food resource. He compiled a wealth of comparative data in 2010 on gum eating across the primate order. These genera stand out. Sixty-seven percent of the *Cebuella* diet comprises gum. For five *Callithrix* species, the highest values are 73% and 83%, and most are above 60%. In contrast, the highest percentages of gum in the diet among tamarin species are 15% and 16%. The Tamarin is not anatomically specialized to procure gum intentionally but eats it when it is found to be already oozing from the bark. The non-callitrichine platyrrhines are trivial consumers of gum. It makes up 4.6% of the diet in squirrel monkeys, 5.9% in woolly monkeys, and only 0.08% in saki monkeys.

In its natural state, gum is a secretion that forms a protective crust to seal an opening in a patch of bark that typically results from mechanical damage. It is a viscous form of dietary fiber, digestible in the same way that cellulose, a related carbohydrate that is a main constituent of leaves is digestible, but only under the right conditions. Both must ferment with the assistance of microbes that are maintained in the gut, because mammals do not produce natural enzymes capable of breaking down these forms of fiber.

A fascinating comparison can be made with the Howler Monkey, *Alouatta*, which is among the largest New World monkeys. Consequently, they have large guts in which the food is fermented after passing through the stomach, and a slow passage rate that moves the food through the gut, which allows time for fermentation, detoxification, and nutrient extraction. The Pygmy Marmoset, *Cebuella pygmaea*, by contrast, is the smallest platyrrhine, yet it does the same thing. Gum passes through its gut very slowly, much slower than in other members of the callitrichine subfamily, giving microbes in the intestine the time needed to digest it. This system appears to be efficient, since pygmy marmosets eat very little fruit in addition to their intake of insects, and they actually consume very small amounts of gum each day. Laboratory studies using gum arabic, an ingredient of chewing gum, used as a proxy for the exudates eaten by callitrichines in nature, indicate that the gum-eating specialists, the Common and Pygmy Marmoset genera, are better able to digest it than the three other callitrichines tested, none of which are obligate gum eaters.

The Pygmy Marmoset and the Marmoset have the gastric adaptations to eat gum, but first they have to harvest it—a Herculean effort for tiny monkeys that weigh roughly 3–9 ounces, 85–255 g. So, they have to put their whole bodies into it. They position themselves on large tree trunks and boughs as they prepare to feed. With a small body size and clawed digits, typical callitrichine traits, *Cebuella* and *Callithrix* have postural adaptations enabling them to cling to large tree trunks. They also have craniodental adaptations specific to the mechanical challenge of scraping and gouging bark to get the secretion started. These features include teeth, the mandible, jaw joint, jaw muscles, and overall cranial morphology. One specialization enables the monkeys to open their mouths in a large gape in order to engage their front teeth, uppers and lowers, in the gouging process.

FIG. 5.3. Scanning electron micrograph of a vertical cross section of the upper portion of a lower central incisor of a tamarin (left) and pygmy marmoset (right) to highlight the thickened outer-side enamel in the latter. From Rosenberger (1978), courtesy of Oxford University Press.

The lower canines are functionally integrated with the incisors to form a gouging battery, six teeth in all when adding right and left sides together (fig. 5.2). Rather than being canted laterally outward like typical platyrrhine canines, the lowers are implanted vertically in their sockets, shaped like the incisors and closely aligned with them. This limits the twisting of the lower jaw as would happen if the canine tips were scraped against the tree directly from a laterally jutting orientation. Spreading the impact of wear across six lower teeth as opposed to four may increase their functional longevity. Crown height and microstructure help as well. The lower incisors are especially tall, which also adds longevity to their useful lifetime, and they are fronted by thick enamel to strengthen the crown and protect it from abrasion while the back sides of these teeth have little or no enamel covering (fig. 5.3). This arrangement apparently serves to maintain the tooth's pointed apex as well, which helps concentrate its penetrating pressure when applied to the bark. Exactly how this works is not clear. It may be that the ultrastructure of the enamel, the microscopic, geometrical arrangement of enamel tissue, is such that during usage the material continually fractures and splits off along the same axis, at an acute angle to the shaft of the crown. Meanwhile, during gouging, as the jaw is raised so the lower teeth scrape against the bark, the upper incisors are embedded into the bark to stabilize the monkey's head.

All these integrated details, like the system as a whole, are composed of de-rived features. Nothing like this appears as a pattern in early primates or in any other modern forms. The rarity of these features and connection to gum eating, an unusual diet, make it clear that the gouging complex is a uniquely derived specialization in the sister genera *Callithrix* and *Cebuella*, another example il-lustrating the power of character analysis for its phylogenetic implications, and feeding adaptations as ecophylogenetic clues to platyrrhine evolution.

Incisors are key to fruit eating

Frugivorous primates are generally interested in eating the fleshy pulp inside the skin of fruits. Their dental adaptations tend to be modest in a mechanical sense. The molars of frugivores are responsible for mashing up the fruit with help from premolars, especially the one adjacent to the first molar, which tends to have the same functional design elements. The chewing teeth work like a linear array of mortars and pestles, a row of blunt cusps pressing and gliding against reciprocally concave basins to crush and grind food. In anthropoid frugivores the incisor teeth, whose main role is to harvest and ingest food, are generally simple affairs. Lower incisors are shaped like chisels with a straight edge, vertical or slightly angled in orientation and aligned in a tight row, making them a good tool set for tearing off a rind and digging out a chunk of fruit. The upper incisors are more complicated in shape because the lateral one must conform to the rounded shaft of the lower canine crown, which slots into position between it and the adjacent upper canine. The canines of most frugivores may or may not be important in harvesting. They often are important in gestural communication. However, the canines of the smaller marmosets, saki monkeys, bearded sakis, and uacaris are important parts of the feeding system.

Variety in the physical properties of fruit interiors and exteriors predicts a range of evolved adaptations among frugivores pertinent to ingestion and mastication. A one-size-fits-all molar model does not go far enough to explain the anatomical variations exhibited by platyrrhines, just as the simple term, frugivory—though botanically faithful—cannot capture the essential source of selective pressures faced by platyrrhine fruit eaters and the diverse evolution-ary responses, which involve teeth and much more.

For wild primates, body size rather than tooth morphology may be the limiting factor in making fruit selections. Smaller species have smaller hands, smaller jaws, and smaller gapes, so fruit must be proportionately small, mean-ing their choices are limited. Larger-bodied primates have more options, and different parts of the masticatory system may be favored by selection depending on the structure of the foods. When confronted with larger, tougher fruits, they must bite off or scrape the dense, fibrous pulp from the rind or the pit before it can be chewed. The incisors are critical for that. The hardest-shelled fruits pose even more mechanical challenges to getting through the exterior. Canines might be used to crack them open, or the premolars immediately behind the canines might be enlisted. When teeth can't do the job, pounding a fruit manually, or with a rock, works for tool-using capuchin monkeys. As the Manu capuchins illustrate, successful frugivory is a lifestyle, and it involves teeth, body, and brain.

Platyrrhine feeding strategies and dentitions reflect all the frugivorous options, in addition to the bark-gouging, gum-eating morphologies. In fact, more than in any other major group of primates, in the New World monkeys the anterior teeth are diversely adapted to the task of harvesting, that initial action to get a morsel into the mouth so it can be eaten. This means that the functional morphology of the anterior teeth holds critical phylogenetic clues, useful for assessing relationships among the extant species and for unraveling the relationships of fossils. Paleontologists have usually given molars higher priority as a source of information because their topographic complexity can be used as a road map to reconstruct phylogenetic history. However, for certain ecophylogenetic groups the incisors and canines provide more important clues when interpreting the platyrrhine fossil record.

The most specialized of the hard-fruit-eating platyrrhines are the sakis, bearded sakis, and uacaris belonging to the pitheciine subfamily. The pitheciines are not hard-fruit specialists in the manner of the capuchin monkeys at Cocha Cashu, Peru, which turn to woody *Astrocaryum* palm nuts in the off-season, when crops of their more accessible, preferred foods are not available. Pitheciines selectively feed on tough fruit all year-round, concentrating on a small number of species in spite of the physical challenges they pose.

Pitheciines are often called seed eaters and seed predators rather than frugivores because, unlike the fig-eating squirrel monkeys, or the ripe-fruit specialist spider monkeys, pitheciines are less interested in eating the carbohydrate-rich flesh of fruit. They are focused, instead, on the protein- and lipid-rich seeds found inside the fruit, which they chew. This is an interesting departure from the coevolutionary role that primates generally play in furthering the success of trees by dispersing their seeds via defecation, dropping, spitting, regurgitating, etc., after which secondary dispersers such as ants and dung beetles may move them even further from a parent tree. It is also a reminder that some scholars maintain that, while primates have certainly evolved adaptations enabling them to use fruit trees as food, it is still difficult to establish a tight coevolutionary link between primates and trees in many cases. That is because many primate and non-primate species utilize the same fruit species, processing and eliminating the seeds in a variety of ways. Dispersed seeds also have a high mortality rate, which makes it difficult to determine if the success of new seedlings is actually dependent on deposition by a particularly adapted primate.

Dietetically, the most extreme pitheciines are the Bearded Saki Monkey and Uacari genera. They rely on seeds, including Brazil nuts, which are well protected from outside interference by a shell, and the nut itself is enveloped by a hard capsule. The fruit does not open naturally when the nuts ripen, at which time they usually fall to the ground. These monkeys also depend on legumes with woody, sheathlike pods that may only split open along a thin seam, or more rotund seed pods that are capped with a structural lid that can be snapped open. In all such cases, accessing the seeds means the monkeys must break open the coverings. That is why the pitheciine seed eaters have evolved a dentition and an entire masticatory system highly specialized for the harvesting phase of feeding. The challenges parallel what the gum-eating Marmoset and Pygmy Marmoset face in breaching a woody barrier to access the food.

Pitheciines have evolved a very unusual morphological system. The lower incisors are tall, narrow, and taper toward their tips, set tightly together to form a wedgelike scoop that is often used to pluck seeds from a fruit that has been opened. The uppers jut out at a steep angle forward from the snout, which positions them to properly occlude with the tall lower incisors. The large lower canines are stout, wide, bluntly pointed shafts with a triangular cross section, splayed sideways away from the toothrow. Of the three premolars, the crown of the first lower one is shaped like a wedge, while the following premolars are designed like mini-molars. The canines and their adjacent premolars are a centerpiece of the shell-breaching system. They act like pliers or lever nutcrackers. The jaws supply the muscular force and leverage to load a shell placed firmly behind the lower canine so it cannot shift in position while the mandible closes on it. Occlusal pressure is transmitted through the adjacent premolars, augmented by their wedged design. To masticate seeds, the last premolars and molar crowns are almost flat and featureless, but with a surface enamel that is textured by a carpet of wrinkles. They are constructed to minimize damage to the tooth crowns. The bumpy enamel surface stabilizes food-breakdown particles from slipping during cyclical crushing and grinding actions.

It is instructive to compare this system with that of capuchin monkeys, which belong to another platyrrhine clade, one historically invested in eating prey. Thus they had a different "starting point." Their capacity to manage tough fruits became augmented through selection of different characteristics. Their most distinctive adaptation in this regard combines thick tooth enamel and cleverness to take advantage of their surroundings—think of the Manu monkeys using pounding maneuvers to crack open hard-shelled nuts. Or, benefiting from the activities of bruchid beetles and the process of organic decomposition influenced by the beetles' feeding strategy, to access tough nuts after they soften.

A fundamental way that species apportion the ecology to allow coexistence is by their position within the spectrum of body sizes. Among the pitheciids, the degree of fruit hardness selected by a species corresponds with body size. The smaller genera feed on the less hard-husked fruits, and the largest genera eat the hardest ones. It is clear that the most divergent, derived anatomy is found in the larger bearded saki monkeys and the uacari. The more primitive condition is that of the smaller forms, the titi and owl monkeys that present a more ordinary anatomy. In evolving the saki-uacari pattern, these monkeys gained a unique ecological position in being able to eat seeds from fruits that are inaccessible to other platyrrhines. For some fruits, like woody legumes, being able to pry open the hull before it bursts open means they can get at the seeds before other monkeys, while some seeds are still relatively soft in their immature stage of growth.

There is an important anatomical and behavioral connection between the two clades of the pitheciid family, the homunculines and pitheciines, that involves the anterior teeth and how they are used in feeding. *Callicebus* and *Aotus*, the homunculines, are distinguished by an incisor-dominant dentition, like that of the pitheciines. They each have tall lower incisors, and *Aotus* has exceptionally broad upper central incisors as well. The lower incisors of *Callicebus* closely resemble those of the pitheciines, which are tall, narrow, and compressed together into an arch. There are other similarities evident in jaw shape relating to large gapes and powerful jaw-closing musculature.

A significant contrast between homunculines and pitheciines involves the canine teeth. Those of *Callicebus* are unusually low-crowned and vertically posted in males and females alike, while they are slightly taller, more pointy, and canted laterally in *Aotus*. Their relatively small size does not mean titi and owl monkeys do not use the canines in harvesting foods. They do. But the role of food harvesting has not strongly influenced the evolution of canine form in homunculines. The more predominant source of selection in this case involves sociality, at least since their last common ancestor split from the other pitheciids. It relates to social monogamy, and a pair-bonded mating system, which is why *Callicebus* and *Aotus* have monomorphic, non-projecting canines. More about this will be discussed in chapter 8.

The incisor-canine tooth sets of *Callicebus* and *Aotus*, on the other hand, are regularly used, and suitably designed in a mechanical sense, for stripping hard-husked fruits and scraping off edibles from hard seeds, although the fruits they bite into are far easier to penetrate than those selected by pitheciines. Another sign of continuity with the pitheciine pattern is that *Callicebus* is also a significant seed eater. In contrast to other New World monkeys of comparable size, the *Callicebus* diet may comprise more than 50% seeds when other fruits are in short supply. Seeds, particularly soft, young seeds, are considered the most important food item for titi monkeys, chosen more often than fruit pulp throughout the year at some sites where they have been studied.

Why did seed eating evolve as a lifestyle strategy in pitheciines? One answer is that seed eating and body mass evolved together because this combination carved out a unique ecological space for pitheciines to exploit. For middle-sized platyrrhines, seeds are a good source of protein, and with the capacity to eat seeds, pitheciines minimize the potential for competition with the squirrel monkeys and capuchins for the other main source of protein used in their size class, insects and small vertebrates. The saki, bearded saki, and uacari post-canine teeth have been selectively engineered in an extreme way to enable crushing and grinding functions. This renders leaves, which must be shredded before swallowing, and tough-bodied insects, which require puncturing and slicing, inefficient alternative sources of protein. Thus the benefits of an intense anatomical commitment to feeding on seeds may have brought with it some resource limitations, though the animals are probably capable of working around them by foraging for younger leaves and softer insects, which are less challenging to masticate. And, as discussed below, the digestive tracts of seed eaters are also compatible with a leafy diet.

Pitheciines have been a major presence in the Neotropics for at least 20 million years, as discoveries of many fossils in Argentina since the 1990s make evident. The saki, bearded saki, and uacari monkeys constitute three of sixteen modern genera, a small percentage of the living South American primate fauna. But if we include fossil genera as well, starting with the early Miocene, the pitheciines are a dominant taxonomic group. If we include the other pitheciid subfamily, the homunculines, living and extinct, the family adds up to about 17 genera, more than the entire platyrrhine radiation that exists today, and they occur across the entire continent. Thus, variations of the hard-fruit and seed-eating feeding strategies have been a prominent, enduring way of life among the platyrrhines.

The fossils tell us that the first major anatomical shifts toward a saki-uacari-like dentition involved a behavioral shift toward ingestion. The incisors, canines, and premolars for picking seeds and breaching and cracking shells are evident in fossils that are 17 and 20 million years old, in *Proteropithecia, Soriacebus*, and *Mazzonicebus*, from Patagonia. At that time, the molars of *Proteropithecia* show incipient signs of being modern but they are designed somewhat differently. By 12–14 million years ago, a fully modern pitheciine dentition is exhibited by the Colombian fossil *Cebupithecia*, found at La Venta, Colombia. It is probably part of the clade formed by *Chiropotes* and *Cacajao*.

Who are the leaf eaters?

Howler Monkeys, *Alouatta*, are the platyrrhines' premier folivore, with extensive craniodental and gastric adaptations for leaf eating. A high proportion of their diet consists of leaves, but they also love to eat fruit and eat it often. A review of 19 studies lasting at least six months and covering all the howler species and a vast territory encompassing many different habitat types, documents leaves and fruit as their most important staples, followed by flowers. No insect-eating was observed. In the combined studies, fruits account for an average of 5% to 56% of their diet. Leaves account for an average of 13% to 81%. But the lower figure belies what the morphology and behavior make clear. *Alouatta* tend to follow the pattern of a classic mammalian herbivore. They are large, slow-moving, and they have capacious fermenting guts, slow passage rates of the digesta, crested shearing teeth, small incisor teeth, and relatively small brains. Similar features are also found among mammalian grass eaters and leaf browsers.

The first field study of monkeys of any kind focused on howler monkeys. It was carried out at the Smithsonian Institution's Barro Colorado field station in Panama in 1934, by the pioneering comparative psychologist Clarence Ray Carpenter. He noted that the leafy diet of howlers was already specified in publications many years prior as "principally leaves and buds, but also fruits and insects," in the most important work on natural history published in the 18th century, Buffon's 36-volume *Histoire naturelle, générale et particulière* (1749–1788). Carpenter confirmed this by recording what the monkeys ate during his daily observations of the animals. He also examined the stomach contents of animals that were sacrificed, contrasting them with the frugivorous spider monkeys he also studied. He wrote, "Howlers consume large quantities of food which is bulky and fibrous. I have removed as much as three pounds of mash from the stomachs of adult howlers, and relatively large amounts of mash from younger animals. They take many times the amount of bulky fibrous foods eaten by spider monkeys of a comparable size."

Carpenter's 3 pounds of mash equates to 17%–20% of the average adult body weight of a Barro Colorado *Alouatta*, though his dissections may have

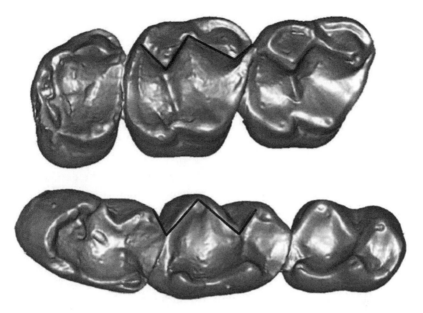

FIG. 5.4. Three-dimensional models of upper and lower molars of a howler monkey with the major shearing crests highlighted by the zigzag line on the second molars. Adapted from Rosenberger et al. (2011).

revealed more than one day's intake. While the howlers eat a bulky, relatively non-nutritious diet, they are nevertheless very successful primates, with a wide geographic distribution today and a fossil record that dates back at least 12–14 million years.

"Semifolivory" may be a more accurate term for categorizing the diet of *Alouatta*. Calling howler monkeys folivores may be too literal because they do actually show a preference for certain fruits when they are available. Leaf eating may thus be seen as a dietary complement rather than a primary staple. Howlers have also been called "behavioral folivores" by Katherine Milton, who pioneered the modern study of howler monkey nutrition and socioecology in the 1970s. This is because they prefer immature leaves that are less toxic and easier to shred. Morphologically, the howler's molars appear to be a functional compromise between a biomechanical design pertinent to shearing, as required in slicing leaves, and for crushing and grinding, as expected in a frugivore (fig. 5.4). And, while there is a significant range in body size, most of the *Alouatta* species are considerably smaller than the other atelids. Other primate folivores and herbivorous mammals in general are normally much bigger than their close relatives that are non-folivorous and non-herbivorous.

That is why I and my colleagues, who were my graduate students at the time, Lauren Halenar and Siobhán Cooke, proposed in 2011 that howlers should be called semifolivores, not folivores. We argued that this was a valuable concept since a comparison with a second platyrrhine genus, the Muriqui, also a prodigious leaf eater, has even fewer of the typical leaf-eating traits than *Alouatta*. The comparison of *Alouatta* and *Brachyteles* lifestyles reveals unique, contrasting, and in some ways unexpected patterns with which platyrrhines have adapted to become leaf eaters. To acknowledge these patterns, the term semifolivores seems apt.

Brachyteles is a biological paradox. Its body build can be described as a large, robust Spider Monkey, as described above and further below in chapter 6. It presents many of the same postcranial adaptations related to quadrupedal climbing, clambering, and high-speed, tail-assisted locomotion, using its proportionately very long limbs, highly mobile shoulders and elbows, and elongated, fully prehensile tail.

The Muriqui seems to move with the same fluid, energetic vigor as *Ateles*. Dentally, however, the Muriqui resembles the Howler Monkey, with large, crested, shearing molars and small incisors, yet its mix of foods falls somewhere between a howler and a spider. Six studies lasting 7–22 months, conducted at several different sites, found that the muriqui's fruit intake composed 35% of the diet, whereas leaves accounted for 47%. These values are comparable to the slow-moving *Alouatta*, not the animated *Ateles*. As in the Spider Monkey, the gut passage rates of digesta are fast, not slow, as in *Alouatta*. The Muriqui brain is as large as the Spider Monkey's, not extremely small relative to body size as in Howler Monkeys. Their speed and distances of foraging travel are faster and lengthier than those of *Alouatta*, though not as extreme as *Ateles*.

In developing our model of atelid evolution, Strier and I interpreted the data to mean that *Ateles* and *Brachyteles*, which are sister genera, evolved from a highly frugivorous ancestor. We hypothesized that the *Brachyteles* lineage diverged from a frugivorous pattern when adapting to the highly seasonal Atlantic Forest where spider monkeys do not live, where it became advantageous to enhance the efficiency of leaf eating in a very large monkey.

This idea is part of a larger hypothesis that Strier and I proposed to explain the radiation of atelids, and how it unfolded as two contrasting lifestyles, slow and fast. At the extremes, the howler's lifestyle is energy-minimizing, while the spider monkey's is energy-maximizing. *Ateles*, a ripe-fruit specialist, is a fast-moving monkey that burns energy quickly. Fruit composes up to 92% of its diet. In more than 12 long-term studies, there was only one case in which less than 72% of the spider monkey diet consisted of ripe fruit. They also have brains of a size expected for their body mass, and a rapid digestion passage rate. In contrast, *Alouatta* is slow, moves little each day, and has a relatively small brain, among the other folivorous behaviors and features noted.

A basic assumption of this model is that the ancestral atelids were frugivore-folivores. But as we have seen, frugivory comes in different forms among platyrrhines: there are ripe-fruit eaters, hard-fruit eaters, and seed eaters. Here, a significant seed-eating component may have been prominent in the last common ancestor's frugivorous diet. Comparative feeding data support this idea by showing the importance of seeds in the Muriqui diet. In a study of *Alouatta* and *Brachyteles* coexisting at the same site, 3.7% of the howler monkey diet consisted of seeds, whereas the muriqui's consumption was 16.5%, even higher than the 12.1% represented by any other types of fruit. Leaves composed 55.3% of the Muriqui diet. Together, leaves and seeds make up almost 72% of the diet. This is important because seed coats are composed of some of the same compounds contained in leaves that require specialized guts to break down and detoxify. Moreover, other primates that had been called folivores are now known to rely on seeds also, like some of the Old World African colobine monkeys, many of which are probably the most specialized primate folivores. Long-term studies of colobine populations show that among the ten highest-ranked seed eaters, seeds compose 32% of the diet. In other words, if a monkey is equipped to digest leaves it can digest seeds, and vice versa.

Among platyrrhines, seed eating may have been a preadaptation to leaf eating. Whole branches of extant New World monkeys ingest seeds in significant proportions, the pitheciids and atelids, in contrast to the frugivorous-insectivorous cebids, which do not. This group includes the medium-sized, hard-fruit eating Titi Monkey genus and all the pitheciines, a group that has specialized heavily on seed eating. The pitheciids and atelids together are monophyletically related, descendants of a single common ancestor. We can triangulate from the present to the past, to seek a dietary common denominator consistent with the non-folivorous dental morphology that might be expected as a preadaptation in that last common ancestor, and that points to a diet that included a significant percentage of seeds. As the atelid clade further evolved and increased in body size, the gut may have already been preadapted to digest both leaves and seeds.

Fossils found at the La Venta site that represent the *Alouatta* lineage show that leaf eating was already established 12–14 million years ago. Remains found there of two species of the genus *Stirtonia* have cheek teeth that are extraordinarily similar to *Alouatta*, and are clearly from a folivorous animal. They demonstrate the longevity of the Howler Monkey lineage and its dietary adaptation. Another fossil from Brazil, *Cartelles coimbrafilhoi*, is related to these alouattines, though it is much younger than *Stirtonia*. It is discussed more fully in chapter 9, and only summarized here because it also suggests an evolutionary pathway from seed eating to folivory. *Cartelles* is known from a relatively complete skeleton with a nearly complete skull and dentition. It weighed more than 20 kg, 44 lb, and is the largest platyrrhine ever discovered. Body size alone suggests this monkey had the capacity to process a bulky, somewhat toxic leafy diet, but it clearly has non-shearing teeth, which fits with seed-eating frugivory. *Cartelles* suggests that seed eating preceded the evolution of semifolivory in the *Alouatta* clade as far back as we can trace it paleontologically at this time. The fossil supports the idea, based on character analysis of the living, that seed eating was most probably an important feature of the common atelid-pitheciid ancestor, and it can serve as a good preadaptive model for the evolution of semifolivory and folivory, a starting point for the transformation from reliance on fruit to dependence on leaves. It would begin with a large fruit eater that has a large gut capable of retaining a large load of seeds as they are digested.

CHAPTER 6

ARBOREAL ACROBATS

Primates are arboreal acrobats. The platyrrhines, among the major radiations of living primates, have evolved the most diversified and unique sets of athletic postures and locomotor styles, which to an important degree reflects the fact that New World monkeys have retained an exclusively arboreal lifestyle. They all live in the trees. This contrasts with the significant terrestrial or semiterrestrial offshoots found among Old World monkeys and apes that have adapted to lifestyles relying on the ground.

New World monkeys have a daily locomotor repertoire that may involve walking, running, leaping, diving, bridging, swinging, clambering, climbing, suspending right-side up or upside down by a forelimb or a hindlimb or a tail. Even their static resting postures vary greatly. They sprawl, sit, crouch, cling, and hang. Some may occasionally go to the ground to briefly feed or travel short distances, but their world is in the trees. The athleticism of New World monkeys and other primates is an outgrowth of the primate order's original arboreal lifestyle that is apparent even among those species that have evolved extensive terrestrial habits and adaptations while remaining agile tree climbers, like Old World baboons, chimpanzees, and gorillas. The extremely complex structure of the tree canopy, the primates' intrinsic genetic milieu, is the setting in which this great variety of positional behaviors, their postural and locomotor styles, evolved.

For some New World monkeys, arboreal dexterity is made possible by unique features that evolved as the fundamental adaptations of major clades and have never evolved in any other primates, such as the prehensile tail and the hands and feet tipped with digital claws, although the latter is mirrored in one living genus belonging to the strepsirhine group, the Madagascan Aye-aye, *Daubentonia*. The capuchin monkeys and the atelid family are the only platyrrhines that evolved tails with true grasping ability. We now understand that this evolved in parallel; however, for centuries zoologists operating within the old gradistic framework suspected a grasping tail represented a connection of some kind that linked these two groups.

Cebus also has the distinction of being the only primate genus that has three biomechanically complex types of grasping organs, or five when all the appendages are added together: the grasping large toes (halluces, plural for hallux) of the feet, the opposable thumbs of the hands, and the semiprehensile tail. Callitrichines are the only adaptive radiation among primates in which the thumb and all the lateral digits of the hands and feet are clawed; the hallux has a flattened nail.

Extant platyrrhines are the only primate adaptive radiation that is not currently represented by a terrestrially adapted genus, yet the potential seems to be there. The reason for this lack remains one of the difficult questions pertaining to New World monkey evolution. Their wide range of body sizes suggests opportunities to exploit small and medium-sized terrestrial niches as other mammalian orders have, and there are vast expanses of unforested terrain in South America that resemble the environments that Old World monkeys and various recently extinct lemurs successfully entered. The smallest terrestrial Old World monkeys are the same size as a squirrel or titi monkey. The fossil record indicates that, like today, platyrrhines have lived in non-inundated forests in the farthest reaches of the continent during the last 20 million years, suggesting ample opportunities to evolve ground-dwelling spinoffs at the edges of wooded areas as these habitats ebbed and flowed in their distribution and extent over time.

Trees are not easy places to navigate and maneuver, so the selective pressures primates experience in adapting to them is varied and intense. The idea of selective pressure goes back directly to Darwin's theory of evolution and the "Struggle for Existence," the title of chapter 3 in *The Origin*. Natural selection is the evolutionary force that results in adaptation. It promotes the widespread occurrence of heritable traits that benefit the survival and reproduction of individuals in a population by differential reproductive success; individuals carrying that trait are likely to have more offspring due to its benefits than others in the population who lack it. The exertion of the natural selection force is what is called selective pressure. Darwin presciently emphasized factors such as "periodical seasons of extreme cold or drought" as a powerful selective force—the Critical Function and Fallback Food Hypotheses, now well supported empirically, are derivatives. He gave an example from his own research, noting that four-fifths of the bird population on his own property was lost after the winter of 1854–1855. Here, weather that was intolerable to many birds was the selective pressure that, in turn, favored those individual birds capable of enduring it, though what traits enabled them to do so Darwin did not determine.

The power of selective pressure and the evolutionary elegance of the effects it yields are manifest precisely at the interface between the animal and its environment, which ultimately determines positional behaviors. When compared to their body sizes, large platyrrhines such as *Ateles* tend to use smaller-diameter substrates for resting, traveling, feeding, and foraging, while the small callitrichines tend to use supports that are larger—often very much larger—for these activities. As a corollary, larger platyrrhines use climbing, clambering, and suspensory locomotion more frequently, while the smallest ones use quadrupedalism and leaping more frequently. These disparities relate to the ways in which the animals must see their world through the proprioceptive sense of their own bodies and limbs in space. To the small Pygmy Marmoset, the treetops are spaces filled with gaps that are relatively large and need to be crossed by leaping or by skirting around them. To the large Spider Monkey, the treetops are composed of branches that serve as a network of walkways that can be reached by an outstretched arm, leg, or tail, which leads to clambering and climbing.

There are two primary levels of natural selection relating to the evolution of positional behavior: the need to forage for food by locomotion, and the means to actually obtain it while stationary. How to maneuver the body from one place to another to get to where food is located; then, how to fix the position and posture of the body in order to secure and consume it. Like the selective impetus to evolve a variety of locomotor patterns in response to the treed environment, the primates' emphasis on the postural aspect of positional behavior is also rather unique among mammals, which again sets the stage for diverse adaptations to evolve. Primates are hand feeders, whereas most mammals are face feeders. Primates tend to manually pluck fruits and leaves from their stems directly, or use a hand or two to pull a food item toward the mouth, while most mammals lack the agility and dexterity to do so. They simply shove their faces into an advantageous position to take a bite.

For a primate, there is also tremendous variety in where exactly food is located. A fruit may hang on a ropelike vine draping down from a branch, or dangle from a stalk fixed in the crown of a 40-foot palm tree or near the top of a 120-foot giant tree that stands alone high above the canopy with no intersecting avenues of entry (fig. 6.1). Desirable foods may commonly be found in the forest understory below the closed canopy, or located at the end of a thin twig in the dense, prodigious growth of a tree's terminal branches. If it is prey, the food may be concealed in the water catchment bowl of a large plant, like a bromeliad, or it may be moving, or hiding motionless on a tree trunk but in plain sight when a sharp-eyed anthropoid comes along. If the primate itself becomes potential prey, another selective factor influencing locomotion is the need to escape. Moving and positioning in all these contexts is locomotion in a large sense. It involves the selective integration of the entre musculoskeletal system.

FIG. 6.1. Structure of a tropical rainforest, with its three-layer arboreal composition. Courtesy of Wet Tropics Management Authority (Cairns, Australia).

A second level of natural selection regarding locomotion and posture relates to the physical points of contact between the animal, its hands, feet, and tail, and the surfaces or substrate it moves upon, the sizes and orientation of tree limbs and trunks, which have their own particular physical properties. Arboreal primates are the mammals that have evolved grasping feet and clutching or grasping hands at these points of contact to keep them from falling off branches and out of trees. The fact that they can physically hold on during static and dynamic activities enables them to contend with a highly complex three-dimensional environment where substrates are discontinuous and arrayed at all angles, gaps are common, and supports are variable in diameter, flexibility, texture, slipperiness, and capacity to support weight. Mobile shoulders and hips amplify the value of grabbing feet and hands to provide the all-angled reach, stability, and agility that enable a primate to work its way through the jumble of branches against the pull of gravity, and sometimes use the kinetic energy stored in bendable, elastic branches creatively, as an underfoot springboard or an overhead slingshot to give its jumps an extra boost when traversing a gap otherwise too big to bridge.

Natural selection has resulted in the diversity of platyrrhine locomotor styles as efficient ways to navigate the physical environment in a community context, where other primates are also sorting out their own locomotor and dietary niches. The importance of locomotion has been understood for a long time, especially in connection with feeding, in the daily life and adaptations of mammals. One of the early discussions of the post-Darwin period was among naturalists who were further elaborating the general principles of the adaptive process and how morphological diversity came about in the origins and evolution of the mammalian orders. In 1902, Henry Fairfield Osborn, a paleontologist and president of the American Museum of Natural History, presented what he called the Law of Adaptive Radiation. It is the underlying idea behind the Ecophylogenetic Hypothesis of platyrrhine evolution, discussed throughout this book. Osborn described how the various clades of mammals tended to innovate new Adaptive Zones and Adaptive Modes by combining different locomotor styles with alternative feeding preferences, all made possible by evolutionary adaptations of the limbs, feet, and teeth.

A student of Osborn, W. K. Gregory, who was to become one of the premier American vertebrate anatomists of his time, demonstrated the profound historical significance of the adaptive interplay of habitat, diet, and locomotion. In completing a project commissioned by his mentor, Gregory summarized the locomotor concepts and characters that had been the historical roots of mammalian classification for millennia. He particularly noted Linnaeus's concept of the "dominating" ordinal character. This would later be recognized as the features defining the great mammalian clades, the orders, where the "nature of the food and mode of obtaining it" were diagnostic. Gregory's synopsis paraphrased Linnaeus in pointing out the main rationales for the organization of mammalian classification by combining features of anatomy, behavior, and environment. The taxonomy was based on the nature of the extremities (hands, clawed feet, hooves, or fins, etc.), the manner of movement (climbing trees, swimming through water, etc.), and the activity involved (prey-snatching, hopping, running, etc.). Critical also to the classification was the presence or absence of a clavicle (collar bone), which relates to shoulder and arm excursion in locomotion.

The indivisible phylogenetic and adaptive elements behind adaptive radiations were recognized in fundamental ways, as naturalists refined their broadly based notions of mammalian diversity and evolution. Yet for lower-level taxa within orders, such as the platyrrhines, where intragroup anatomical variations were far more subtle than the decisive features that distinguish the ways of life among the orders of horses, whales, and bats, for example, it took decades longer to comprehend their interactive role in generating diversity, and it required an inventory of direct observations of behavior in the wild.

Locomotor types

Clingers, climbers, leapers, and more

To document the locomotor behavior of wild primates and examine the morphology of locomotion, primatologists have devised a set of descriptive categories, knowing full well that there are many variations on each of these themes. Four major locomotor types were proposed by John R. Napier and Prudence H. Napier, the British husband and wife team instrumental in establishing the modern study of primate functional morphology. In their landmark 1967 volume, *A Handbook of Living Primates*, the foundation of modern primatology, they describe vertical clinging and leaping, quadrupedalism, brachiation, and bipedalism, and discuss how forest structure relates to arboreal locomotion. The Napiers were largely concerned with the role of the forelimbs and hindlimbs in effecting a locomotor style. The monkeys' fifth appendage, the tail, figured less importantly in their analysis. Because of this, spiders and other atelines were at that time called brachiators or semibrachiators because of their resemblance to apes in the upper body, in aspects of chest, vertebral, shoulder, and forelimb anatomy.

That view has changed after decades of methodological field studies in the wild and in the lab which made it possible to relate morphology more explicitly

TABLE 6.1. Principal modes and frequencies of locomotion among New World monkeys. Outstanding patterns are highlighted.

	Quadrupedalism (%)	Climbing and Clambering (%)	Leaping (%)	Clawed locomotion (%)
Callimico	18	4	62	16
Saguinus	48.9	11.5	32.4	9.2
Leontopithecus	51.8	5.8	26.1	15.3
Callithrix	32	13	23	32
Cebuella	27.1	5.8	24.4	42.7
				Suspensory locomotion
Cebus	55.1	17.2	21.3	6.4
Saimiri	65.3	9.2	23.7	3.7
Callicebus	56.8	16.1	26.6	1.5
Pithecia	32.5	16.6	50.3	1.9
Chiropotes	53.7	15.9	29.3	1.7
Alouatta	52.0	32.8	4.8	11.8
Lagothrix	35.4	41.7	7.4	15.7
Ateles	31.9	22.1	7.5	32.7

Source: Youlatos and Meldrum 2011

to behaviors that were well documented and quantified. The new approaches also accounted for body posture as a full part of the animal's positional behavior repertoire and adaptive profile. As a consequence, a fifth locomotor type, climbing, or climbing and clambering, was identified as being a vital part of the behavioral pattern. In addition, the crucial role played by the prehensile tail in atelids now fit better with the description of clambering and climbing, rather than trying to relate their locomotion to the specialized form of arm-suspension practiced by Old World gibbons and siamangs, the original models for the brachiation locomotor concept. The atelids move quite differently. Relatively large and heavy, they use their long, mobile arms and legs—and tail—to reach in all directions for handholds in the trees in order to distribute their weight among several supports for stability as they clamber through a network of branches.

The predominant locomotor mode among platyrrhines is arboreal quadrupedalism (table 6.1), four-footed walking and running along branches. On the rare occasions when they come to the ground they also walk and run quadrupedally. In the trees, variations of quadrupedalism occur as a function of body size. Leaping between branches occurs commonly among the small and medium-size classes of monkeys that are essentially quadrupeds, while climbing, clambering, and suspensory locomotion prevail among the largest genera, but they are nonetheless essentially quadrupedal in design. These behavioral differences relate to body proportions. The smaller and middle-sized species are longer-legged quadrupeds and that facilitates leaping, while the largest species tend to be long-armed quadrupeds that also use various modes and degrees of forelimb-dominated suspensory movements and postures. However, at the body mass extremes, these generalized patterns are integrated with unique, transformational specializations, as exemplified by the clawed callitrichines and the prehensile-tailed atelids. In each of these groups, these novelties enable the animals to extend their opportunities in the forest by opening up gateways to opposite sides of the habitat's vertical axis, below and above the canopy, affording access to the understory for callitrichines and to the emergent trees for atelines.

Of the four major locomotor categories described by Napier and Napier, obligate bipedalism exists only in humans. Vertical clinging and leaping is not common among platyrrhines. Holding fast while clinging to one vertical or oblique tree trunk and then jumping from it to another, with the trunk held upright and propulsion almost exclusively supplied by the legs, is a specialized means of traveling across gaps between branches, and it occurs mainly among some lemurs and galagos and in tarsiers. More common are the horizontal leaps made as extensions of the four-footed striding patterns of a small quadrupedal monkey like the Squirrel Monkey, Titi Monkey, or Owl Monkey, which consists of an extra effort to push beyond a space in the canopy that cannot be negotiated by extending the forelimbs. The Saki Monkey, *Pithecia*, of the same size class, uses a style of vertical clinging and leaping locomotion while it is active in the understory, where there are few interlacing branches to provide quadrupedal pathways—but not powered by specialized, long hindlimbs or feet, as in prodigious vertical clingers and leapers, like some strepsirhines and tarsiers.

The most interesting platyrrhines that have been called vertical clingers and leapers are the small callitrichines, though their specialized mode is actually a claw-clinging postural adaptation to facilitate food acquisition, and not a locomotor adaptation based on leaping leg-power that benefits traveling. These are among the conceptual, anatomical, and adaptive differences from the basic models described by the Napiers that have been recognized by the tremendous growth of field studies since their synthesis on locomotion. As with studies of feeding, they underscore the importance of observing behavior in the wild in order to understand the adaptive significance of a form-function anatomical system.

Claws are advantageous to callitrichines as small primates with small hands and feet, while negotiating large-diameter supports that are too wide to be grasped, especially when the animals are oriented vertically or obliquely against the pull of gravity. An additional reason such supports, in a broad range of diameters, pose a challenge to them, is that the callitrichines have essentially lost the capacity of the hallux to produce a strong pedal grasp, which is the most fundamental way primates hold fast to a branch. This is highly unusual and it has occurred only one other time in primate history, as bipedalism evolved in hominins. This basic feature of all the nonhuman primates has been a key to their arboreal success since they all shared a common ancestor more than 55 million years ago.

Foot-grasping is made possible by the evolution of a robust large toe with a saddle-shaped joint at the base that connects it to the rest of the foot, enabling forceful flexion of the large toe toward the sole of the foot while it is applied to the substrate in a clasping position. Pedal grasping is a prolonged action that takes place throughout a walking or running cycle as the hind feet alternate their individual holds on a branch, or as the animal extends its torso forward while jumping or downward while hanging, for example. But unlike the patterns of other nonhuman primates, callitrichines have unusual digital proportions and a relatively feeble hallux with a simplified basal joint that greatly limits the effectiveness of muscular, hallucial grasping. Lacking an independently moving or opposable thumb as well, the fanlike arrangement of the callitrichines' long fingers also limits their manual grasping capacity. Callitrichine claws, an array of tiny grapnels, are an alternative method of maintaining traction in the absence of true pedal or manual grasping.

Because they have claws, callitrichines are not limited by substrate diameter or inclination. They can rest and travel in the canopy and below it, where horizontal supports can be infrequent. Most importantly, claws enable the callitrichines to position themselves while feeding on freely oozing gum or while scraping bark to stimulate a flow. That is a crucial advantage for the Marmoset and Pygmy Marmoset. They not only cling to broad tree trunks while lapping a trickle of gum but also leverage upper-body muscle power in this way to enhance the efficiency of bark-gouging.

Other platyrrhines, such as saki monkeys and capuchins, also employ unique postures in critical feeding situations, hanging by the feet, or by feet and semiprehensile tail, in order to free both hands while handling difficult foods. These maneuvers are reminders that even the most dynamic behaviors that have a significant biomechanical impact on the skeleton are not always selected for traveling locomotion per se. They must be evaluated in context to determine their evolutionary significance. The most acrobatic locomotor behaviors are not always the adaptive keys to the success and survival of a species. The simpler

Tree shrew	Goeldi's monkey	Marmoset	Squirrel monkey	Capuchin monkey
Tupaia	*Callimico*	*Callithrix*	*Saimiri*	*Cebus*

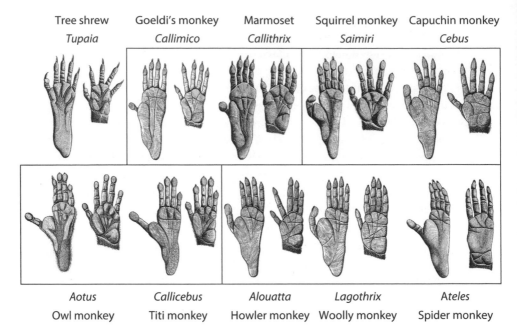

Aotus	*Callicebus*	*Alouatta*	*Lagothrix*	*Ateles*
Owl monkey	Titi monkey	Howler monkey	Woolly monkey	Spider monkey

FIG. 6.2. Varied feet and hands of platyrrhine primates and the Tree Shrew, all from the left side and brought to the same foot length. Rectangles emphasize the similarities of monophyletic groups. Adapted from Biegert (1961).

postures—hanging in place by feet, tail, or claws—can determine how a primate actually manages to eat and obtain its energy from the environment.

Feet and hands tell the story of platyrrhine evolution

Teeth and skulls have been the traditional source of information for reconstructing the most important outlines of evolutionary history of the various primate adaptive radiations. But for platyrrhines, the feet and the hands tell key parts of their story as well, in fine resolution. The morphological and functional diversity of New World monkey cheiridia—the anatomical term applied to both hands and feet—is remarkable (fig. 6.2). They range from a structure that approximates the look and behavior of a paw to one that resembles the human hand. The pawlike form is conducive to flexing actions of the fingers and toes, and to spreading them apart and contracting them together in a fanlike fashion—but not to being molded to conform to and wrap around a shaped object, or to adjusting the entire foot or hand in a posture that places the animal at varied angles relative to the substrate. Primate feet and hands are able to maneuver as a whole unit in three planes, to pitch, roll, and yaw in relation to the long bones, and the foot is designed particularly to be set down on a branch situated in the cleft between the hallux and the sole of the foot. No other mammals have evolved this level of maneuverability. It is a highly evolved arboreal form-function system.

The anatomy and behavior of callitrichines has posed many perplexing problems. With regard to their hands and feet, the presence of claws on all the fingers and the four lateral toes—the hallux is flatly nailed—has been one source of debate: is this pattern primitive or derived? The fossil record clearly shows that all the early primates of the Eocene epoch had nailed fingers and toes, as well as a strong, grasping hallux, and this history places the callitrichines in sharp relief. Their claws are not as prominent as the mammals on which we rely as models of the primitive, pre-primate anatomical condition, such as the tree shrews. Their halluces are fully offset from the other digital rays in a way that benefits the pedal grasp, also in contrast to tree shrews, in which the large toe is integrated with the others (fig. 6.2). The pulpy, terminal touch pads of the callitrichine fingers and toes are also better developed than in tree shrews, indicating that the fingertips play a more important tactile role in locomotion and in manipulating objects.

On the other hand, the callitrichines' large toe is feeble when compared with the early primates and with all other New World monkeys. This indicates the hallux has a reduced capacity for grasping, even when it is deployed on the side of a branch opposite the sole of the foot. The most likely explanation for this combination of features is that the callitrichine foot has been redesigned to rely on grappling relatively wide substrates with clawed digital rays, which callitrichines encounter with high frequency because of their small size, while also maintaining the tactile acuity of the excellent, soft-touch fingertips shared by all living primates, a boon to precise locomotion on slender branches and twigs and to sensitive object manipulation. In reconstructing how these characteristics may have evolved, a comparison with close relatives offers insight. The nails of the Squirrel Monkey, *Saimiri*, tend to be prolonged beyond the fingertip, narrow, and pointy, and those of the Capuchin Monkey, *Cebus*, are comparable. This does not mean they are literally the "starting points" toward callitrichine morphology; rather, they are ancestral models of what the pre-callitrichine pattern might have been like.

The array of pedal forms among the living cebids is thus instructive about callitrichine origins and adaptations. Within cebines, *Saimiri* and *Cebus*, the hands are informative for other reasons. Both genera are excellent manual probers and object manipulators, especially the capuchin monkeys, which have evolved significant thumb prehensility. This means that their clade may be defined in part by these assets, which are a fundamental part of their predation tool kit.

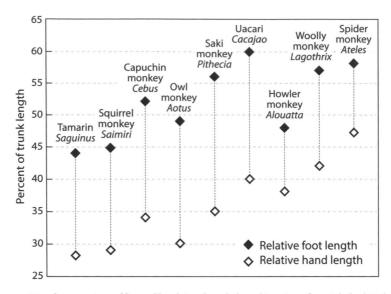

FIG. 6.3. Length proportions of feet and hands in selected platyrrhines. Data from Schultz (1956).

Among the pitheciids, the form and function of hands and feet also reveal distinctive evolutionary patterns. For example, the Owl Monkey, *Aotus*, and Titi Monkey, *Callicebus*, share two distinctive traits not seen elsewhere among New World monkeys. For reasons that are obscure, their fifth pedal digits are markedly short by comparison with the other toes. In *Callicebus*, manual digit V is notably short as well.

A second trait well illustrated in the image of the *Aotus* foot in figure 6.2 is a prolonged, rather delicate, clawlike nail on the second digit, which is comparable to the condition of *Callicebus* and a host of strepsirhines. These are grooming claws, used to comb the fur. The morphology is distinctly different from the nails found in other platyrrhines and the claws of callitrichines. In *Aotus* the grooming claw is upwardly canted, pointing away from the fingertip and the underlying bone to which it attaches, which has a distinctive, complementary shape to support that orientation. In callitrichines, all claws are downwardly curved, toward the palm and the sole. It is likely that mutual grooming, like the tail-twining ritual discussed later, is part of the intensely tactile behavioral repertoire that helps maintain social bonds in these monogamous pitheciids, and additional evidence of their close phylogenetic relationship.

The cheiridial proportions of pitheciids also appear to be distinctive, although the available data are incomplete and based on limited sample sizes (fig. 6.3). With reference to trunk length, the feet of *Pithecia* and *Cacajao* are long, especially for their body size class. Behaviorally, they are also notable for foot-hanging behaviors. The proportions of *Aotus* are similar. Relative foot length is exaggerated when compared with relative hand length, especially when owl monkey proportions are compared with cebids and atelids.

Among atelids, there are several patterns evident in the hands that also add to knowledge of their evolutionary history. Among the atelines, hand:foot proportions are distinctive (fig. 6.3). Both are relatively long compared with trunk length, that is, body size, especially in *Ateles*, the Spider Monkey, and in *Brachyteles*, the Muriqui. Also, there is less of a length differential between the cheiridia in the atelids, particularly in *Alouatta*, though neither its hands nor its feet are as elongate as in the more specialized, acrobatic atelines. The larger feet and hands provide mechanical advantages in supporting the body and, particularly, in effecting handholds when locomoting, as the positional behaviors of *Ateles* and *Brachyteles* are in many respects forelimb-dominated. They use a hooklike, hand-grasping pattern, which is associated with curved hand bones, a pattern that is evidence they are sister genera. They also share another interesting, novel feature. Though reportedly variably expressed in different Muriqui populations, both genera have thumbs that are either outwardly vestigial or missing entirely: no manual I phalanx can be seen extending from the palm of the hand. Actually, within the palm, *Ateles* and *Brachyteles* do have a first metacarpal bone, but it is trivially small relative to the others and to the typical platyrrhine hand (fig. 6.4; plate 2.12). Another feature of interest is prominent in *Alouatta*, the Howler Monkey, and is seen to a lesser extent in other atelids and in pitheciids. They show a natural functional split of the fingers between the second and third digit. Digits II and III are used as a pincer. The natural set of the hand is not exclusively a position in which the thumb is recruited to oppose the palm and other digits in grasping mode. Instead, the pitheciids and atelids have a tendency to grasp a branch with its axis falling closer to the middle of the palm. This would seem an obvious precondition to the loss of the external thumb in *Ateles* and *Brachyteles*.

Hanging, clambering, and locomoting with a prehensile tail

Atelines have reinvented what it means to be a quadruped. With limbs that are long relative to trunk length, long hands and feet, mobile shoulders, and stiff backs, they resemble the great apes. What they have in common behaviorally is that they all climb. But atelines are much smaller than these apes, and they spend all their time in the trees, unlike the comparatively gigantic, semiterrestrial chimpanzees and gorillas. Atelines also have a fully prehensile tail; apes have no tail. While in the trees, the atelines' body type effects a climbing and

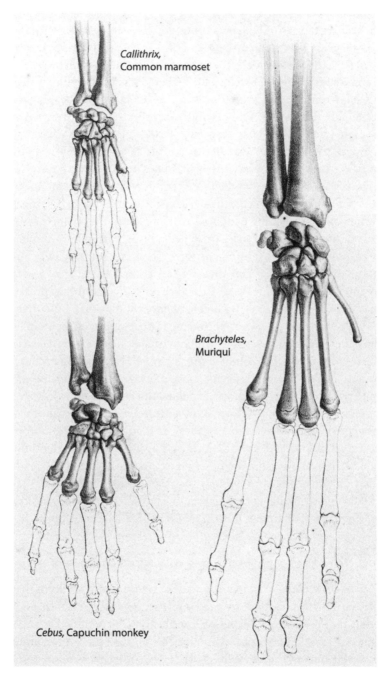

FIG. 6.4. Right hand and wrist skeletons of (counterclockwise from top left) a marmoset, capuchin, and muriqui. From Blainville (1839).

clambering version of quadrupedal locomotion. The apes are capable of this as well, but their large size limits how often they engage in this sort of behavior.

The fully prehensile tail of the four ateline genera is their most obvious shared locomotor innovation, though its involvement in positional behaviors varies among them. In dynamic situations the tail can be used in tight coordination with the arms to facilitate above-branch quadrupedal travel. In resting postures it can provide an auxiliary support, wrapped around a nearby branch. While feeding, it may be used alone or in combination with a single arm to let the animal suspend itself beneath a branch. The well-studied, highly agile *Ateles* is the genus most adept and innovative in using the tail, relying on it extensively. Field reports indicate that the spider monkey's prehensile tail is used together with its arms in up to 62% of locomotor bouts and 54% of feeding bouts. The spider's reliance on arm-tail, coordinated handholds during traveling enables it to move easily through the interlaced, bendable twigs that form the closed canopy of the rainforest with little concern for maintaining balance, as would happen if they were restricted to locomoting on the top surfaces of branches. To the Spider Monkey, there is no practical distinction between above-branch or below-branch locomotion, which is a great convenience to a relatively heavy arborealist. In contrast, *Alouatta*, the Howler Monkey, is a slow, cautious, above-branch quadruped that uses the tail more for stability, as when reaching to bridge a gap between branches. Many observational studies confirm that howlers, unlike atelines, do not employ tail-arm combinations while feeding or moving.

The Spider Monkey can use its prehensile tail to pick up an object as small as a peanut. This display of control and sensitivity in an organ other than the primate hand or foot is remarkable. The manner in which the tail is used in dynamic locomotion is equally impressive, as the monkeys move hand-over-hand and foot-over-foot, pulling and pushing their way through a network of branches with the tail bent up from the base of the vertebral column to grasp an overhead branch, acting as if it is a third hand.

In Woolly Monkeys it is less extraordinary. There is a significant difference in the tail's role that distinguishes woolly and spider monkeys. Tail-assisted, below-branch locomotion in *Lagothrix* appears less coordinated and efficient than in *Ateles*. It involves choppy, short strides as opposed to the fluid manner in which spider monkeys move. Not surprisingly, *Ateles* in the wild use varieties of suspensory locomotion twice as much as *Lagothrix*, and travel at least twice as fast. The detailed kinesiology is even more telling, as laboratory studies of video recordings have shown. In general, the woolly's tail acts like a safety device during quadrupedalism, sliding along a branch above the body as the monkey moves forward. In spiders, the prehensile tail, longer and more bendable, especially toward the end, plays a more dynamic role. During phases of the quadrupedal step cycle, the tip of the *Ateles* tail floats above a branch until the moment when it grasps the support in coordination with every other handhold, landing alongside the leading hand. The precision involved in placing the tail this way demonstrates an uncanny ability to control its position in space. When the tail is released, the center of mass shifts downward and the monkey swings below the fixed hand like a pendulum, taking advantage of the kinetic energy of a body already in motion.

There is a more distinct contrast between the staid, quadrupedal locomotion of *Alouatta* and the three other atelids, each of which relies more on arm-swinging locomotion with tail involvement. The Woolly Monkey seems to be an intermediate between the extremes. The Spider Monkey and Muriqui share many detailed postcranial traits in common and they more closely resemble one another in postural and locomotor behavior than either genus resembles *Lagothrix*. *Ateles* is the more slender version, and *Brachyteles* the more thickset, of a common form-function skeletal and neuromuscular system.

Platyrrhines are the only primates that evolved grasping tails

Among platyrrhines, the only primates that have evolved grasping tails, this feature has evolved twice, in parallel in different clades; once in the last common ancestor of the four atelids and once in the Capuchin Monkey lineage. The constellation of differences distinguishing *Cebus*, with its semiprehensile tail, and atelids, with the fully prehensile tail, represents two different pathways to the evolution of tail grasping, for overlapping yet quite distinct reasons. For one, in *Cebus*, tail grasping is closely connected with foraging and feeding, as a postural adaptation. In *Alouatta* the tail's biological role may be similar, but in the atelines it is related more to traveling, a locomotor adaptation.

That the tails of Capuchin Monkeys and atelids are morphologically different is a fact that has been understood at least since 1862. At that time, the physician and natural historian J. H. Slack pointed out that the capuchin tail tends to be lax and has hair covering it fully to the tip, while the atelid tail is prehensile, functioning like a fifth hand, and has a patch of skin on its underside. His work was well known, published in the premier American scientific journal of the time, the *Proceedings of the Academy of Natural Sciences of Philadelphia*, which was started by Benjamin Franklin. Knowledge of the functional capacity of the atelid prehensile tail goes back even further. In 1700, the British explorer and naturalist whom history also remembers as a pirate, Captain William Dampier, vividly described the behavior of howler monkeys and remarked, "The Tails of these Monkeys are as good to them as their Hands; and they will hold fast by them."

These outward anatomical distinctions were extended by later anatomical and behavioral studies to show that, as had been suspected, the Spider Monkey tail is more sensitive to touch, with skin better designed even at the cellular level to limit slippage when it grasps than the furry Capuchin Monkey tail. This sensitivity is made possible, in part, by the evolution of the microscopic substructure in the long, narrow patch of friction skin, which is finely corrugated like the palm of a hand and stretches for roughly one-third its length along the underside of the tail toward the tip.

There are other differences as well. There is a strong correlation between body size and tail length in New World monkeys (fig. 6.5), but there are several outliers. The short-tailed *Cebus* is one of them. Its tail is shorter than predicted based on it body mass. The atelids, in contrast, have tails that are longer than is expected based on the trendline. This contrast is another reason why the semiprehensile and fully prehensile tails have different functional capabilities. Each outlier, but of very different kinds, contributes to the evidence that tail-grasping evolved independently. In other words, the semiprehensile tail is not an intermediate stage or forerunner of the prehensile tail, as one might intuitively expect. Each specialization evolved by different pathways and plays different roles in order to gain different biological advantages. One gains benefits in postural contexts and the other in its potential for locomotion, keeping in mind that in the least elaborate atelid version, seen in *Alouatta*, feeding and foraging is the predominant context in which the tail is used.

Figure 6.5 also highlights the case of the Uacari, *Cacajao*, another outlier with curious proportions: a relatively very short tail. Its nearest relatives, *Pithecia* and *Chiropotes*, also have tails that are shorter than other monkeys of their size class. In the other pitheciid clade, the homunculines, the tail of the Owl Monkey, *Aotus*, is also relatively short, falling below the Squirrel Monkey, *Saimiri*, in the plot, whereas the Titi Monkey, *Callicebus*, exhibits a relatively long tail. The varied, overall pattern of relative tail size among the platyrrhines suggests that its length has both increased and decreased evolutionarily, in ways that make it difficult to sort its status in each genus. It is intriguing, however, that four of the five pitheciid genera have relatively short tails, suggesting this is a specialized characteristic of the family that may have been transformed in the one outlier, *Callicebus*. Perhaps not coincidentally, titi monkeys are the ones that use the tail most conspicuously in tail-twining, a social behavior discussed in depth later. Rather than being short, a long, ropelike tail can be more easily wound around another monkey's tail.

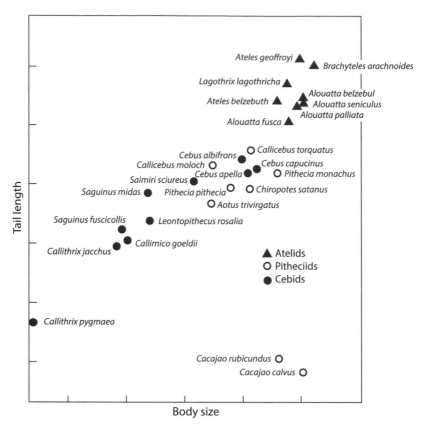

FIG. 6.5. Plot of tail length against body size in platyrrhines. Adapted from Rosenberger, 1983.

Alouatta belzebul	Red-handed howler	*Callithrix jacchus*	Common marmoset
Alouatta fusca	Brown howler	*Cebuella pygmaea*	Pygmy marmoset
Alouatta palliata	Mantled howler monkey	*Cebus albifrons*	White-fronted capuchin
Alouatta seniculus	Venezuelan red howler	*Cebus apella*	Tufted capuchin
Aotus trivirgatus	Northern owl monkey	*Cebus capucinus*	Panamanian white-faced capuchin
Ateles belzebuth	White-bellied spider monkey	*Chiropotes satanus*	Black bearded saki
Ateles geoffroyi	Black-handed spider monkey	*Lagothrix lagothricha*	Brown woolly monkey
Brachyteles arachnoides	Muriqui	*Leontopithecus rosalia*	Lion marmoset
Cacajao calvus	Bald uacari	*Pithecia pithecia*	White-faced saki
Cacajao rubicundus	Red uacari	*Pithecia monachus*	Monk saki
Callicebus moloch	Red-bellied titi	*Saguinus fuscicollis*	Brown-mantled tamarin
Callicebus torquatus	Collared titi	*Saguinus midas*	Red-handed tamarin
Callimico goeldii	Goeldi's monkey	*Saimiri sciureus*	Common squirrel monkey

The hypothesis that the prehensile abilities of capuchin and atelid tails evolved in parallel is consistent with the findings of researchers studying these animals in the wild. Their observations demonstrate that there are many differences in form, function, and biological role. We now know that their tails are used in very different ways. For example, *Cebus* generally carries its tail stretched out behind the body during quadrupedal locomotion and it is used infrequently in weight-bearing mode while traveling. It is, however, used as a brake following a leap and is quite important in feeding activities. When hanging upside down or reaching out to grab food, *Cebus* commonly adopts an inverted tripod posture, with two feet planted and the tail serving as a hook to anchor the body. Similarly, the tail stabilizes the body by wrapping around a branch when two hands are used in manipulating or pounding objects. Among adults, the tail is rarely used alone in suspension, but young capuchins do hang by it without additional support.

Anthony Rylands succinctly described how and why the capuchin prehensile tail is used: "I have always related its prehensile tail to the forceful, destructive, manipulative foraging that is so prevalent. When bashing things about and pulling and tugging and having your hands fully occupied you need more than just two feet to keep yourself from falling out of the tree."

Ateles, in contrast, frequently engage tail-arm combinations during quadrupedalism and while hanging freely. Their tail-assisted, clambering locomotor specialty is a style of rapid directional travel within the canopy that enables them to easily traverse gaps in footing. Quantitative studies of behavior bear out the contrast. Tail-assisted quadrupedal travel in *Cebus* accounts for about 13% of locomotor time, while in *Ateles* it can involve 62%. The large amount of time invested in tail-assisted traveling in *Ateles* is related to their feeding strategy, which involves frequent, rapid travel to far-off sources of ripe fruits.

In contrast to morphological and behavioral studies showing that *Cebus* and the atelids have become specialized in different ways, the skeletal basis of prehensility in the least agile, most doggedly quadrupedal atelid, the Howler Monkey, *Alouatta*, precisely matches the patterns of the other atelids. This is consistent with the hypothesis that the capuchin semiprehensile tail evolved independently and the brain, further considered below, also supports this idea. For example, in *Ateles* the area of the parietal lobe dedicated to the tail is expanded and the brain's external surface has been reshaped as a consequence. The same cortical enlargement and rearrangement of grooves and folds occurs in howlers, too, though their brain is far smaller relative to body size, poorly folded, and contoured quite differently from a spider monkey's brain. In *Cebus*, which has a highly enlarged and convoluted cortex, the tail representation takes up a smaller fraction of the parietal lobe, and the genus does not exhibit atelid-like changes in the arrangement of brain convolutions.

Deconstructing the elements that contribute to tail functional morphology makes it quite clear that the grasping tail evolved twice among platyrrhines. In other words, the capuchin system is not homologous with the atelid system. They do not share the same genetic, evolutionary origin. This approach tells us how it happened and what the dual selective advantages may be. But it says nothing about why only the New World monkeys have experienced this evolutionary coincidence when it would seem that other arboreal primates could also benefit from having a prehensile tail to serve as a fifth hand to secure them in an arboreal environment. Is there some underlying evolutionary process, or pattern, that may explain this phenomenon? Is there something in the nature of platyrrhine morphology and behavior that is widespread and conducive, or preadaptive, to the evolution of tail prehensility? To address those questions, it is useful to consider what several other platyrrhines do with their tails.

Tails for balancing, embracing, and coiling for social bonding

Titi Monkeys, *Callicebus*, have an easily observable tail specialization that is a ritualized form of behavior central to their social system: two or more monkeys coil, or twine, their tails together when resting or sleeping, probably to enhance the bonds between mates and within a nuclear family. The Owl Monkey, *Aotus*, which is socially pair-bonded as are Titi Monkeys, also engages in tail-twining (plate 12). Tail-twining is not a remarkable physical feat when compared with the weight-bearing or dynamic capacities of capuchin or atelid tails, but it is significant in the context of the question: Why did tail prehensility evolve only in platyrrhines—twice? Looking elsewhere, there is another member of the monophyletic pitheciid group that includes *Callicebus* and *Aotus*, the Black Bearded Saki, *Chiropotes satanas*, which also uses its tail in a specialized way. It is recruited as a brace positioned between the grasping feet while the animal hangs upside down to procure, bite into, or feed on large or hard-shelled fruits that require a lot of manual handling time. Therefore, three of the five pitheciid genera employ the tail in highly particular ways. The only ones that don't are the very short-tailed Uacari and the Saki Monkey.

In another family, in the Squirrel Monkey genus, the nearest living relatives of *Cebus*, any special features that might be shared in common with Capuchin Monkeys are traits that potentially existed in their last common ancestor. The data suggest that the last common ancestor of these cebines had already evolved an advanced degree of tail control and prehension, though not quite the type that eventually came to characterize *Cebus*. Evidence for this earlier evolution was presented in one of the first books dedicated to the study of a platyrrhine, *The Squirrel Monkey*, edited by the comparative psychologist Leonard A. Rosenblum and the research veterinarian Robert W. Cooper, published in 1968. The book included a chapter summarizing aspects of the earliest field study of wild *Saimiri*, by Richard W. Thorington Jr., then curator of primates at the Smithsonian Institution's National Museum of Natural History, and a chapter by Frank DuMond on squirrel monkeys living in a seminatural habitat in Florida. These papers by Thorington and DuMond, and another by Rosenblum, made several key observations about tail use in squirrel monkeys.

Thorington observed that for the first few weeks of life, *Saimiri* infants, when carried on the backs of their mothers, wrap their long tails tightly around her torso to avoid falling off. When adults are feeding and foraging, he described tail use as follows: in a "quasi-prehensile manner . . . hanging by their feet to reach some item of food . . . they looped their tails over a branch for additional support." DuMond states that while the tail is non-prehensile, it is used for balance and stability in moving, as a postural brace, and that it can also be wrapped around a tree limb for support. Rosenblum says that the "infant's tail is somewhat prehensile . . . capable of supporting the neonate's entire weight for brief periods." Photographs of squirrel monkeys frequently show them with their tails wrapped around their own bodies while resting; with a mother's tail wrapped around a baby when it is sleeping on her back; or with an infant's tail tip coiled into a loop.

Evidently, examples of unusual forms of tail control and use occur in three of the major platyrrhine clades, cebids, pitheciids, and atelids. This suggests that ancestral platyrrhines were also able to use the tail in comparable ways outside its more generalized role as a balancing organ that is active during locomotion. If, for example, a strong, manipulable tail was beneficial to secure neonates from falling off a mother in the earliest New World monkeys, there is a chance that by natural selection this capability could have become enhanced and extended into adulthood under the right conditions. Ancestral platyrrhine tail form and function may have been a preadaptive precursor to the varied, and independently evolved, tail-use patterns now found in the three families. There may have been a variety of social, developmental, postural, or locomotor traits that would have benefited from a shared anatomy that predisposed groups of New World monkeys to evolve semiprehensile and prehensile tails and other, more subtle behavioral variations like tail-twining.

Another factor that may have disposed platyrrhines to enhancing tail manipulability relates to body posture, how they sit. Small and medium-sized New World monkeys are haunch-sitters (fig. 6.6). They tend to sit on or across branches in a hunched posture, resting their torso on folded hindlimbs. The tail is draped over the same branch or an adjacent one where it plays a significant

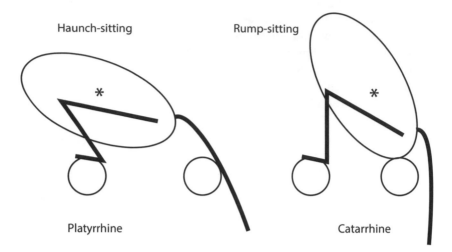

FIG. 6.6. Schematic model of the sitting postures of a haunch-sitting platyrrhine and a rump-sitting catarrhine. The center of gravity is represented by the asterisk.

proprioceptive role, and it may be actively recruited for balance and support when the branches on which they rest are thin and flexible. As body weight increases, the tail's potential role as an additional support, under more control, becomes advantageous. Contrasting this with catarrhine resting postures clarifies this point. Catarrhines are upright rump-sitters. They transfer body weight through the buttocks while sitting, specifically to one of the pelvic bones, the ischium. This is why the skin covering the ischium becomes hardened and calloused, and ischial callosities develop in all Old World monkeys and some apes. With the center of gravity of a sitting catarrhine situated near the rump, the tail plays no significant supportive role.

CHAPTER 7

MANY KINDS OF PLATYRRHINE BRAINS

As with other systems and traits, the evolution of the brain in platyrrhines is influenced by the balance of ecology, behavior, and phylogeny. When it comes to evaluating brain-to-body size relationships, the most commonly used point of reference for making evolutionary comparisons, we look at standards that describe this relationship within and between taxonomic groups, and we focus on the departures from that pattern that identify groups for special consideration. As in other primate adaptive radiations, some New World monkeys have brains that are large relative to body size and others have brains that are small relative to body size, and there is no single, uniform explanation that covers all instances consistently. Component parts of the brain are also influenced by natural selection, and several monophyletic groups are noteworthy in exhibiting specializations of the cortex tied to adaptations in morphology and behavior that characterize these groups, such as those relating to feeding, foraging, and locomotion.

Increase in relative brain size is a pattern seen in the evolution of primates as well as other mammal groups, but it is not an inevitable trend. Primate evolution is not channeled by directional selection to evolve larger and ever-larger brains, although on average primates as a group have larger brains than most other orders of mammals. The brains of various primate taxonomic groups have gotten relatively larger or smaller, as adaptive responses to circumstances, and, over time, many have probably remained about the same size since their clades originated.

Selection for intelligence has played a role in shaping brain size evolution in the Capuchin Monkey, *Cebus*, known as the cleverest New World monkey. It is hard to define and measure intelligence in a standard way among species with very different lifestyles and behavioral configurations that are often so alien to our own. Still, relative to body size, *Cebus* has by far the biggest brain of all the living platyrrhines, even exceeding the relative brain size of all living apes. Its predilection for using tools is one of the striking indicators of high intelligence, but tool use per se is not the singular driving force that influenced the evolution of the capacious *Cebus* brain. Selection for advanced cognitive abilities would be highly advantageous for other fundamental aspects of their daily life as well.

This proposition was evaluated by behavioral ecologist Charles Janson in a series of experiments carried out in the wild in 1983. He hung eight feeding stations in trees more than 160 meters apart and baited them with bananas. During the study, Janson increased or decreased the size of the food reward to mimic the ripening patterns of fruits the monkeys normally encountered while foraging, and he analyzed the capuchins' movements, including their repeat visits to the sites. He found that in addition to the monkeys' ability to remember the existence, location, and richness of a feeding site, already demonstrated in other studies, the capuchins in this experiment also displayed knowledge of when they last used each feeding site. Their movements suggested flexibility in making travel decisions based on their expectation of which site was likely to be the most rewarding, given its distance from the feeding site they had been visiting. Janson suggested that this unusual ability to integrate knowledge of elapsed time with the memory of the "who, what, and where" that underpins their cognitive food reference system is a remarkable ability that evolved as a foraging adaptation, which may reflect selection for intelligence in all primates. The hyperintelligence of capuchins probably relates to a specific type of foraging—the searching, testing, manipulative, branch-breaking and nut-pounding manner described above—and other aspects of their lifestyle, like their complex social systems and communication patterns discussed below.

Other New World monkeys have evolved relatively large brains independently and are not tool users, such as the Bearded Saki, Uacari, and the Spider Monkey. For them, selection to enhance general intelligence may not be the best explanation. In these cases, natural selection for encephalization, enlargement of the brain, may be a response to more specific features of their ecological and/or social situations, rather than the promotion of benefits that accrue from just being smart. Dietary quality is also a significant, and rather explicit, factor in shaping brain size evolution among platyrrhines. The folivorous *Alouatta* is a primary example. It has an exceptionally small brain as an adaptation to eating low-energy, leafy food. And, as we shall see, there are other direct and indirect effects related to body size evolution in several taxa that must be taken into account when assessing relative brain size in the New World monkey radiation.

Studying brain size and shape

Three methods to examine brain size in platyrrhines are presented here. One takes a graphic look at the relationship between relative brain size and body size in platyrrhines among the extant primates (fig. 7.1). The other two document relative brain size in New World monkeys by comparing them with other mammals, using measurements to examine the diversity of platyrrhine brain size and how much larger or smaller their brains are by comparison (table 7.1). Each of the major platyrrhine clades is represented in fig. 7.1, which shows the scaling relationship between brain size and body size, the degree to which an increase in brain size is biologically coupled with an increase in body size. The four trendlines represent the scaling relationships of non-platyrrhine primates to provide standards of comparison. They illustrate that for a given body mass, New World monkeys and other anthropoids have larger brains than the strepsirhines. This difference is especially evident if one prolongs the trendlines for the Old World apes (Hominoidea) and monkeys (Cercopithecinae and Colobinae) toward the left side of the chart. All three extensions plot well above the strepsirhine regression line; their brains are more encephalized.

Scanning the chart shows that each of the three platyrrhine families exhibits a range of relative brain sizes; that is, their respective circles and ellipses fall at varying levels above the strepsirhine line. Among cebids, callitrichines have relatively small brains, falling closest to that line, while cebines, *Saimiri* and *Cebus*, the Squirrel Monkey and the Capuchin Monkey, have relatively large brains. Among pitheciids, the Owl Monkey, *Aotus*, and the Titi Monkey, *Callicebus*, have the smallest brains; within the pitheciine subgroup, the Saki Monkey, *Pithecia*, has the smallest relative brain size while in the Bearded Saki, *Chiropotes*, and the Uacari, *Cacajao*, the brain is quite enlarged. Among atelids, *Alouatta*, the Howler Monkey, has the smallest relative brain size, and the atelines are encephalized. Taken together, it is clear that platyrrhines have evolved relatively large brains independently in these three clades.

Several outliers are important to consider as they illustrate other principles guiding the direction of brain size evolution and its possible determinants. For example, the relatively small brain of the Howler Monkey is striking. The strepsirhine regression line intersects with the ellipse that encompasses the variation in brain:body size proportions seen in the sample of *Alouatta* individuals, which plot well below the cluster of points representing its closest relatives, the atelines. This is consistent with the measurements provided in table 7.1, which are explained below. As mentioned, howlers have anomalously small brains for their body size as a consequence associated with their leafy diet. This reflects a well-known correlation that links a folivorous or herbivorous diet with unencephalized brain across modern mammals. On the chart, the generality of this relationship in primates is illustrated by comparing the elevations of the two Old World monkey regression lines. The trendline for Colobinae, also known as Leaf-eating Monkeys, falls below their sister group, the more omnivorous and frugivorous Cercopithecinae, the Cheek-pouched Monkeys.

The particularly large-brained platyrrhines compose another set of outliers when compared with the strepsirhine standard and their nearest New World monkey relatives (table 7.1). The two cebine genera, the Squirrel Monkey and Capuchin Monkey, are markedly encephalized. One case seems easy to explain: *Cebus* by all accounts is an exceptionally smart monkey. But are *Saimiri* also smart, even if they do not seem, using our observations and our human-centric perspective, to be as smart a monkey as a capuchin? Or, are there other factors that need to be considered? Perhaps both are relatively encephalized, at least in a modest way, as part of their phylogenetic heritage, possibly as a foraging adaptation relating to the way they actively seek out food by manually manipulating materials in their environment. Another explanation for the encephalized squirrel monkey brain—neither are mutually exclusive—is that it is a byproduct of the process involved in having become phyletic dwarfs. Their anatomical growth patterns resemble those of a capuchin monkey except that bodily growth is halted earlier in *Saimiri*, and adulthood is reached at a smaller absolute body size. One effect of that is the production of a large brain in a small body. In a parallel development, *Cebuella*, the Pygmy Marmoset, represented by the small outlier within the callitrichine ellipse that plots to the far left (fig. 7.1; table 7.1), may be an effect of dwarfing as well, but through a different mechanism of growth and development.

Among pitheciids, the relatively large brains of the sister genera *Chiropotes* and *Cacajao*, the Bearded Saki and Uacari, also stand out. These animals are not well known, but they share in common several features that may be related to their jointly advanced level of encephalization. Like atelines, they live in fission-fusion social groups, as discussed in the next chapter. These groups are generally quite large in size, and the smaller subgroups that forage as temporary units roam widely in large home ranges. *Chiropotes* and *Cacajao* also have highly selective diets that require detailed monitoring of tree phenology across a vast area. Thus it is possible that ecological and social complexity may be factors that have selected for a relatively large brain in their last common ancestor. An additional factor that may be related to encephalization is the previously mentioned fact that males of these genera (the female conditions have not been described) exhibit delayed maturation, as discovered by field primatologist Thomas Defler, who raised several in captivity and was able to observe this pattern. It resembles the slow growth rate that is well established

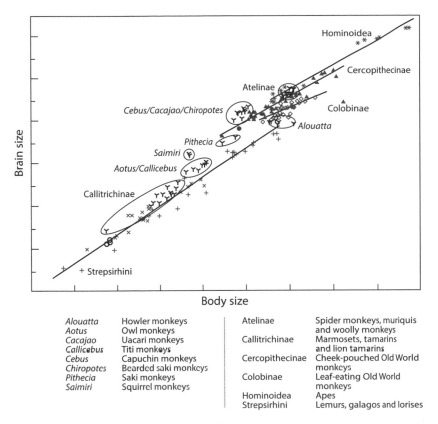

FIG. 7.1. Relationship between body mass and cranial capacity in living platyrrhines and other primates. Adapted from Isler et al. (2008).

in *Cebus*, another large-brained platyrrhine in the same size class. In humans, apart from being able to benefit from learning prior to achieving adulthood, there is another connection between slow maturation (i.e., extended childhood and prolonged adolescence) and very large brain size, and a similar phenomenon may be occurring among these platyrrhines. To maximize the energy necessary to support a prodigious amount of brain growth during early postnatal phases of life history, weight gain is deferred until adulthood is reached.

Relative brain size overall is only part of the story of platyrrhine brain evolution. Complexity and within-brain proportions are others. Complexity comes in several forms, including the brain's external shape, its internal wiring, and the proportional variations reflecting evolutionary emphases on different functional areas. Since different roles are attributable to localized parts of the brain, one important field of study examines the morphology of the brain's outermost surface, the convolutions of the cerebral cortex, which can reveal which of the brain's functional units may have expanded or contracted in size. These convolutions are actually folds or ridges of the cortex, called gyri (singular, gyrus) that are each offset from one another by a crease, a sulcus (plural, sulci). Like a crumpled piece of paper, folding of the cortex allows the brain to maintain a relatively small volume in proportion to its surface area. Folding means that more brain cells can be packed efficiently into a small space. In addition to general construction principles that benefit by not being overbuilt, tightening the spatial relationships among cells via convolutions may enhance interconnectivity, providing an advantage to the brain's wiring system, thus to brain performance.

The degree to which the cortex is convoluted depends on the size of the brain as well as species body mass. Platyrrhines with large brains, such as squirrel and spider monkeys, have many convolutions exposed on the surface of the cortex (fig. 7.2). Howler Monkeys, in contrast, although among the largest New World monkeys, have brains that are small relative to body mass, and the external surface lacks some of the fissures seen in other atelids of comparable body mass. *Alouatta* also has a brain of unusual design. Rather than being rounded like a soccer ball, it is shaped more like an American football, an oblate spheroid, that is, an elliptical sphere that is compressed from top to bottom, at a right angle to its long axis (fig. 9.11). This reflects the brain's relatively small size and the highly modified cranial morphology of Howler Monkeys, one of the many anatomical consequences related to the evolution of the extraordinarily large vocalization apparatus. At the smallest body sizes, in callitrichines, few fissures are evident. They have relatively smooth brains. That has suggested to some that their brains are primitive, but the lack of convolutions may relate mostly to the fact that they are small mammals with small brain volumes.

Because the inside of the cranial cavity mirrors the surface contours of the brain accurately, there are several ways that the brain's convolutions can be reconstructed from fossils and mapped. For example, sediment may fill in and harden against the inner shell of the bony braincase, making a natural internal cast, or endocast, that replicates its surface anatomy. Even if skull bones are later lost during fossilization, or become broken, the endocast can remain intact as an accurate model. If the braincase is not filled with natural sediment, an artificial endocast of a fossil can be made by molding the inside of the clean braincase. Liquid latex is poured through the large foramen magnum at the base of the skull and allowed to harden to form a flexible, balloonlike reproduction of the cavity that can be physically withdrawn through the same opening. Filling that with a solid material, such as semihardened plaster, will

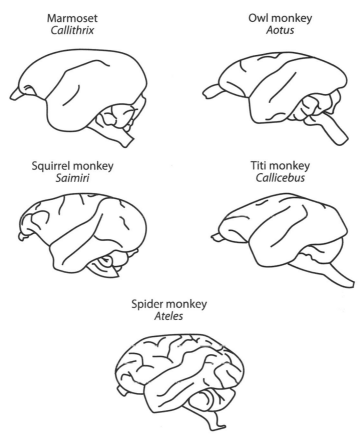

Marmoset
Callithrix

Owl monkey
Aotus

Squirrel monkey
Saimiri

Titi monkey
Callicebus

Spider monkey
Ateles

FIG. 7.2. Left-side views of five platyrrhine brains brought to the same length. From *Comparative Mammalian Brain Collections.* http://neurosciencelibrary.org/.

then hold the authentic volume and shape of the brainbox, turning it into an anatomical surrogate of the cerebrum itself. Endocasts are also useful for measuring brain size by water displacement, though the simpler method of filling an empty braincase with small inert material, plastic beads for example, and then measuring their volume by pouring them off into a graduated cylinder is the more common approach. The advent of CT scanning has also been a boon to this type of qualitative and quantitative research, especially for fossils. It allows the development of digital, virtual brains as three-dimensional models. This is a safe, non-contact process—the specimens may not even need to be cleaned of adhering matrix—and the digital model can be easily shared among scientists as research-quality material.

Brain-to-body-size relationships

The full storage and processing power of the brain is related to its size, even if this is only a crude measure of performance. Mammals vary enormously in body mass, so actual size metrics such as brain weight or volume cannot be independent measures of the brain's properties or functional potential: big animals will have big brains; little animals will have little brains. That principle was illustrated with reference to the non-platyrrhine primates. But how do platyrrhine brains stack up against those of other mammals?

One study by R. D. Martin, a British zoologist and biological anthropologist and a leading figure in the area of primate brain research as well as primate evolution, examined this question using the brain sizes of non-primate mammals as a basis for comparison. Martin asked how much each platyrrhine genus, and those of other primate genera, deviated from the proportions of living mammals generally regarded as being rather primitive, a group that includes ecologically and behaviorally unspecialized insectivores, shrews and moles. The sample included 11 genera of living platyrrhines (table 7.1). The method involved calculation of a trendline that describes the brain-to-body-size relationship common to these basal insectivorans. Then, Martin computed the Index of Cranial Capacity (ICC), which quantifies the degree to which the observed brain size of a primate genus deviates from the insectivoran trendline. If the ICC value is larger than 1, it means the brain is relatively larger than expected in basal insectivorans of the same size. Anything below a value of 1 would mean that the brain is smaller than expected.

Martin's sample was drawn from all three platyrrhine families, so it is an effective way of examining the sweep of the New World monkey radiation. From table 7.1 it is evident that the platyrrhines consistently have brains larger than the basal insectivoran norm. The two highest ICC measures are found in *Cebus* and *Cebuella*, at 11.7 and 10.2, respectively. Their brains are more than 10 times the size that is expected if they scaled like a presumptive primitive mammal. The two smallest values are in *Saguinus* and *Alouatta*, at 4.3 and 4.7.

Table 7.1 also provides results from a different analysis arrived at by a comparable method, presented as EQ, the Encephalization Quotient, developed by psychologist Harry Jerison in his pioneering 1975 book on the vertebrate brain, *Evolution of the Brain and Intelligence*. It is conceptually similar to the ICC method but generates smaller figures because of a difference in the way Jerison selected his reference model, which is based on a broad sampling of mammals rather than a select group. In this case, a value of 1 approximates the average EQ for a mammal, so capuchins have a brain that is more than three times the size of an "average" mammal of comparable body weight. As

TABLE 7.1. Comparison of absolute and relative brain sizes among living platyrrhines. Outstanding patterns are highlighted.

Genus	Cranial Capacity (cc)	Index of Cranial Capacity	Encephalization Quotient
Callimico	11.1	5.2	1.5
Saguinus	9.9	4.3	1.3–1.5
Callithrix	7.2	6	1.4–1.5
Cebuella	6.1	10.2	1.8
Aotus	16.9	4.8	1.4–1.7
Callicebus	18.3	4.9	1.5
Saimiri	23.6	7	2.4–2.6
Cebus	76.2	11.7	2.8–3.1
Pithecia	N/A	N/A	1.8
Chiropotes	N/A	N/A	2.3–2.7
Cacajao	N/A	N/A	2.7–3.1
Alouatta	60.3	4.7	1.4–1.5
Ateles	108.8	7.3	2.2–2.5
Lagothrix	97.2	7.8	2.4

Sources: Hartwig et al. 2011, Martin 1990.

to where *Cebus* stands with regard to other primates, Jerison found that its EQ of 2.8–3.1 is even larger than the metrics of the three great apes (1.4–2.5). Overall, Martin's and Jerison's results and conclusions for the platyrrhines are quite comparable.

These studies clearly demonstrate that Capuchin Monkeys have very large brains. At least part of the evolutionary explanation for this must be connected with their manual dexterity and their inherent inclination to manipulate the environment. The higher order of intelligence displayed by tool use extends from there. Tool use is the manifestation where the cognitive cascade of highly complex thoughts unfolds as an impulse to obtain an object—a non-vegetable object, not simply something separated from a tree, like a fruit or a broken branch, or an animal—to examine it, learn its properties, employ or discard it, store and retrieve memories about it. Many brain parts must be working together as the capuchin manipulates its environment or works with a tool. Yet other factors, such as the cognitive requirements associated with a complex social organization, explored in the next chapter, cannot be discounted as a major selective factor responsible for genetically building such a bright brain in the first place.

Phylogeny has a large role in shaping adaptation, and there may be an underlying phylogenetic aspect that gives historical context to the potential of the *Cebus* lineage to amplify brain size as their body size increased. This is suggested by the presence of a relatively large brain in *Saimiri*. The Capuchin Monkey's living sister genus, the Squirrel Monkey, presents an ICC of 7.0, well above the Titi and Owl Monkeys of comparable body size. This means that both cebine genera are prone to benefitting from selection for brain size increase above the platyrrhine norm, that it probably derives from the predaceous ecological context in which the clade originated, and that they inherited the genetic potential for advanced encephalization from their last common ancestor. In other words, they both have relatively large brains because their "starting point" may have been advanced in this regard by comparison with other New World monkeys. Following that, the baseline cebine brain was influenced by selection to evolve an even larger size in the *Cebus* lineage in support of high intelligence, whereas the degree of encephalization in *Saimiri* may be less correlated with native, acute smartness than with factors associated with dwarfism and sensory specializations.

The monkey stole my keys

Intelligence and dexterity are tightly correlated

Cebus and *Saimiri* share capacities, and a derived set of behaviors, associated with foraging for prey that have been inherited from their last common ancestor. This is borne out by novel structural features of the cebine brain, in the organization of the visual cortex located in the occipital lobe, at the back of the brain. It suggests they may see the world in much the same way, and differently from other platyrrhines. In effect, both are prodigious, visually directed faunivores that search for concealed prey, unlike most New World monkey species that pluck insects from places where they are visible. Capuchin and squirrel monkeys unroll leaves and probe crevasses where an invertebrate arthropod or small vertebrate may be hidden. Although squirrel monkeys lack the precision thumb-grip of the capuchins and are not known to use tools, they are manual manipulators with a keen sense of sight, foreshadowing the capuchin's skills.

As Jerison pointed out, increasing reliance on vision in predaceous mammals is associated not only with enlargement of the brain's visual areas, it also correlates with relative enlargement of the whole brain. The amount of information taken in by the eyes and stored and processed as data is large and complex. These two patterns are uniquely evident in the cebines. For *Cebus*, the intense innate curiosity seen in their interest in physically exploring novel objects and using tools is an undeniable sign of a broad intelligence. As Jacob

Bronowski wrote in 1973, in *The Ascent of Man*, "The world can only be grasped by action . . . the hand is the cutting edge of the mind."

How intelligent and dexterous are capuchin monkeys? I learned for myself in the early 1980s in an incident on my way to a field trip in the forests of Northeast Brazil. On the way, I stopped at the now defunct Zoological Garden of Niterói, near Rio de Janeiro, to see what kinds of monkeys were on exhibit, to better recognize the species I might encounter fleetingly in the field. There, I saw a Buffy-headed Capuchin, and to get a better look at it I dangled my key ring to lure the monkey nearer to the wire front of the cage. He came closer and, in a flash, stuck his slender hand through the wire, snatched the keys from me, and scampered to a perch in the back corner of his cage. As I watched, dumbstruck, he proceeded to wedge the end of my car key into a slot between the wooden boards of the platform where he was sitting and began pulling on it, bending the key. I and my colleagues pleaded with the monkey, in English and Portuguese, to stop. A zookeeper came to the rescue. He grabbed a banana, stuck it onto the end of a long branch, and with this he coaxed the monkey toward the front of the cage again. For this small-handed capuchin, taking the banana off the stick required a two-handed grasp so he had to let go of his new toy. He dropped the keys and the zookeeper was able to retrieve them. Fortunately, the bent ignition key still worked. To be outsmarted by a monkey was an unforgettable lesson about capuchin brains.

I learned more about capuchin tool use while chasing monkeys in a forest fragment on an enormous Brazilian ranch. One morning, I heard a thud-thud-thud coming from the forest edge. My first thought was that someone was cutting firewood, which was prohibited. But the repetitive sound seemed odd. An axe- or machete-wielding person would take an occasional break, and the timing between strikes would be more random and spaced out, not like the rapid, rhythmic thud-thud-thud I was hearing. The sound also had a hollow tone I did not recognize. After a few minutes of walking through the bush I was standing under a very tall Jatobá tree, and there I saw a capuchin sitting hunched in the canopy, pounding a fruit held in two hands against the wide branch. Jatobá fruits are large, with a hard, podlike casing several inches long, and they are difficult to breach. Each pod contains a beige or greenish powdery pulp, and that is what the capuchins eat. I watched as it eventually smashed and broke open the pod. It sat flicking flakes of shell to the ground and munching on the pulp, its furry brown face dusted with Jatobá residue.

Of all the platyrrhines, only *Cebus* has ever been seen to behave this way. No other New World monkey is known to be a tool user and none have brains as encephalized as this genus. The cognitive psychologist and primate conservationist Benjamin Beck, author of the classic reference book *Animal Tool*

Behavior: The Use and Manufacture of Tools by Animals, defined tool use as a behavior in which an external object is enlisted to achieve a physical goal that is not possible by using one's own body, a hand, foot, or mouth. An extensive repertoire of tool-use behaviors of capuchins has been carefully detailed by comparative psychologists. It includes probing actions like those of the Niterói monkey; creating compound tools, that is, pushing one object to effect movement of another; breaking and intentionally dropping branches near human intruders and throwing stones at potential predators; using a stone to hammer a nut selectively placed on a hard, anvil-like surface to maximize the effective power of the striking force; and bringing rocks to dry spots on the ground and using them as shovels to dig for water.

Fingertips, precision grips, and tool use

Recognizing the crucial role of manual skills in human evolution—the use and manufacture of tools—in the 1960s, groundbreaking primatologists John and Prudence Napier organized two seminal classifications to describe behaviors at the vital interface between animal and environment: the locomotor categories described in chapter 6, and use of the primate hand in grasping. They identified the two most important prehensile maneuvers of the hand, the power and precision grips. A power grip clutches an object between the fingers and the palm of the hand with the thumb applying counter pressure to stabilize it, the way a human holds a hammer. A precision grip delicately, or forcefully, pinches an object, between the fingertips of the thumb and another digit, usually the index finger, the way a human steadies a nail to be hammered. This exacting hand posture brings the tips of thumb and index together in the shape of an O, which maximizes the surface area of contact between the fingers' pulpy touch pads. It allows, in John Napier's words, "an unlimited potential for fine pressure adjustments or minute directional corrections."

Among New World monkeys, only big-brained capuchins have the musculoskeletal architecture that can produce precision grips. And only they have the neural mechanism that facilitates the advanced level of precise control that is required. Experiments have shown that they employ several variations of the grip depending on the task, but most often they do it by opposing the thumb and index finger in some fashion.

Area 2 and Area 5 are labels used by neuroscientists for two fields of brain tissue visible on the outer layer of the cerebral cortex, in the parietal lobe, which covers about one-quarter of the brain's surface in the area above and behind the external ear. The parietal lobe is where the brain processes mechanical and thermal information that comes from the skin—the sensations of touch, pressure, temperature, pain, and proprioception—which means sensing bodily self-awareness, how one recognizes and modulates the spatial relationship, positions, and orientation of limbs by knowing the angular relationships of the joints connecting them.

Parietal Areas 2 and 5 collect input from various parts of the body, but our interest here concerns the forearm and hand, because a monkey or a tool-using civilization depends on them. This is where extraordinary manual dexterity and a precision grip come together in the exact coordination of eyes, arms, hands, and fingers. It is where precise motor actions of the arm and hand are planned and monitored by the brain in milliseconds. It is where the brain manages the interface between the hand and an external object, where finger position is controlled and monitored to reflect an awareness of an object's size and shape. Capuchins rely on Areas 2 and 5 for their tool use.

A number of New World monkeys have been studied by neuroscientists interested in comparing and contrasting the development of Areas 2 and 5 among species in order to relate the gross anatomy of the hands with the manner in which the hands and fingers are guided by the brain. These studies examined the extent to which parallel evolution in the brain may facilitate the expression of precision grips in capuchins, rhesus macaques (an Old World monkey), and humans.

Area 2 is absent from the brains of the Tamarin, *Saguinus*, the Owl Monkey, *Aotus*, the Titi Monkey, *Callicebus*, and the Squirrel Monkey, *Saimiri*. Their hands tend to work as grabbers and they lack versatility in thumb movements. However, Area 2 is well defined in capuchins, where inputs from the face and hands predominate. We know less about Area 5 in platyrrhines. It is either entirely absent or very poorly developed in the titi but, again, it is well defined in capuchin monkeys. Its inputs are almost exclusively restricted to the forelimb and hand. Its importance has been shown by studying people with injuries to Area 5. It impacts the coordination of the hand with the arm, wrist, and shoulder that is necessary for accurate reaching movements. In rhesus macaques, Area 5 manages "preshaping" of the hand to prepare it for grasping an object. Of the various anthropoids studied, the capuchins and rhesus come closest to humans in the development of the same brain regions and in their hand use. The capacity evolved in the three groups independently, via parallel evolution. In addition to the motor control necessary in a tool user, keen eyesight is also fundamental to performing precision grips, and *Cebus* has been shown experimentally to resemble humans in visual acuity and other measures, such as sensitivity to brightness and adjustment to intermittent light stimulation, though they do not have the same type of color vision.

The visual cortex of capuchins and squirrel monkeys is not only larger than in other platyrrhines, it is also organized differently. In cebines, groups of cells are arranged to receive input from the right and left eyes together, while in other New World monkeys the right and left inputs are isolated from one another. The wiring of the visual cortex also differs in the way it connects with another complex unit of the brain, the limbic system, which encompasses a large territory of the cerebral cortex. Learning, memory, motivation, problem solving, emotions like pleasure, and many other roles have been associated with the limbic system. The visual cortex and the limbic system appear to be more highly integrated in *Cebus* and *Saimiri* than in other New World monkeys. That may explain their persistence as predators, *how* and *why* they may both relish discovery of a bug or a bird's egg as a morsel of food. It may explain why capuchins are so focused on toying with and using objects in their daily lives. They may be motivated to do so because their reward system becomes engaged, so they derive pleasure from the activity.

The sensorimotor strip in the brain controls tail use

Capuchins are not the only New World monkeys that have evolved a remarkable neurological capacity for interpreting sensation, effecting motion control, and applying dexterity through an appendage. An entire clade accomplished something similar in the form of a prehensile tail. This, too, is imprinted on the external morphology of the brain, in the arrangement of the gyri and sulci of the cerebral cortex found in atelids, in an area called the sensorimotor strip (fig. 7.3). The morphology of the strip is distinctive in these prehensile-tailed monkeys. They present a subdivision that coordinates the tail that is unique in that it is easily seen on the surface of the brain. In other monkeys, even the semiprehensile-tailed *Cebus*, the homologous region is positioned differently. It is not visible on the lateral surface of the brain because it is only exposed in the deep midline cleft that divides the right and left hemispheres. Enlarged and shifted from the midline to a more lateral position, the atelid tail unit may be packed with more neurons, and it may also provide a more advantageous wiring system that coordinates the tail with the limbs, which are also controlled by the sensorimotor strip. Only one of the atelids, *Ateles*, has been studied electrophysiologically in the lab in order to document the role of this area. But because the anatomy is uniform among all four, we assume that the same region of the cortex in the others is responsible for producing the same behaviors. This is true even in the decidedly smaller brain of *Alouatta*.

The atelid prehensile tail is hypersensitive, multiaxial, superbendable, and ends in a strong tip. As in other primates, the tail is furry. But, as previously discussed, on its underside, for the last third of its length, there is a hairless,

PLATE 1. *Cebus nigritus*, Black Capuchin Monkey (courtesy of Mark Bowler).

PLATE 2. *Saimiri sciureus*, Common Squirrel Monkey (courtesy of Mark Bowler).

PLATE 3. *Callimico goeldii*, Goeldi's Monkey (courtesy of Mark Bowler).

PLATE 4. *Saguinus imperator*, Emperor Tamarin (courtesy of Mark Bowler).

PLATE 5. *Leontopithecus rosalia*, Golden Lion Marmoset (courtesy of Andreida Martins, Associação Mico-Leão-Dourado).

PLATE 6. *Callithrix flaviceps*, Buffy-headed Marmoset (courtesy of Carla B. Possamai, Projeto Muriqui de Caratinga).

PLATE 7. *Cebuella pygmaea*, Pygmy Marmoset (courtesy of Mark Bowler).

PLATE 8. *Pithecia pithecia,* White-faced Saki (courtesy of Marilyn Norconk).

PLATE 9. *Chiropotes chiropotes*, Red-backed Bearded Saki (courtesy of Mark Bowler).

PLATE 10. *Cacajao calvus,* Red Uacari (courtesy of Mark Bowler).

PLATE 11. *Callicebus torquatus,* Collared Titi Monkey (courtesy of Mark Bowler).

PLATE 14. *Brachyteles arachnoides,* Muriqui (courtesy of Carla B. Possamai, Projeto Muriqui de Caratinga).

PLATE 15. *Lagothrix lagothricha,* Woolly Monkey (courtesy of Mark Bowler).

PLATE 16. *Alouatta seniculus*, Red Howler Monkey (courtesy of Mark Bowler).

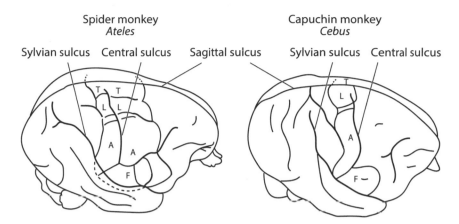

FIG. 7.3. Lateral views of the brains of a prehensile-tailed monkey, left, and a semi-prehensile-tailed monkey, right. Abbreviations for areas representing anatomical regions: T, tail; L, leg; A, arm; F, face. Adapted from Radinsky (1972).

soft strip of friction skin, embossed with the same kinds of miniscule, patterned dermal ridges that all primates, including humans, have on fingertips, palms, and soles. Because of its specialized, multijointed anatomy, an atelid tail can coil around a branch and twist about its own long axis while carrying the monkey's full weight. It is strong enough to fling a 20 lb, 9 kg, spider monkey from one perch to another. Most amazingly, the spider monkey's tail acts like a fifth limb, fully incorporated into locomotor movements during a hand-over-hand, suspensory-style locomotor cycle or during the quadrupedal step cycle that is used in walking and bounding on branches. It is completely functional at all times rather than just being an accessory of occasional value, acting as an emergency brake as it is does in howlers, or in woolly monkeys, where it acts as a stabilizer.

The semiprehensile tail of *Cebus* lacks a comparably specialized representation in the brain. Their tails are adroit and strong enough to prop up the body in a head-down climb or while hanging to feed in suspension. It is very handy as an anchor when the capuchin is manipulating and banging on things. But its structural evolution took a different turn from that of the Spider Monkey, as capuchins use their tails differently. Its tail lacks the hyper-maneuverability seen in *Ateles* and other atelids. There is no naked skin to enable comparable tactile sensitivity, and it is actually short for the body length of the animal, which makes it more stout than flexible. Significantly, it is never used in phasic coordination with the forelimb or hindlimb during locomotion. Once again, teasing out such distinctions in behavior, form and function, in systems ranging from the musculoskeletal to the neurological, provides powerful evidence that grasping tails evolved twice among platyrrhines, in parallel.

Evolution of the brain in platyrrhines is shaped by phylogeny, ecology, and social behavior

Each of the three major clades of platyrrhines includes genera that have relatively small brains and genera in which the brain is relatively large, which suggests that a relatively small brain was the ancestral condition in all three groups and that selection for brain size enlargement has also occurred three times in parallel. Among atelids, Howler Monkeys are noteworthy because they have relatively very small brains in a clade that is large-bodied and also includes relatively large-brained animals, like Spider Monkeys. However, an exceedingly small brain like a howler's is unlikely to be primitive in this group. It is a specialization, as are many other unusual features of *Alouatta*, adaptations to minimizing energy expenditure in connection with the low-quality, semifolivorous diet of the genus. On the other hand, a leafy diet does not always correlate with a reduction in brain size in atelids. The Muriqui is semifolivorous and it is not de-encephalized. *Brachyteles* is also a high-energy locomotor and lives in large social groups that are more complex, as discussed in the next chapter. A relatively large brain may be a specific advantage in this context, as the history of encephalization in this genus may also be a definitive function of phylogeny if these factors—the ecology of locomotion, social organization and encephalization—have been coupled since the common ancestry that Muriquis share with Spider Monkeys.

In pitheciids, the Saki Monkey, *Pithecia*, is less encephalized than the Uacari, *Cacajao*, and the Bearded Saki, *Chiropotes*. *Pithecia* is also more primitive than the other two in dentition and cranial morphology, and presents a less complex form of social organization, as discussed in chapter 8. The other two genera of this clade, *Callicebus* and *Aotus*, the Titi Monkey and Owl Monkey, also have relatively small brains. Both are pair-bonded like some *Pithecia*, and live in small groups. It seems likely that among these five genera a relatively small brain, as seen in the homunculines and the Saki Monkey, is the ancestral condition, whereas the dietetically hyper-specialized Bearded Saki and Uacari, with very large, complex social groups that roam over expansive territories to feed, as a consequence have experienced selection for brain enlargement.

In the third case, cebids show that relative brain size increased in the Capuchin Monkey and the Squirrel Monkey—the two genera with the largest brains by far of almost all living platyrrhines. *Cebus* and *Saimiri* also share numerous specialized ecological and behavioral adaptations not seen elsewhere. The data indicate that one other cebid genus, the Pygmy Marmoset, has a relatively large brain in comparison with the other callitrichines. This may not be a case where selection for encephalization is the explanation. The relatively large brain of *Cebuella* may be a byproduct of significant body size dwarfism. It is also worth noting that relative brain size in *Cebuella* may be somewhat overestimated in the studies cited, which were done when the available body mass measurements for the genus were not particularly robust and may have underestimated their weight. Nevertheless, this taxonomic distribution among cebids, coupled with the particularities of the large-brained genera, also suggests that a relatively small brain was the ancestral condition in the callitrichine clade.

No single factor explains why selection favored increased brain size in New World monkeys; however, there appears to be a common denominator shared by the three groups of platyrrhines that have evolved noteworthy degrees of encephalization. They live in groups that are socially complex. Conversely, genera with relatively small brains live in simpler groups, smaller in size and lacking the behavioral diversity and individual flexibility that would accompany the daily mixture of males, females, varied kinships, and subgroups of varied ages.

CHAPTER 8

THE VARIETIES AND MEANS OF SOCIAL ORGANIZATION

Platyrrhines, like all anthropoids and nearly all primates, live in relatively stable social groups that are reproductive units. The individuals sleep together, rest together, travel together, feed together, and communicate information about the environment, such as the presence of a predator or the location of a sleeping hole or a fruit tree. Group living provides mating opportunities and an environment for rearing offspring. It provides a situation where young can develop, learn and mature in relative safety. The sizes of these groups, their structures and social rules vary among New World monkeys. Their social groups can remain relatively cohesive units for generations, as witnessed by those observing them for decades. Karen Strier, who has studied the Muriqui in the field for more than 35 years, has known some females from the beginning of that time that became grandmothers and great grandmothers. However, no group lives as an isolate. Troops in the same forest interact at various levels with others belonging to the same species. One band may signal to another, vocalizing about their presence to an adjacent group, and two interacting groups may even get involved in a vocal, face-to-face altercation. There are also exchanges of individuals between troops, which happens when animals that have reached reproductive age transfer to another social group to mate and to resituate themselves. Attacks also occur, sometimes involving infanticide, when a few roaming adult males oust the resident male of a troop as a way to overtake the females and establish their own reproductive entourage. Though acts of violence and aggression may occur as a pattern between individuals or social groups belonging to the same species, aggression between primate species living sympatrically is rare.

Among the 16 extant genera of New World monkeys, the various social organizations and mating strategies embedded in them are as diverse as other facets of their biology. This suggests that such variations are also the consequences of the evolutionary imperative for each genus to find its niche in an exclusively arboreal community (table 8.1). The radiation's diverse social systems are phylogenetically patterned, as are the adaptations in body size, diet, and locomotion, a balance wherein closely related genera tend to have comparable systems as well as specialized deviations. The factors that explain both the similarities and differences of social organization are complex. They relate to ecological parameters such as body size and dietary profiles, and to reproductive parameters like litter size, gestation length, and the age of onset of puberty. This complexity is compounded because of our increasing awareness that interspecific differences in grouping patterns exist among some of the genera, which can make it difficult to generalize genus-specific norms. In any case, a core Darwinian principle relating to fitness helps explain the evolution of New World monkey social systems. They each provide the context in which an individual female or male best maximizes its genetic contribution to the next generation by producing offspring or assisting in the successful maturation of close kin.

An examination of the variations in platyrrhine sociality begins with some of the special forms of communication that have evolved in support of those systems. Communication is at the root of all social relationships and it is the mechanism that regulates interactions within and between social groups. Platyrrhines have evolved an extensive array of communication strategies in a complex environment that poses challenges to the exchange of information, both aurally and visually. While the forest may be relatively noiseless much of the time, arboreal monkey locomotion can be noisy, especially when many are on the move. Smaller species are quiet as they move through the trees without disturbing branches very much, but the heavier animals inevitably bend branches when traveling, producing a whooshing sound when they rebound and the copious leaves rub against one another. The ubiquitous foliage is a visual barrier and the monkeys cannot easily see each other. It also muffles and reflects vocalizations. Achieving efficient communication through the modalities of sight, scent, and sound is, therefore, of high selective value to social, arboreal animals like the platyrrhines.

Group size and foraging patterns determine to what extent members become spatially dispersed during the day, which influences the methods of communication they use. Vocalization, a method of communicating over long and short distances, is one of these, but sounds that are broadcast widely can be heard by other animals, including competitors and predators. The capuchins and squirrel monkeys, and some of the callitrichines, are examples of the gregarious kinds, constantly in contact with each other through chirping, contact-call vocalizations. Some platyrrhines are preternaturally quiet in order to avoid

TABLE 8.1. Summary of social organization, mating systems, and social dispersal patterns among New World monkeys

	Social organization[a]	Group size	Group dynamics	Parental care	Mating system[b]	Dispersal[c]
Cebus	**MM**/MF	13–35	M/F HIERARCHY	SEMI-COMMUNAL	PGAMY(PGYNY)	M
Saimiri	**MM**/MF	15–54	M/F HIERARCHY	SEMI-COMMUNAL	PGAMY(PGYNY)	M,F,B
Callimico	**MF**/MM	4–12	F HIERARCHY	COMMUNAL CARE	PANDRY	B
Saguinus	**MF**/MM	2–13	F HIERARCHY	COMMUNAL CARE	PANDRY	B
Leontopithecus	**MF**/MM	2–11	F HIERARCHY	COMMUNAL CARE	PANDRY	B
Callithrix	**MF**/MM	2–16	F HIERARCHY	COMMUNAL CARE	PANDRY	B
Cebuella	**MF**/MM	2–9	F HIERARCHY	COMMUNAL CARE	PANDRY	B
Callicebus	P	2–8	M/F COOPERATION	MALES	MGAMY	B
Aotus	P	2–4	M/F COOPERATION	MALES	MGAMY	B
Pithecia	P (MM/MF)	3–9	M/F COOPERATION	MATERNAL	MGAMY	B
Chiropotes	MM/MF	9–33	FISSION-FUSION	?	PGAMY(?)	F?
Cacajao	MM/MF	16–44	FISSION-FUSION	?	PGAMY(?)	F(B?)
Alouatta	**1M**(MM/MF)	3–22	M HIERARCHY	MATERNAL	PGYNY(PGAMY)	B
Lagothrix	**MM**/MF	4–43	FISSION-FUSION	MATERNAL	PGAMY	F(B)
Ateles	MM/MF	16–55	FISSION-FUSION	MATERNAL	PGAMY	F
Brachyteles	MM/MF	11–80	FISSION-FUSION	MATERNAL	PGAMY	F

[a] Bold font indicates strong orientation; MM = multi-male; MF = multi-female; 1M = one-male/multi-female group structure; P = pair-bonded; terms in parentheses = observed variations

[b] MGAM = monogamy; PANDRY = polyandry; PGAMY = polygamy; PGYNY = polygyny; terms in parentheses = observed variations

[c] M = male; F = female; B = both

detection and have evolved a communications strategy that incorporates visual and olfactory signals consistent with a camouflaged lifestyle. Others are not so muted, as we shall see.

Olfactory signals are a short-distance communication method able to convey a specific kind of social information that cannot be transmitted through other modalities, such as the reproductive state of females or the identity of the natal family to which an individual belongs. In some very specialized cases, the chemical content of a sender's scent has a direct effect on the physiology of the recipients, in a way that deeply influences not only their behavior but also their lives as a whole. When messages are delivered in the form of scent, there are also anatomical adaptations, skin glands, relating to odor production and, often, stylized patterns of behavior that are enacted when odor is deposited. Species that make particular use of visual gestures have also evolved unique anatomies and behaviors that give shape to the signals.

A day in the life of a platyrrhine

The social life of New World monkeys begins in the morning, when they wake, carries on throughout their daily activities, and continues into the night, when they go to sleep. The one exception to this rule involves the Owl Monkey, *Aotus*, which is fundamentally nocturnal, so the day-night cycle is reversed. Its pattern changes, however, among the Owl Monkey species living far to the south in South America. When moonlight is dull and temperatures are cool, they shift their activities. Then, they are active for periods during the day as well as at night.

It is possible to outline a monkey's day by examining its activity rhythm, the temporal patterns it follows, and its activity budget, which is a measure of the investment made in different types of behaviors. This information is the foundation of any ecological and behavioral profile of a platyrrhine species. While a variety of methods may be used to collect the data, all tend to count events or situations: where they are in the trees, when, how often, and how long an activity takes place. The aim is to develop a descriptive catalog of a group's behavior by recording what adults and non-adults do each day, and averaging each of the classes of behaviors across identified individuals to build up an outline of species-specific activity. Long-term observations of more than one troop of a target species that cover all seasons of the year is the baseline approach because their profiles can change from one season to the next as food availability fluctuates. Some research projects have gone on for decades, and they reveal longer-term variations due to other factors, including significant changes in climate and troop size due to intrinsic population growth rates or loss to disease, famine, and other factors. Differences in behavior may also be influenced by natural variations in troop size and composition, that is, by adult male to adult female ratios, and by proportions of young at various ages that occur in the wild. It is also beneficial to collect data in a realistic ecological context, accounting for the presence of other species living in the same area with overlapping requirements, which may influence what the subject monkeys do day-to-day.

Recording this extensive catalog of observations to document the behavior of a group of arboreal monkeys moving throughout the day is a challenge. Teamwork is an asset, and good tools are a requirement, including binoculars, compass, stopwatch, camera, measuring tape, a clinometer to measure tree height, waterproof notepaper and pencils, and plastic bags to hold specimens of animals' feces and the plants they eat. That was the tool kit developed by primatologists during the 1960s and 1970s, what Warren Kinzey and I used when we studied titi monkeys in the Peruvian Amazon.

Methods have evolved since then, though none can substitute for the experience of being in the field and watching monkeys, because something new and unexpected happens every day. Now, there are handheld, programmable devices to record a field-worker's observations in real time. Tracking monkeys in the forest is done more easily, and their movement measured more accurately, using lightweight radio collars worn around a monkey's neck, or transponders inserted under the skin. Setting that up means a monkey has to be trapped, or darted, and anesthetized, then released. In the process, an animal's age, health, and reproductive status can be assessed, blood samples can be drawn, hair samples can be taken, dental molds can be made, and an identifying marker can be deployed to distinguish one animal from another, a dyed rump or tail, a collar of colored beads, or a microchip with GPS capability. Motion-sensitive camera traps can be installed in the field to collect visual information without an observer being present. It is also now possible to collect fecal and urine samples of individually identified platyrrhines in order to research their hormonal activity, and how it influences their behavior.

The introduction of low-cost methods to test the genetic relatedness of individual troop members, as well as the relatedness of separate groups living in proximity, is an important innovation that contributes to our understanding of interpersonal relationships, group dynamics, and patterns of offspring care. In 2016, Paul Garber, who has done pioneering research on callitrichine behavior, locomotion, and ecology and has written extensively on platyrrhines and their conservation, published such a study, with colleagues, on Weddell's Saddle-backed Tamarin, *Saguinus weddelli*. It is a model of how modern fieldwork is done, on both small and large platyrrhines, and the quality of the data that can be collected. The work was carried out in two phases, beginning in 2008 and then followed up in 2016. Using banana-baited traps, Garber's team caught, marked, and released 67 tamarins belonging to 12 different social groups living near one field site in Bolivia. When possible, under anesthesia, individuals were tagged with a miniature, subcutaneous radio transponder.

All individuals were weighed, measured, and fitted with lightweight identification collars. Tooth impressions were made in order to assess their relative ages based on the species-specific dental eruption pattern and tooth wear. A dental exam recorded the age-related degree of dental staining. The status of females, whether currently nursing or had been previously, was inferred by measuring nipple length. Hair follicles were removed to genotype each animal in order to establish the parentage of adults, assess the genetic relatedness of troop members, and evaluate whether one or both sexes disperse from the natal group. The results confirmed that the mating system of these tamarin groups is most frequently polyandrous, one female mating with more than one male. Both sexes tend to leave the natal group but they sometimes delay their departure and stay on as adults, presumably to assist in caregiving. The data also suggested that same-sex group members are closely related, but the parents of juveniles were not always part of each troop.

The daily routines of platyrrhine monkeys tend to be rather standard, and when not resting or sleeping the monkeys seem to be always moving and eating. However, there are notable variations among the genera because of the contrasting needs and lifestyles of the major ecological divisions. At the extremes, the frugivore-faunivores are constantly on alert and actively collect prey while travelling between fruit trees. The committed frugivore-folivores, in contrast, can be more sedentary. They are able to acquire leaves in bulk without moving much, but they need long bouts of rest in order to conserve energy, since leaves do not provide much.

Upon waking, after roughly 12 hours of sleep, monkeys move away from their communal sleeping site, which typically changes from night to night. The smallest species, the callitrichines, may sleep in a tangle of branches or a tree hole. Troops of the medium-sized monkeys, like titis, sleep on one or a few boughs of a tree, and the larger atelid species use one or more trees with large, sturdy branches. Social group size and body mass dictate the specifics of how large the sleeping tree or sleeping hole must be to accommodate its members. They also determine how many trees will be visited during the group's first feeding bouts of the day, which tend to favor fruit that is available nearby. Traveling and feeding on fruit, and perhaps catching a few insects encountered along the way, make up the first hours of activity. By midmorning, as the day heats up, activity slows. Midday hours, when it is hottest, involve limited, if any, traveling. The monkeys tend to take a siesta. No longer on the move, it is a time for juveniles and young adults to play while adults rest. Adults may quietly socialize, spending time in physical contact with one another. The afternoon is again spent foraging, feeding, and traveling before evening sets in.

In general, the first hour or two in the morning and in the late afternoon are the most active periods. The Collared Titi Monkeys, *Callicebus torquatus*, that we studied in northern Peru usually arose at or before dawn. The family group had three activity peaks during the day, at around 8 a.m., 11 a.m., and 2 p.m. Fruit was by far the most important food eaten in the morning, and fruits and leaves were preferred in the afternoon. This meant the titi monkeys ate high-energy, easily accessible food first, when their energy stores were depleted after a long sleep and they may have felt hunger. Later in the day, they bulked up on a combination of rapid- and slow-release energy sources before retiring. For frugivorous-faunivorous capuchin monkeys, fruits also come first. Though they are very skilled foragers, it has been suggested that insect-hunting may be inefficient in the early morning hours under dim light. During all of these activities, most monkeys interact and remain in relatively close proximity to one another. The exceptions are among species that have evolved social systems designed to allow subunits to disperse widely, sometimes for days at a time.

Leila M. Porter conducted the first year-long research project on Goeldi's Monkey, *Callimico goeldii*, in the late 1990s, in Bolivia. Hers is a case study that illustrates the activity budgets of monkeys living sympatrically, as she also collected comparable data on two species of Tamarins, the Red-bellied Tamarin, *Saguinus labiatus*, and the Saddle-backed Tamarin, *Saguinus fuscicollis*, that lived in the same forest as her study troop of Goeldi's. During Porter's study, the sizes and compositions of all the groups changed as the result of births, deaths, and intertroop transfers, but the monthly averages remained consistent. Considering adults only, the Goeldi's Monkeys had the smallest groups of the three species, consisting of two or three adult males and females. The lesser known Red-bellies had as many as five to eight adults in the troop at a time.

All three monkey groups began their day one or two hours after sunrise throughout the year, but there was a distinct difference in when they retired (fig. 8.1). The Goeldi's tended to wait until sunset was near before bedding down, which was always in vine or leaf tangles at a height of more than 10 m. The tamarins, in contrast, gathered at their sleeping sites one or two hours before sunset. They habitually slept in tree holes scattered about their territories, hardly ever using the same one during successive nights. Occasionally, they slept in the dense fronds of palm trees.

As a species that is normally jet-black and somewhat larger than the tamarins, the Goeldi's did not seek the protection of sleeping holes as various cal-

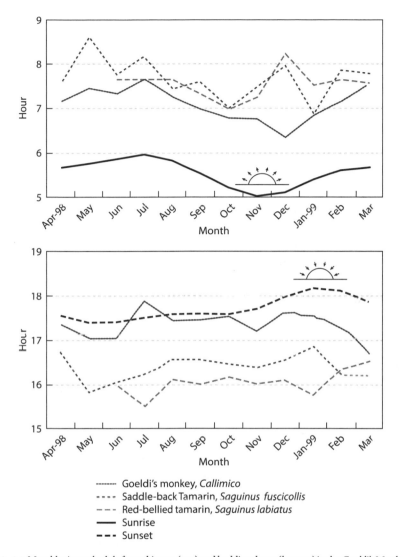

FIG. 8.1. Monthly time schedule for waking up (top) and bedding down (bottom) in the Goeldi's Monkey and Saddle-back and Red-bellied Tamarins. Adapted from Porter (2007).

litrichines do. Camouflaging coat color may be enough to make them safe from predation as nightfall approaches, and being in the open for a longer period at dusk may also provide useful time to forage and feed. Indeed, the peak feeding time when the Goeldi's ate their food specialty, gelatinous fungi, low to the ground and in the canopy's shadow, occurred in the final hours before they stopped being active. In contrast, among the tamarins, late afternoon

TABLE 8.2. Daily activities as proportion of total behavioral observations of three species of callitrichines

	Daily Activities (%)					Resting Behaviors (%)		
	Feeding	Foraging	Resting	Traveling	Other	Scanning	Grooming	Sitting
Goeldi's Monkey	9	6	66	17	2	87	11	1
Saddle-backed Tamarin	8	10	54	22	6	71	25	3
Red-bellied Tamarin	10	12	46	31	1	66	33	1

Source: Porter 2007.

foraging and feeding may be more constrained if their movements during that period are tied to the location of sleeping sites, or if inadequate light makes foraging for moving prey inefficient. If they scatter, it may be more difficult to coordinate meeting up at a common sleeping hole, and it may also attract the unwanted attention of nocturnal predators.

Of the three species of callitrichines living sympatrically at the same location in Bolivia, the breakdown of their daily activities, calculated as a percent of total behavioral observations, is shown in table 8.2. During the day, the species each present a different activity pattern, which separates them ecologically, though the tamarins are most similar to one another, as would be expected. Belonging to the same genus, they have very similar requirements. Most of the Goeldi's resting activities involve scanning the surroundings, perhaps for food, predators, or other groups of monkeys with which they habitually associate, such as the tamarins. They spend the least amount of time grooming, an intense form of social behavior, while the tamarins spend 25%–33% of their resting time grooming themselves and one another. Inspecting and cleaning the fur of dirt and parasites is an important part of their health ritual, but it is also a basic form of amicable, tactile communication between troop members.

Communicating through visual displays

Perhaps the simplest, least intrusive form of interpersonal communication is visual. This may be especially effective in platyrrhines because, as anthropoids, they have a highly developed visual system and all but one genus are diurnal, active when light is available, even if the light is dimmed by the foliage, which is their milieu. Visual signaling also requires the least expenditure of energy, if any. Once a feature that serves as a visual gesture or signal becomes manifest in adulthood, the energy costs to maintain and employ it are negligible. One exception involves male squirrel monkeys of some species who get "fatted" during the mating season and become more visually impressive, as discussed in chapter 2. Their weight gain probably does produce an extra physiological load.

The outward appearance of many platyrrhines is striking, with fur that is distinctively colored and patterned. Some of their faces are highlighted by leathery patches, marked with stripes, framed by beards and by patches around the eyes. Their tails can be especially bushy or silky smooth; some are distinctly short or long. Features like these may serve as species identifiers, a common mammalian method to avoid crossbreeding between closely related species living nearby in the same habitats. Some forms of appearance are gender-specific. The sexual dichromatism seen in the White-faced Saki Monkey, *Pithecia pithecia*, where males and females have very differently colored and patterned faces, is an example. It is an obvious way for the same-sized males and females to identify each other's sex immediately, without concern for interpersonal aggression. This is a case where natural selection favors the absence of a body size difference between the sexes, perhaps for ecological reasons and/or social reasons, such as equalizing resource needs and minimizing the potential dominance of one sex over the other. Because pelage coloration and pattern are permanent adornments, intrinsic parts of the body, they can perform their roles in communication passively.

Some bodily features, on the other hand, are dynamically employed as visual signals in sending a message. The Emperor Tamarin, *Saguinus imperator*, tosses its head in display, and the intensity of this behavior is presumably heightened by the animal's bright mustache (see plate 4). To signal arousal, fur at the top of the head in Cottontop Tamarins, *Saguinus oedipus*, and along the tail of the Pygmy Marmoset, *Cebuella pygmaea*, is raised in piloerection. In Geoffroy's Tamarin, *Saguinus geoffroyi*, the tail is coiled in a sexual display. Goeldi's Monkeys grimace in fear. The Capuchin Monkey and Squirrel Monkey, *Cebus* and *Saimiri*, bare their canines to show aggression using a stereotypical open-mouth/bare-tooth threat display. Male squirrel monkeys spread their thighs and expose their genitals as a show of dominance, or self-esteem according to some researchers.

The Arch Display by the Owl Monkey, *Aotus*, and the Titi Monkey, *Callicebus*, during which the animals bend their vertebral columns downward as if they are deeply bowing bipeds, is an aggressive signal of agitation. The Saki Monkey, *Pithecia*, also uses this visual display. It is homologous in all three genera, that is, genetically determined and inherited from their last shared ancestor. The Bearded Saki and Uacari, *Chiropotes* and *Cacajao*, perform ritualized tail-wagging as a sign of excitement, or as an intergroup visual signal, even though the latter has a shortened tail. Woolly Monkeys of the genus *Lagothrix* show hostility by stiffening their limbs in a quadrupedal stance, raising and curling their tails, and thrusting their faces forward. Female Howler Monkeys, genus *Alouatta*, curve their lips into a circle and flick their tongues to solicit copulation. The Spider Monkey, *Ateles*, capuchins, and saki monkeys shake and/or break branches as a reaction to predators and intruders.

The upper and lower canine teeth play an important role in visual signaling in many New World monkeys. Their evolution is strongly influenced in complex ways by social organization. When the canines are large and conspicuous, baring them by opening the mouth in a yawnlike expression is often a sign of aggression. Aggressive behavior is often employed in regulating interpersonal relationships is some taxa. Thus the canines are also used in fighting, and for protection from predators as well. Canines may also be exposed while vocalizing, as when a callitrichine opens its mouth wide to emit a high-pitched call.

In social systems where intratroop or intertroop competition is an important factor, the canines tend to be tusklike, tall and conspicuous because they project vertically beyond the row of the postcanine teeth. The crowns of upper canines are typically taller than the crowns of lower canines. Size differentials are exaggerated in males. This is a common pattern in all three platyrrhine families. For example, among atelids, in howler monkeys the length and breadth dimensions of male canines (taken at the base of the crown) can be 30% to 40% larger than those of females. Among cebids, squirrel monkey males have canines that are 20% to 30% larger than those of females. In male bearded sakis, a pitheciid, the canines are 20% larger than the female canines. It is worth noting that in some taxonomic groups, the canines' social roles are balanced evolutionarily with dietary adaptations to determine details of the morphology. For example, in sakis, bearded sakis, and uacaris, all with exceptionally prominent canines, the shapes and orientations of lower crowns are especially designed to facilitate opening the woody shells of heavily protected fruits. Evolution typically generates elegant solutions so that structures are designed to perform more than one role.

Canines are also sexually monomorphic, the same size and shape in males and females, in several New World monkey genera, although there are marked differences among them in the essentials of canine morphology and in correlated social organizations as well, suggesting the pattern evolved three times in parallel. Canine monomorphism is associated with social groupings in which there is limited aggression and/or significant male-female cooperation within troops. For example, in pair-bonded monogamous genera, such as the Titi Monkey and Owl Monkey, *Callicebus* and *Aotus*, the canines are small and monomorphic. There is one owl monkey species, in Argentina, in which males have slightly taller upper canines than the females that can be distinguished by formal statistics—though what this may mean to the monkeys is hard to know. As a rule, *Aotus* is monomorphic. The titi canines barely project beyond the adjacent teeth and are exceptional in having bluntly tipped crowns. The owl monkey canines are more tusklike and pointed. Small, non-projecting, non-dimorphic canines also occur in Muriqui, the atelid *Brachyteles*, which live in large social groups. A third contrasting pattern of sexually monomorphic canines occurs in callitrichines, where the canines of both sexes are of the same size and shape but the uppers and lowers are both high-crowned, vertical tusks.

The two different patterns of monomorphically small canines are examples of parallel evolution, as indicated by their fundamental morphological differences, contrasting social contexts, and different diets as well—only in titi and owl monkeys are the lower canines crucially important in feeding. Small canines evolved as a shared derived trait in titi and owl monkeys in connection with a pair-bonded, monogamous mating system, and where these teeth work alongside the incisors in harvesting hard-husked fruit. Small canines evolved in muriquis in connection with a multi-male/multi-female social system and polygamous mating system in which individuals are non-aggressive. There does not appear to be any feeding advantage to their small size, and cooperation among Muriqui males is an important aspect of their mating system. Males do not impede one another's mating opportunities. They also cooperate in territorial defense.

The monomorphic variant found among callitrichines, where the canines are large and conspicuous as opposed to being small and discrete in the other examples, is a third case of parallelism. The only other anthropoid primates to exhibit this pattern of large-canine monomorphism are the Old World lesser apes, gibbons and siamang; however, in contrast to callitrichines, gibbons and siamang live in monogamous pairs, while the callitrichines often live in multi-female/multi-male troops that are also polyandrous. As mentioned, upper canines are normally taller than lower canines when the canine complex is designed to be eye-catching. This suggests that the absence of a gender-based, upper-lower size differential in callitrichine canines, as in gibbons and siamang, evolved because selection favored large canines in females. A likely explanation for this condition is that competition exists among the females in callitrichine troops. The underlying adaptation relates to enhanced female aggression. It is a factor in their unusual form of female-oriented social organization, which involves female-female dominance and the reproductive inhibition of subordinate females by alpha females.

Tail-twining in Titi and Owl Monkeys as tactile communication

Tail-twining, winding the tails of two or more individuals together, is a form of tactile communication that is a crucial diagnostic behavior exhibited by the monogamous Titi Monkey, complementing their blaring dawn calls. It is commonly seen when the adult pair sit side by side touching one another, facing in the same direction, or facing away but with their backs touching, with one tail twisted around the other. When a juvenile is included in the set, all three tails are entwined. This behavior is a highly attentive form of social contact. Long periods of tail-twining among all troop members occur before the animals go to sleep. It probably serves to maintain and solidify the bond between mates and provide assurance to offspring. Tail-twining has also been seen while the pair-bonded titis are involved in intergroup encounters and while they cooperatively perform long calls as duets. A highlight of my early research experiences as a primatologist was seeing and hearing this daily ritual in the field.

It has also been documented that tail-twining occurs in Owl Monkeys. Reports of this are rare, perhaps because the nocturnal *Aotus* is difficult to see in the dark in the wild, and because of the way in which they are typically housed in captivity, with nest boxes to sleep in. It may also be the case that this behavior occurs infrequently in this genus. Nevertheless, it has been regarded as an important tactile interaction between adult males and females.

Martin Moynihan, a communications specialist and the first scientist to write a monographic treatment of platyrrhine diversity, behavior, and ecology from personal experience in his 1976 book, *The New World Primates*, added this footnote to his 1966 study of communication in captive *Callicebus*: "Since writing my earlier account [in 1964] of *Aotus*, I have seen Night monkeys intertwine their tails in just the same way as Titis. This was done by a captive pair in the National Zoo in Washington, as they slept side by side on a perch, in the open. (I overlooked this pattern before simply because all the other captive Night monkeys observed were provided with boxes, into which they retired to sleep. There is, however, a drawing of South American Night monkeys with intertwined tails in Cruz Lima, 1945)." In 1980, Dennis Meritt Jr. described this behavior in a pair of captive animals while he was Director of Animal Collections at Lincoln Park Zoological Gardens. He wrote to me in 2018: "When we saw tail-twining in our captive animals we knew that the pair was not only compatible but would indeed be one of our breeding pairs and the start of a new lineage. I have also seen this behavior in the wild in Paraguay." Observers working with the long-term, ongoing "Owl Monkey Project" in Formosa, Argentina, where the animals are active during the day, have also documented tail-twining (see plate 12).

Tail-twining, found only in these two genera, is a derived feature they share because they are closely related and inherited the behavior from their last common ancestor. If tail-twining in the comparatively well-studied *Callicebus* is explicable as a ritual that helps maintain a rare mating system and social organization, this interpretation applies to less well-studied *Aotus*, too, where the same general parameters of reproductive and social life are found. A striking implication of this hypothesis relates to the antiquity of this behavioral pattern. Both the fossil record and the molecular clock indicate their last common ancestor existed at least 20 million years ago. That means tail-twining is a gene-based social ritual, an instinctive fixed action pattern, that has been sustained in each of these lineages for roughly 20 million years.

Vocalizing with roars and duets

The roaring calls of the polygamous Howler Monkeys are aggressive signals for internal and external audiences. These are the most impressive vocalizations produced by a platyrrhine, or any primate or land mammal other than elephants and lions. An alpha male's calls advertise physiological and social status, a preemptive show of prowess to fend off other howlers that may have designs on his females as potential mates. In a very different, monogamous social system, another type of long-distance call is the impressive, staccato dawn call of Titi Monkeys, a complex set of loud notes that flow into a pulsating blast of sound, which can continue for up to seven minutes. It is a between-group territorial call, a long-distance alert that helps maintain spatial separation among densely packed social units of male-female pairs and their offspring.

There are several variations to the howler's powerful call and various circumstances in which they are produced. The calls may occur at any time of day or night, but they are often produced at dawn. The roars, and all their vocalizations, are low-frequency, which is ideal as a long-distance signal that must carry through a teeming, acoustically dense forest. It is very common for the bellows of one group to be answered by others, setting off a reverberating chain reaction through the forest, a crowd of choruses led by adult males and joined by females and even juveniles. This makes it clear that howler roars are designed for intertroop communication. The vocal mechanism that produces and projects the deep howler voice appears to limit their ability to emit high-pitched sounds, which are commonly used in short-distance signals in other New World monkeys but are not part of the *Alouatta* vocal repertoire.

Howler Monkeys live mainly in cohesive single-male/multi-female social groups, though troops comprising a male and female pair have also been encountered, as have bands with several males and females. Rarely are there

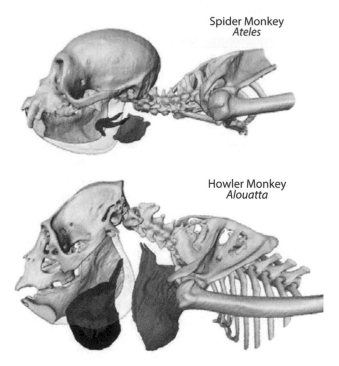

Spider Monkey
Ateles

Howler Monkey
Alouatta

FIG. 8.2. Comparative anatomy of the head, neck, and shoulders of two monkeys illustrating the exceedingly enlarged bony hyoid (beneath the skull) and thyroid cartilage (in the throat) of howlers. Adapted from Dunn et al. (2015). Courtesy of Jacob Dunn.

more than 10–15 individuals in a social group. Howlers have been studied intensively at many locations over several decades. They range over an enormous expanse of South and Central America and there are multiple species, so variations in recorded group size and composition are to be expected. Still, these documented variations leave unanswered many questions about the structure and evolution of such a non-rigid social system and the demographics of *Alouatta* across their distribution.

Howlers are sexually dimorphic, aggressive monkeys despite their inherently low-exertion demeanor. Groups rest for hours during the day, individuals sprawled out on large branches as they slowly digest low-energy, leafy food. Their roaring vocalizations are produced by a greatly enlarged hyoid-laryngeal apparatus situated in the neck (fig. 8.2). The mechanism is sexually dimorphic, much larger in males than in females. The shape of the *Alouatta* cranium and mandible that reflects this remarkable structure is another measure of the evolutionary importance of vocalization in the howler lifestyle and a striking example of the role played by social behavior in shaping morphological evolution. Despite their bellicose ways, howlers are not territorial as are the duetting titi monkeys, the other loud-calling platyrrhines. Troops of howlers may actually intermingle while feeding in a large tree. This is why some researchers stress that the howlers' specialized vocal behaviors are also an important means of regulating intratroop interactions, an aspect of a controlling, male-oriented social system.

Another facet of the *Alouatta* social system relates to out-of-group dispersal and troop formation. In howlers, subadult males and females that are nearing sexual maturity both leave their natal group to form, or join up with, another troop. To a resident adult male, a departing female that is not his daughter may represent the loss of a potential breeding opportunity, but the withdrawal of a male may be a benefit, the loss of a potential rival.

Male dispersal is associated with violent levels of aggression. An emigrating male or two may form a new breeding group by hostile takeover. They try to establish their own troop by dislodging and replacing an alpha male and commandeering his females. This is often accompanied by infanticide, as the new male or set of males may kill the nursing offspring of the females he aims to control. Doing so may give a new alpha male an advantage in what is called Darwinian fitness, the direct contribution to his reproductive output—offspring—as the females become receptive reproductively when they are no longer suckling young. Infanticide may also benefit the intruding male indirectly by reducing competition for food and care among the new offspring that he will sire, which also contributes to his fitness. Though females have surely lost their own prior investment in progeny as a consequence, it is to their own evolutionary advantage to resume reproduction as well. This dynamic underscores the importance of females' recognizing the protective role that an alpha male can play in their lives, hence their attention to his displays of mettle, his roaring calls.

The long-distance calls of titi monkeys, usually produced first thing in the morning, are higher pitched than a howler roar. They are more easily deflected by forest cover and do not carry as far. Yet these vocalizations are instantly recognizable by their beat. Titi calls do not have to carry as far because their groups are not as widely separated as are howler groups. These monkeys are more densely packed into smaller territories. Their dawn calls are often duets performed by the mated pair soon after they wake up and leave their sleeping tree, which in some localities may happen before sunrise. Titi calls are a mechanism that serves to mark territorial boundaries and repulse neighboring troops, to protect crucial fruit sources. When neighboring groups meet at the borders of their territories, both may engage in dueling duets and other visual displays, such as running vigorously back and forth in the trees.

Social groups of titi monkeys invariably comprise a single male-female pair of monomorphic adults with no visible differences in appearance, and their offspring. They are enduring, potentially lifelong in cases, sociosexual units, without any signs of dominance or aggression between the partners, and with many signs of cooperation. For example, males defer to females, letting them go first when departing a sleeping tree, initiating troop movements, and entering a feeding tree, and males tend to the care of offspring to a significant degree. In rare instances a third adult troop member has been seen to associate with a mated pair in the wild. But these all seem to result from unusual circumstances, which attest to the fixed evolutionary nature of this system. *Callicebus* social groups remain small, with no more than three young. When approaching maturity, the offspring are forced by the parents to disperse in order for them to form their own groups.

The social mechanisms that maintain the pair-bond have been studied in the wild and in captivity. Laboratory studies show that hormonally mediated female-female intolerance, male-male jealousy, and a very intense, mutually beneficial cooperation between mates appear to be involved. The ritualized vocal duet is one sign of cooperation, and the assistance of males in rearing offspring is another. The social bond between mates is also strengthened by a high frequency of grooming. Grooming has been observed to account for 10% of their daily time budgets given to travel and maintenance activity, with an additional 9% added once the group retires to the sleeping tree. Then, a grooming session may last as long as three hours before sleep. It is also when the animals devote a lot of time to tail-twining, the most striking behavior connected with social bonding in titis.

Paternal care is an important element of the pair-bonded social system in the Titi Monkey. The male carries their infant from a few days after its birth, returning it to the female for suckling. He also does most of its grooming and supports it with food. This was borne out in a study by primatologist Dawn Starin, who was part of our titi monkey team in Peru in 1975 and documented food transfers in the group. She showed that a titi infant and juvenile will beg for food from a male at a much higher frequency (67%) than from a female (11%). All the food transfers that resulted from infant prompts came from the pair's male. The types of food provided, whether fruits or insects, were items the young were not able to acquire at their stage of immaturity. The male also played with the juvenile, tolerated food-taking, and acted in an important protective role. In one incident observed by Starin, the father attended to the juvenile when it was startled by a tree falling close by. In another that we all witnessed, he rushed to comfort the juvenile when it fell out of a tree and landed on the ground.

Significant levels of paternal care, fathers caring for offspring and cooperating with females, is found in two subfamilies of two major platyrrhine clades, though it is rarely a prominent feature among other New World monkeys or any other primates. Among cebids, it occurs in callitrichines, and among pitheciids it occurs in titis as well as in owl monkeys. Monomorphism is one anatomical feature with clear-cut behavioral implications that is found in all the seven genera that exhibit this pattern of paternal care to a significant degree. It enables the development of a balanced relationship between males and females, which must exist if males undertake roles otherwise typically associated with motherhood.

It is important to recall that monomorphism also occurs in two other platyrrhine genera, in the spider monkey and muriqui, where it is associated with a fission-fusion social structure involving large groups in which no significant paternal care is exhibited. In these cases interpersonal aggression is also low. This makes the fission-fusion system possible, particularly the flexible compositions of the small foraging parties that may change from day to day and involve mixed-sex and mixed-age groupings.

Aspects of the titi mating system and social organization are exhibited by saki monkeys, which also use duetting calls during intertroop encounters. Owl monkeys do this as well, even though these nocturnal platyrrhines live a much quieter lifestyle, with no daily dawn calls, so the acoustics of the calls coordinated by the mated pair are more muted. The calls of all three genera serve varying roles relating to the regulation of troop spacing at spatial boundaries. Calling may be used to defend fruit trees and/or to defend space per se, perhaps to exclude potential sexual competitors from entering a home range.

The roaring calls of Howler Monkeys are a unique case that involves a variety of rather extreme behavioral and anatomical adaptations. Still, the fact that loud calls of three other platyrrhine genera, *Callicebus*, *Aotus*, and *Pithecia*, are important parts of their vocal repertoires requires more explanation, especially as there is a phylogenetic connection among them. They are all members of the pitheciid family clade and two, the Titi and Owl Monkeys, belong to the same subfamily. All three genera are monogamous—although some species of *Pithecia* may not be—and monogamy is very rare in primates, as it is in other mammals, too. Yet, it occurs in these three Neotropical monkeys in a classic pattern, which involves vocal territorial defense and duetting. Anatomically, they are all sexually monomorphic in body size, canine size, and, except for the saki, outward appearance. This suggests that the last common ancestor of *Callicebus*, *Aotus*, and *Pithecia* was similarly adapted to loud calling and duetting, for territorial defense in order to protect crucial resources and to support a small-troop, monogamous lifestyle. Again, this differs from the highly sexually dimorphic howlers, which are not territorial, and where the male roaring may play an important role in mate-guarding, keeping other troops from interfering with an alpha male's females while also maintaining order within the group by an aggressive display.

The roars and duets of howlers and titis are not the only form of vocal cooperation and interaction that may be pertinent to within-group cohesion. The leading figure in the study of platyrrhine vocal communication, Charles Snowden, and his collaborators have conducted long-term studies of callitrichines, mostly in the lab via experimentation. They showed that Pygmy Marmosets have highly flexible vocal repertoires which are influenced by social and environmental variables, more so than is typical of other primate species. These differences are referred to as dialects, which develop as the babbling calls of infants are molded by responses uttered by adults and other troop members. The Pygmy Marmosets also coordinate vocalizations by turn-taking. Snowden attributed these sensitive nuances in vocal communication to the cooperative form of social organization found among callitrichines. These platyrrhines live in social groups called a cooperative breeding and communal rearing system that involve intense infant care given by fathers and other troop members.

Sending scent signals

A primate's scent can be produced in two ways, via secretions from specialized skin glands and by urination. Both are modalities of chemical, olfactory communication used by platyrrhines. Scent signals are long-lasting, so they can perform a communicative function even after the individual that deposited the odor has moved on. As various studies have shown, they are highly complex chemically, and can therefore transmit a lot of personal information, such as the identity of the species of the depositor, the identity of the individual, its natal group, and the animal's sex and reproductive status. Ascertaining a female's sexual receptivity is valuable in many social situations. In New World monkeys there are no outward physical signs of ovulation. This is unlike some Old World anthropoids like baboons and chimpanzees where females present sexual swellings that become exaggerated when she is maximally fertile. Leaving an identity sign via odor may also be advantageous among groups whose members become widely dispersed and out of visual contact for prolonged periods, as occurs in spider monkeys, with their fission-fusion type of social organization.

The monogamous Owl Monkey, living in the nocturnal world where visual signals are less effective, relies heavily on olfactory communication to regulate interpersonal interactions. Among callitrichines, communal rearing offsets the competitive social demands of a high potential reproductive rate, and they employ scent in a dramatic way. The alpha female emits scented signals that are implicated in suppressing the breeding of lower-ranking females so that these non-reproducing females, and the resident males, are free to assist in caring for her infants. Infant care is a heavy burden, as a reproductively active female may be raising and gestating twin litters at the same time.

Given this last remarkable social adaptation, it is not surprising that among New World monkeys the twin-bearing callitrichines are known to use odor most intensively for communication. They have specialized skin glands in three areas of the body that give off scent. One field occurs in the anogenital region. Another is near the pubis, called the suprapubic gland. The third is near the breastbone or sternum, the sternal gland. By rubbing one or more of these padlike patches of glandular skin on a surface, the animals scent-mark their environment, or each other. The social significance of these actions is sometimes difficult to discern. They are often staged, adding a visual performance to the action. For example, a marmoset or tamarin will rub its sternal and suprapubic glands on a branch while lying down with their legs dangling on either side as they pull themselves forward pressing against the bark. The gum-eating Marmosets and Pygmy Marmosets rub their anogenital glands against trees to mark holes that have been gouged when obtaining gum. In the latter case, where the flow of gum may take time to exude, the scent-marks may be important in intertroop communication: one group leaves a message for another, letting it be known that it had been there and might be coming back soon. The intensity of odor deposited at gouged holes may provide important temporal information. In one study of marmosets, troops occupying the same area took turns visiting gum sites but managed to avoid interacting.

Being nocturnal, the Owl Monkeys have an extensive repertoire involving olfactory communication, mostly for intragroup communication. The adults live monogamously as pair-bonded mates in close quarters, where interpersonal interactions involve only two adult individuals. Social interactions between mates often consist of scent-marking and sniffing, rather than visual or tactile gestures, or vocalizations.

The most informative, detailed research on olfactory communication in *Aotus* is based on 12 months of data coming from an ongoing study of a captive colony of male-female pairs at the DuMond Conservancy in southern Florida, which was conducted by primatologists Christy Wolovich and Sian Evans, two leading experts devoted to studying owl monkey behavior. They described a wide range of sociosexual behaviors mitigated by scent. Like callitrichines and several other platyrrhines, owl monkeys have sternal glands near the breastbone and anogenital glands. In addition, they also have a specialized pad at the base of the tail called the subcaudal gland, and glands on the cheek. Several of these areas are groomed by a partner, or rubbed against them, in sexual contexts or during pair formation. Sternal marking coincides with male penile erection and mounting, and the subcaudal gland is used by females to scent-mark when new male-female pairings are established. It is thought to convey distinct chemical signatures offering information about age, sex, and the natal group into which the animal was born. Males also overmark, rubbing their subcaudal gland on surfaces that have been marked by their mate.

Urine plays a significant role in owl monkey olfactory communication, too. Males ingest droplets of female urine as it is excreted and they will lick surfaces where a female has urinated. They urine-wash by wetting the hands with urine and rubbing them on the soles of the hindfeet. This behavior has been interpreted as a method to mark travel pathways and/or territories in the wild. When paired, males engage in urine-marking and other scent-marking behaviors like anogenital rubbing more often than females, but, in general, females urinate more often than males. To Wolovich and Evans, this suggests that males intensely monitor the reproductive status of their female partner by decoding signals relating to the ovulatory cycle.

The Spider Monkey is another genus, with a different social organization, in which urine plays an important role, and it is associated with a specialized anatomy. *Ateles* females have an elongate, pendulous clitoris, whose size, shape, and color varies among individuals. Because the genus is sexually monomorphic, field observers normally use the clitoris to distinguish females from males and to identify them individually. That may also be the case with the monkeys themselves. Urine, and possible vaginal secretions that drip from the clitoris, are used in communication. Adult males smell branches where a female urinated, more often than other adult females do. Immature individuals grasp the clitoris and smell their hands, and lick urine where it is deposited. In forming mating dyads and during copulation, males intensely sniff and handle the female's clitoris, which suggests it may serve as an attractant.

One hypothesis for the evolution of the specialized clitoris emphasizes its olfactory potential as a long-distance means for males to identify females. Spider monkeys live in large, multi-male/multi-female, fission-fusion societies, where the troop splits up into small, daily foraging parties, whose composition and duration vary. The parties may roam widely to find food, the ripe fruits and young leaves that make up the *Ateles* diet. Working as a visual and olfactory identifier whose odor may change with time, the elongate clitoris may provide locational signals useful in tracking the movements of individuals with known group affiliations. Also, in spider monkey society, subadult females disperse from their natal group but males remain. An enhanced set of visual and olfactory cues may play an important role in identifying dispersing females that could become absorbed into a new troop.

Social dispersal is a mechanism that helps avoid incest and inbreeding, which would have deleterious genetic consequences and can complicate mating systems. When juveniles approach reproductive age they have to leave the natal group to establish their own mating opportunities. In the majority of platyrrhines, both males and females disperse, but there are exceptions. Females disperse in the fission-fusion society. This happens in the three ateline genera, *Lagothrix*, the Woolly Monkey, *Ateles*, the Spider Monkey, and *Brachyteles*, the Muriqui, and also possibly in two pitheciids, the Bearded Saki, *Chiropotes*, and Uacari, *Cacajao*. The latter pair have not been studied well enough yet, but we do know that their large social groups split up to form smaller foraging parties, indicative of the fission-fusion system. The dispersal of subadult females in ateline society may prefigure their roaming behavior as reproductively active adults, as mothers, as a phenomenon related to securing food. In spider monkeys, a mother and offspring may forage alone or become the nucleus of a small foraging party, the size of which depends on fruit availability. In a marked contrast, *Cebus* is only one clear example among New World monkeys of a genus where males disperse from the natal group. This pattern has major behavioral consequences in capuchin society that will be discussed further below.

The fission-fusion system of spider monkeys is an adaptation that allows large, cohesive groups, some including more than 50 animals, to be maintained while reducing what may otherwise become intense intratroop competition for resources. These are large monkeys with high energetic demands associated with travel. The foods *Ateles* prefer are ripe fruits, widely dispersed as clusters in small growing patches, and different fruits have different ripening schedules during the year. Moving a large troop through the forest to such sites, where only a small number of individuals can feed at any one time, could impose large energy costs but a relatively small return for each individual, so small foraging units can advantage each of them by ensuring the likelihood of an adequate supply of foods and limited competition from conspecifics. This hypothesis is supported by the observation that foraging parties tend to be larger when food is more abundant locally, when home ranges are very large and there is no competition from neighboring troops. The increased competition among individuals that would be experienced if spider monkey troops were to forage together would be heavily borne for a long time by subadults in this large, slow-growing platyrrhine. The large-sized aggregate troop, in turn, may benefit individuals by enabling the exchange of knowledge about food sources and by offering collective vigilance. Spider monkey adults and juveniles are preyed upon by eagles and large cats, like puma and jaguar.

A fission-fusion social system also occurs in the Uacari, *Cacajao*, where groups of 200 individuals have been recorded, though in a number of studies 20–50 individuals are the more frequent counts, and foraging units of three to five have been documented. Foraging parties may travel apart over a distance of 2 km, 1.2 miles. Like *Ateles*, *Cacajao* foods are widely dispersed and temporally limited, consistent with the explanation that a fission-fusion grouping pattern is a correlate of foods' temporospatial distribution. As with *Ateles*, the large-group aggregation of the uacaris may serve as a defense against predators, while small foraging parties make feeding more efficient by reducing intratroop competition. *Cacajao* are alert to the presence of large birds overhead, to which they respond by emitting alarm calls and dropping to safer places below the canopy. They are also wary of large cats, which trigger a different alarm call.

The odoriferous callitrichines

The cooperative breeding of two-molared callitrichins is associated with the most intriguing form of olfactory communication among platyrrhines. Their groups range in size from 2 to 16 individuals, and although there may be several adult females, only one breeds. The single breeding female releases pheromones, chemical signaling substances that influence the physiology and behavior of others, that block the reproduction of the other females by suppressing ovulation. In the rare instances where two females breed, they are mother-daughter pairs. The number of adult males in these groups ranges from two to several.

Reproductive suppression in callitrichins is connected with several characteristics that determine a high reproductive potential and output. A female produces twin litters and begins ovulating again shortly afterward, two to four weeks after giving birth, thus enabling her to conceive and gestate a new set of offspring while nursing a dependent set of twins. This happens twice a year in marmosets. They also become sexually mature at a young age. Females can conceive as early as 12–17 months of age, and males can father offspring at 15–25 months. Compared with all other anthropoids, individual callitrichins have the highest reproductive potential, though it is dampened by several real-life factors, including a female's limited tenure as the lone breeding female in a troop, high infant mortality, and the effects of pheromonal suppression.

A consequence of this pattern is the high energetic load assumed by a reproducing female as she lactates and gestates simultaneously. Her energy costs are considerable also because neonates are relatively large, together weighing 20%–27% of the mass of adult females in the Common Marmoset, *Callithrix jacchus*, for example. By being able to curb reproduction in other females, the alpha female may attract more helpers to care for her offspring, because they are not obligated to care for their own, and resident males will not divert their attention elsewhere. This social system provides helpers. In Common Marmosets a breeding male carries and cares for his twin offspring almost immediately, the day when they are born. He periodically returns them to the female to nurse until they are weaned. Additionally, each subsequent set of twins has an older set of siblings, who also help care for the infants. From an evolutionary perspective, which is based on the notion that each individual gains by adaptations that maximize their genetic output, the system works. This is because the activities of all the participants, who are close familial relatives, parents and siblings, assist in the propagation of the genes they share with their many brothers and sisters. Individuals may also gain experience in child rearing that benefits them later in life. This collaboration is the communal care system.

Always giving birth to twins is believed to be a unique pattern that evolved from a single-litter reproductive system in the ancestor of the two-molared callitrichin genera, *Cebuella, Callithrix, Leontopithecus*, and *Saguinus*. Many facets of the behaviors and cooperative breeding arrangement seen among them are tied to this unusual evolutionary strategy. They are predicated fundamentally on genetic changes in their ancestry that influenced the release of two fertilizable ova in females that would develop within a uterus designed for single-birth pregnancies. This type of uterus, present in all anthropoids, has a single-chamber structure for growing a single fetus. In most mammals, strepsirhine primates, and tarsiers, the uterus has two compartments, which is associated with multiple fertilizations. The cross-cutting, counterintuitive correspondence between litter size and womb design is one of the reasons the callitrichins' reproductive system and social organization is thought to be derived rather than primitive among platyrrhines. It is also derived among callitrichines. The three-molared Goeldi's Monkey, the first-branching lineage of the subfamily, gives birth to singletons.

There is a constellation of unusual phenomena characterizing these genera that are integral parts of a novel evolutionary strategy. They include maintaining a dwarfed body size, patterns of placentation, high neonatal-maternal size ratios, interrupted rates of bodily growth and development, sexual monomorphism, cooperative rearing, reproductive suppression, and occupying an ecological niche in which insectivorous or faunivorous tendencies are fundamental to their feeding ecology. An important breakthrough in our understanding of these phenomena came in 2014, when the entire genome of the Common

Marmoset was sequenced. This allowed researchers to identify the genetic basis of twinning, and to correlate it with an unusual pattern of intrauterine growth resulting in slowed development that delays the onset of significant organ growth and continues after parturition. It has the effect of limiting adult body size, meaning its result is dwarfism.

The unique uterine environment in which twin callitrichins develop also has consequences for behavior and social organization. As fraternal twins, the two fetuses derive from two different ova which, at fetal stage, develop two placentas that become joined, allowing the exchange of blood, cells, and genes between them. This means each individual develops from its own embryonic tissues, its own germline, as well as from the tissues generated by the stem cells of its brother or sister. In other words, genes are not inherited only from parents, but some are also passed horizontally, between the siblings. On a genetic level, individual callitrichins can thus be more than brothers and sisters, or normal fraternal twins. Technically, this pattern is a reproductive syndrome that produces offspring called chimeras, named after a mythical Greek creature that combines body parts taken from different animal species.

Chimerism in callitrichins appears to be beneficial, because fusing placentas and exchanging genes may initiate the developmental and hormonal processes that produce adult monomorphism and the homogenization of adult behaviors, like the enhanced parental care provided by siblings and fathers. The rarity of physical aggression between and within the sexes is probably a correlated effect.

Another advantage that relates to communal caregiving is the attractiveness of males to offspring. The complex ways in which individuals are genetically related may influence how well they recognize kin and cooperate. To test this idea, researchers collected epithelial cells from 30 infant Wied's Marmosets, *Callithrix kuhlii*. Some of these cells proved to be chimeric and others were non-chimeric. Thus some of the infants had mixed germ lines and were, genetically speaking, more broadly related to the other animals in this sample. Because epithelial cells develop into skin, which also means glands, they may be associated with olfactory signals that enable kin to recognize one another. In the experiment, when adult males were exposed to various infants, they carried chimeric babies more often than non-chimeric babies, possibly because the males responded more favorably to offspring with which they may have shared a more intense pattern of genetic kinship cues. In the communally breeding, non-monogamous callitrichin social system, where knowledge of paternity is ambiguous in the presence of other potentially copulating males, sensitivity to chimeric babies may in some way increase a male's feeling that he might have sired the infants with which he has a special connection.

Foraging parties

The platyrrhines that specialize by feeding on carbohydrate-rich, ripe fruits whenever possible are the atelines, *Ateles*, *Brachyteles*, and *Lagothrix*, although the Muriqui is also adapted craniodentally to ingest leaves, and it does so prodigiously as a fallback food when living in forests that have been degraded. Because ripe fruit is distributed in the rainforest in widely separated small patches, the atelines spend considerable amounts of their day traveling from one patch to another. Also as documented in *Ateles*, they may travel widely to locate fresh, immature leaves that are easier for them to digest because they lack the shredding, specialized, shearing cheek teeth and detoxifying guts, like howlers. Immature leaves give them an additional source of protein. Their social systems are aligned with the temporospatial requirements of this diet. Like chimpanzees, which are also ripe-fruit specialists, atelines live in a bifurcated society that comprises large social groups, roughly 12 to 50 individuals, that break up into smaller foraging and feeding parties of varying composition, often along gender lines. These smaller units may last for several days until their configurations are reshuffled.

The feeding, foraging, and social behaviors of the three ateline genera differ from one another in several subtle ways, but the fission-fusion pattern is a consistent feature of their social lives. On one level, the division of a large social group into smaller units seems like an inevitable consequence of the arboreal environment. Larger monkeys need more space than smaller monkeys, and coordinating the activities of a multitude of group members that are spread out in the forest cannot be done effectively without attracting attention from predators, and given the barriers to communication. Another explanation for this arrangement is that it minimizes the potential for competition at feeding sites, especially during lean periods when ripe fruit is scarce. In view of the heavy energetic cost involved in roaming widely each day to feed, this makes sense as well. There is also another factor that suggests why ateline males and females tend to travel and feed separately, which adds credence to the idea that avoiding feeding competition is a crucial benefit. It has to do with their body size and life history parameters.

Following general rules pertaining to primates and other mammals, as large platyrrhines, the atelines reproduce more slowly than small monkeys, have longer interbirth intervals lasting about three years, grow more slowly, and have a prolonged phase of dependence and semi-independence before becoming sexually mature adults. As they approach maturity, when they are between four and six years old, females disperse from the natal group. This places an exceptional burden on mothers, who must nurse, wean, and feed their offspring for years, and teach them how to forage on a very large number of plant species inside a large home range. Thus the small-unit foraging subsystem may provide special benefits to females by sparing them and their offspring from having to contest with other group members that require the same diet. In other words, a fission-fusion system may provide an advantageous selective edge biased toward females when they are large, slow-growing, and forage widely on a daily basis, whether they are arboreal, semiterrestrial, or even terrestrial primates. The temporary, small-party provision is balanced against the safety benefits of living in enduring larger groups.

It is interesting that the other closely related, large platyrrhines, the Howler Monkeys, present a starkly contrasting form of sociality even though they experience similar biological conditions associated with large size. They live in much smaller social systems where troop members spend all their time in close proximity. Diet makes all the difference. Here the maternal burden to ensure that offspring are fed is simplified by the physiological and behavioral adaptations to folivory that evolved in *Alouatta*. By comparison with atelines, food is everywhere and there is no imperative for the type of long-distance travel that could be an impediment to group cohesion.

The other New World monkeys that are organized in fission-fusion societies are the pitheciines, *Cacajao* and *Chiropotes*, although we are only in the beginning stages of describing and documenting their social lives, and their life history patterns. A basic difference between the Bearded Saki, Uacari, and atelines is that the pitheciines are far smaller in size, and are thus likely to have more compressed subadult life stages that require maternal support and oversight. However, in addition to the energetic expense of long-distance travel, *Chiropotes* and *Cacajao* expend significant energy costs in the handling time needed to dismantle the well-protected, hard-shelled fruits on which they feed. Thus the same model may apply to them. Small foraging parties centered on females and offspring may be beneficial because they are less likely to be inhibited by feeding competition from other troop members.

Capuchin gestural language

In the realm of sociality, Capuchin Monkeys are naturally distinguished, as they are with other systems like locomotion and feeding and brainpower. They are highly social, exhibit cultural traditions and complex interpersonal relationships, such as multi-individual coalitions, and their social systems, which vary among the species, sometimes defy standard definitions. So it is not surprising that all capuchins are highly communicative, employing a large repertoire of signals across the modalities. An account by Freese and Oppenheimer published in 1981 summarizes the gestures reported in the literature and lists 33 visual signals, including 20 bodily movements and 13 facial expressions, 23

FIG. 8.3. Some facial expressions of the Tufted Capuchin, *Cebus apella*. From Fragaszy et al. (2004), reproduced with permission of the licensor through PLS clear.

vocalizations, and 17 tactile patterns. A study of wild capuchins in 2001 that recorded calls and used sound spectrograms to distinguish their frequency and tempo detected 30 different calls.

Their visual signaling behavior is highly varied and sophisticated. Capuchins have more facial expressions than any other platyrrhine. They have better-differentiated facial muscles that make slight variations possible, and they are highly attentive visually, a likely byproduct of their acutely visual method of prey detection. The inner mouths of *Cebus* are also amenable to evolving varied visual signals preformed by the lips. For example, the males have large canines, which they use in threat displays; the canines may be bared whole or in part while the jaws are open or clenched, with the lips framing them in the round or by a horizontal slit. Different meanings are attached to each expression, and research findings within and across *Cebus* species demonstrate that the faces they make, which may be accompanied by vocalizations, are encoded accurately by the recipients.

Among the Tufted Capuchin's expressions (fig. 8.3) are the Silent Bared Teeth Display in adults and non-adults; Open Mouth Silent Bare Teeth

Display; Open Mouth Threat Face; Relaxed Open Mouth Display; and Scalp Lifting. All but the scalp lift are accompanied by vocalizations. The Silent Bare Teeth Display may occur during courtship. The Relaxed Open Mouth Display occurs during play. The Open Mouth Threat Face occurs in a dangerous or threatening situation and is accompanied by barking calls.

Always on the move and frequently obscured by foliage as they forage, individual capuchins vocalize constantly throughout the day, publicizing information about location, food, predators, and their personal state, whether they are feeling threatened or content. Some researchers maintain that food-related calls reflect their willingness to share food based on the social rank of other group members, how far away, and how many there are. Other researchers consider food-related calls to be simpler signals, notifying other monkeys about the caller's location in order to keep the individuals spatially separated, to enhance their foraging efficiency. This would be advantageous if an animal is foraging for invertebrate or vertebrate prey, a highly individualized activity that would be hampered by interference. In another aspect of their lives, there is a specific call used by adult female White-faced Capuchins, *Cebus capucinus*, to initiate and arrange troop movements. It is sometimes used cooperatively by several females to influence and coordinate travel.

Several leading experts on capuchin behavior and evolution, Dorothy Fragaszy, Elisabeth Visalberghi, and Linda Fedigan, have questioned whether there is a common form of the social organization that characterizes the various species of the sexually dimorphic Capuchin Monkey genus. They concluded that female bonding is at the root of their social systems, unlike the situation in most other platyrrhines. One major piece of evidence supporting this hypothesis is that it is the male that disperses from the natal group—in other platyrrhines it is either sex or just the females. Also, in capuchins, affiliative interpersonal relationships between females, and between females and males, tend to be stronger than those between males, which are fiercely competitive. Yet there are variations in how males interact. In the multi-male groups of White-fronted or White-faced Capuchins, males are codominant, tolerating one another as they feed on fruits in large-canopy trees. In contrast, in Tufted Capuchins, which are palm fruit specialists, dominant males monopolize these narrow-crowned trees. An alpha male is aggressive toward subordinate males, females that do not mate with him, and offspring which are not his, keeping meddlers from accessing the palm fruits he wishes to control.

Females, possibly as matrilines, form alliances that share food resources, assist in caring for offspring, establish same-sex dominance hierarchies, and organize troop movements. There are several reasons a system organized around female bonding is advantageous. One is that capuchins mature slowly and adult females have long interbirth intervals, so a mother and her close kin benefit one another by being helpers. But in forming coalitions, females also benefit because capuchin society is harsh. Working together, females can mount a mutual defense against rogue males during hostile group takeovers, when females may be wounded and their offspring are sometimes killed. This may happen when two or three marauding, non-resident males seek to appropriate a troop's females by assaulting and evicting the existing dominant male. The fighting involves biting and slashing with trenchant canines.

Strong alliances among females, and between females and the alpha male, may afford some protection in such circumstances. During the years of a male's tenure as the alpha, there are likely to be several episodes of invasions by bachelor males. This tendency may be why adult capuchin males tolerate one another as troop members, as they become allies in fighting off aggressors. It requires a complex balance involving a high degree of social intelligence for an alpha male to weigh the imperative to maintain the fealty of his female consorts while relying on his male comrades, potential competitors for the females, to help safeguard against intruders, should the time arrive.

Over decades of study, Fragaszy, Visalberghi, and Fedigan documented such takeovers every three to four years in White-faced Capuchins, and the severity of social upheaval upon females that often results. They wrote that adult females "lose the resident males with whom they have formed affiliative bonds, they suffer wounding and they may lose their young infants. Their group and typical foraging patterns are thrown into disarray as males rampage through the forest, looking for one another. No wonder that females and juveniles stay away from inter-troop encounters!"

When such violence occurs, female capuchins are known to exercise different options, another sign of their intelligence. In some troops, high-ranking females lead all the group's females and their young away from the fighting and from males who may be chasing after them. Or, they may team up with a resident male to ward off intruders. In some cases, once the fighting is over and the alpha male is deposed, females may begin to affiliate with the invaders, who are potentially infanticidal, presumably to protect their young. These patterns of high-level male aggressiveness, female-female collaboration, and readiness of females to consort with invading males are characteristics associated with the unusual pattern of male-only dispersal found in capuchins.

An evolutionary model of platyrrhine sociality

What form of social organization might we reconstruct as the ancestral, or most primitive, pattern of the living platyrrhines? We can apply the method of character analysis to look for common denominators among the component parts of social organizations in order to build a model pertinent to the radiation's origins, which may be a logical predicate relevant to the evolution of other traits and systems. First, we can eliminate the outlier characteristics that have no bearing on the reconstruction, specializations that are rare or relate to unusual ecological or behavioral conditions that most likely would not have been building blocks in the last common ancestor. There are four such outliers: female-only dispersal, male-only dispersal, reproductive suppression, and monogamy.

Organizations in which only females disperse occur in fission-fusion systems in two socially specialized clades: atelines and, possibly, the pitheciine sister genera *Chiropotes* and *Cacajao*. It appears to have evolved in parallel in these clades in connection with similar, specialized foraging patterns involving long-distance travel to find fruits at the right stage of ripeness, for which particular locomotor, behavioral, and feeding adaptations have also evolved as correlates. None of these elements are likely to be primitive for platyrrhines.

A male-only dispersal pattern is even rarer among New World monkeys, and unlikely to have been present in the last common ancestor. It definitively occurs only among capuchins. There is a strong female-centered aspect to capuchin sociality that is also unusual and probably linked with male dispersal.

A derivative of this pattern may be part of squirrel monkey society, which in some species is sexually segregated, with males forming bachelor groups apart from the females, who band together, until mating begins. This may be important because *Cebus* and *Saimiri* are sister genera that share many novel ecological, anatomical, and behavioral resemblances that may have evolved when the clade originated. Several unusual, characteristic factors are consistent with this hypothesis: significant degrees of sexual dimorphism in canine size in both genera; permanent body-size dimorphism in adult capuchins and temporary, simulated body-size dimorphism in some squirrel monkeys during the fatted-male period; high levels of aggression exhibited by males; capacity for females to coordinate and live cooperatively; the separateness of bachelor capuchin males traveling together in small coalitions looking for mating opportunities; and the formation of months-long bachelor male groups in squirrel monkeys. These appear to comprise an integrated pattern of behaviors and morphologies that characterize a cebine style of multi-male/multi-female social organization. None of the elements are likely to have been ancestral in platyrrhines.

Reproductive suppression is restricted to callitrichines and involves many correlated specializations, no elements of which occur in other platyrrhines. This argues against its being part of the ancestral social or mating systems of New World monkeys.

Monogamous pair-bonding is another extremely rare mating system that also involves numerous unique behaviors and morphologies expressed in only two closely related genera, *Callicebus* and *Aotus*. The constraints of this system minimize the possibility that a similar model of monogamy was present in the last common ancestor of platyrrhines.

With outliers eliminated, what remains are the traits and patterns that may have been inherited from a common ancestor, the commonly shared components of sociality found in relatively distant platyrrhine clades with different types of social systems.

There are two commonly shared features and societal arrangements: dispersal by both males and females out of the natal groups, and multi-male/multi-female systems that are male oriented. Dispersal by both males and females out of the natal group when they approach sexual maturity is widespread, occurring in most platyrrhine genera and in all three family-level clades. As such, it is likely to have also been present in the platyrrhine common ancestor. Multi-male/multi-female social systems that are male oriented are similarly widespread, existing under different conditions of social dynamics, in several mating configurations and different parental-care arrangements.

The foregoing suggests the general outlines of ancestral platyrrhine sociality. The groups were probably of modest size and consisted of multiple males and females that mated polygamously. Sexual dimorphism was moderate, and the cooperation provided by males toward rearing offspring was limited. Both males and females dispersed from the natal group. Olfactory and visual forms of communication were favored over loud, long-distance calls. However that social system is defined, it must have corresponded to the requirements of local ecology, especially feeding habits. Because it makes sense to reconstruct the ancestral platyrrhine as a monkey of modest size, a mixed, non-specialized diet based on fruit supplemented by insects and/or a modicum of leaves would have been likely.

Under this hypothesis, platyrrhines subsequently evolved several notable, divergent variants of social systems and lifestyle behaviors. The pair-bonded, monogamous, intensely cooperative system shared by *Callicebus* and *Aotus*, and in less extreme fashion by *Pithecia*, is a derived phenomenon that evolved from a multi-male/multi-female system. It probably began by extending the temporal duration and spatial separation of consort pairings, or small feeding parties, foraging for an alternative type of food, fruits with relatively tough skins, perhaps because they were not yet ripe, with more easily edible seeds. Because the core element, pair-living, occurs in two pitheciid clades, it follows that this may be a primitive feature of the socioecology of the family. In turn, this suggests that the larger social groups evolved as their own shared specialty in the other pitheciines, the little-known Bearded Saki and the Uacari, which are undoubtedly sister genera. They evolved features paralleling the atelines, the fission-fusion subsystem, which in both clades is an adaptation to feeding on fruits that are widely distributed in patches of forest, and are exploited at specific stages of ripening. It is possible that the large-group construct of fission-fusion in pitheciines evolved by accreting small units into a larger whole, while the ateline pattern evolved by splitting a formatively large group into subunits as large body size evolved and a roaming style of foraging was emphasized.

The specifics of the cebid social system departed from the ancestral platyrrhine pattern as a degree of female-oriented, cooperative offspring care was introduced before the two subfamily-level clades, the cebines and callitrichines, split from the cebid common ancestor. In addition to essential care provided by mothers, other females—mothers, sisters, and aunts—of the troop may have optionally provided extra care for dependent infants, carrying them occasionally, grooming them, protecting them, and sometimes even nursing them. These behaviors are exhibited sporadically by squirrel and capuchin monkeys today.

In callitrichins, such behaviors evolved to become obligatory, as a necessity for successful rearing of multiple offspring simultaneously, in connection with twinning. The efficacy of assisting females, that is, reducing the energetic load relating to reproducing, nursing, gestating, and caring for infants, was enhanced by evolving from a male-oriented polygamous system to a female-oriented system that provides helpers. In this system, multiple males have the potential to sire offspring, once or twice a year for several consecutive years, which assures their genetic contribution to the next generation, a benefit for transforming socially into a female-oriented system. The help given by other closely related female group members is reciprocated when each of them gives birth. The collective effort in this cooperative arrangement encompasses infant carrying, food sharing, vigilance, predator protection, and resource defense.

This big-picture model of the evolution of New World monkey social systems is neither comprehensive nor simple. It involves many interconnected hypotheses concerning behavioral parameters and links to ecology, and it lacks basic information about the species-specific hormonal regulation of behavior that is the foundation of interpersonal conduct at the root of sociality, and directly influenced by selection. A principal feature of the model emphasizes the ecophylogenetic aspect, that behaviors and social arrangements of the living monkeys are patterned and integrated phylogenetically, just like anatomical traits, feeding preferences, locomotion, and hand use. This means a temporal dimension can be added to the hypotheses, based on fossils and the molecular clock.

For example, the entire configuration of pair-bonded monogamy, involving behavior, anatomy, communication, etc., as it is played out exclusively and in similar ways in the tail-twining titi and owl monkeys, must reflect their shared genetic heritage, carried forward when these two lineages split from their common ancestor. Both the molecules and the fossils place the split at more than 20 million years ago. This implies that the core behaviors we see today are at least 20 million years old. The behaviors that underpin the female-centered systems of squirrel monkeys and capuchins, lines that diverged at the same time, are equally ancient. Similar logic applies to other platyrrhine clades. The multi-individual embrace of tail-hanging atelines dates back more than 10 million years. The fission-fusion social systems of *Chiropotes* and *Cacajao* is roughly 7–14 million years old. Reproductive suppression in callitrichins may go back 15 million years.

Many of the essential behaviors we see today, all integrated by evolution, that shape the lives of platyrrhines, whether in locomotor style, dietary choices, interpersonal behavior, methods of communication, or organizations of social groupings, have had very long, continuous histories, beginning at least 20 million years ago. Many of the New World monkeys are, therefore, living fossils in their own right.

CHAPTER 9

20 MILLION YEARS

Every Fossil Tells a Story

The paleontological record of New World monkeys grew slowly at first but the pace of discovery accelerated rapidly near the end of the 20th century (fig. 9.1). The first specimens were named in 1838 when paleontology had not yet gained the stature of an authentic scientific discipline. While the material was referred to in 1859 by Darwin in *The Origin*, it wasn't until the 1890s that any fossil Neotropical monkey figured into the 19th century's fledgling notions of primate evolution. Only three legitimate, new fossil platyrrhine genera were named during the first 100 years since that initial discovery of a few postcrania, none of whose cranial bones or teeth were ever found. It was *Protopithecus brasiliensis*, from a cave in the Minas Gerais State, Brazil, about 130 miles from the Muriqui Project of Caratinga, the location of the most important research and conservation initiative devoted to *Brachyteles*, a potential close relative. Only one more legitimate platyrrhine genus had been named when the 1800s came to a close. A pulse of discovery occurred in the late 1940s, and then the pace increased rapidly beginning in the 1980s, driven largely by international efforts led by Marcelo Tejedor, John Fleagle, Richard Kay, Takeshi Setoguchi, Masahito Natori, Laurent Marivaux, and their many colleagues in Argentina, Colombia, and Peru. There are now more than 30 established genera known from the South American mainland and the Greater Antilles in the Caribbean Sea (fig. 9.1; table 9.1). The ones discussed in this chapter have special places in the history of New World monkey evolution, and in the evolution of our knowledge about their history.

This represents a remarkable, rapid increase in knowledge. Only eight genera, half the current biodiversity as measured at that taxonomic level, were recognized in 1979 in Szalay and Delson's classic review of primate paleontology, *Evolutionary History of the Primates*. When the number of known fossil genera eclipsed the number of extant genera in the 1990s, a turning point in perspective was reached. Still, while recent decades have witnessed exponential taxonomic growth of the platyrrhine record, the amount of information

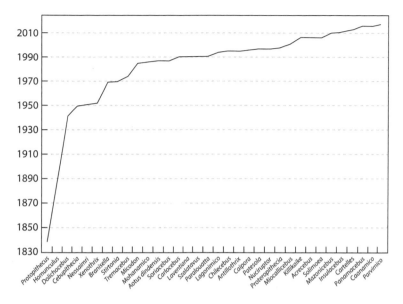

FIG. 9.1. Increase over time in the number of unique, named fossil genera. *Aotus dindensis* is also included for its historic significance. *Ucayalipithecus* was discovered in 2020 (see end of chapter).

that is actually available pales in comparison to the fossil record of Old World anthropoids, covering the same periods of Earth history.

What has been gained with these discoveries has allowed a new understanding of platyrrhine evolution that goes beyond what can be gleaned from studies of extant forms alone. It became possible to discern patterns of stability and differentiation over geological time, in the phylogeny, morphology, and ecology of the New World monkey radiation whose members still live today. The Long-Lineage Hypothesis is based on the discovery that 7 of the 16 extant platyrrhine genera, *Cebuella*, *Callimico*, *Saimiri*, *Cebus*, *Aotus*, *Callicebus*, and *Alouatta*, are represented by fossils very closely related to them that are 10–20 million years of age. The Long-Lineage Hypothesis has become an important model describing the pattern of evolution pertaining to the last 20–25 million years of New World monkey history—the existence of multimillion-year-old lineages that have occupied distinct, enduring ecological niches without much change.

As the fossil record increases, it is likely that more such lineages of similar character will also be found. This prediction is suggested by the molecular evidence that demonstrates, according to leaders in the field such as Horacio Schneider and Iracilda Sampaio, that it was "in the late Oligocene (~25 Ma) and early Miocene (~23.8 Ma) that the cladogenesis producing the crown families Pitheciidae, Atelidae and Cebidae occurred." More may also be found

TABLE 9.1. Paleontological record of New World monkeys

Age	Epoch	Genus	Geography
11,000 Yrs	Holocene	*Xenothrix*	Jamaica
1.4 Mya	Pleistocene	*Antillothrix*	Hispaniola
		Insulacebus	Hispaniola
		Protopithecus Caipora Cartelles Alouatta mauroi	Brazil
		Paralouatta varonai	Cuba
2.6 Mya			
6.8 Mya	Pliocene		
9		*Solimoea Acrecebus*	Brazil
11		*Cebuella Cebus*	Peru
12–14		*Micodon Patasola Dolichocebus*	Colombia
		Neosaimiri Mohanamico	Colombia
		Lagonimico Nuciruptor Miocallicebus	Colombia
		Cebupithecia Aotus dindensis	Colombia
		Stirtonia	Colombia
15.7	Miocene	*Proteropithecia*	Argentina
16.4		*Homunculus Killikaike*	Argentina
17		*Parvimico*	Peru
		Soriacebus Carlocebus	Argentina
		?Paralouatta marianae	Cuba
20		*Tremacebus Dolichocebus Mazzonicebus*	Argentina
		Chilecebus	Chile
21		*Panamacebus*	Panama
23 Mya			
26	Oligocene	*Branisella Szalatavus*	Bolivia
29 Mya			
36–40	Eocene	*Perupithecus Ucayalipithecus*	Peru
56 Mya			

with additional study of fossils that are already in hand. Whether or not these predictions turn out to materialize, the density of long-lived, genus-level lineages and suprageneric clades that we currently understand with high or reasonably high confidence, especially those involving fossils that coexist in space and time, paints a coherent picture, an illustration that they are all interconnected evolutionarily because the primate ecosystem has been evolving as a unit for millions of years in South America.

Among the 30-plus extinct genera that now make up the platyrrhine fossil record, three extinct species older than 10 million years are allocated to genera still living: *Aotus*, *Cebus*, and *Cebuella* (table 9.1). *Alouatta* is also represented by an extinct species of unknown age. It is probably much younger, probably from the Pleistocene Epoch, between 12,000 and 2.6 million years old.

Many platyrrhine fossils consist of only a single specimen from one individual, usually a few isolated teeth or a partial set held together by a small piece of lower jaw. This paucity of fossils markedly contrasts with the voluminous collections of fossils documenting the evolution of Old World monkeys and apes. Fossil New World monkey cranial remains are precious few, and postcranial fossils are even rarer. Fossil limb bones, which may amount in total to no more than an ankle bone or the end of a humerus or tibia, are associated with about a dozen platyrrhine genera, and only five—all very much younger than the rest by many millions of years—are represented by what we might call partial skeletons. These are sometimes called subfossils, like those found in dry cave sites in Brazil, because they have not become fully mineralized and are of relatively recent vintage, belonging mostly to the Pleistocene.

There is a geographic bias in our fossil sample, too, because it is difficult to find productive localities in the Amazon where Neotropical monkeys now live in relative taxonomic abundance and would likely have been very diverse in the past as well. Exposures of fossil-bearing Cenozoic Era rocks are rare there. The vast majority of monkey fossils have been found in or near the drier, comparatively barren regions of the Andes, where erosion has exposed geological deposits from periods when primates existed there under different climatic conditions prior to extensive mountainous uplift.

These regional differences exemplify to some extent the variety of ways that paleontologists have searched for and discovered the fossils. In Patagonia, where mammal paleontology originated in South America and many remains of large fossil mammals are rather common, paleontologically speaking, surface collecting, visually scouring exposures where fossils can be seen once they erode through the sediment, has been the major method—walking the terrain and then getting a closer look at its surface by crawling along on hands and knees, face to the ground. In the Amazon, intense collecting efforts in recent years proceed by digging mud from riverbanks at age-appropriate geological levels when water levels recede, then screen-washing the sediment through a graduated series of mesh filters to separate teeth and bone from the wash. Such a laborious approach sometimes pays off handsomely, as happened when a Peruvian-French team discovered exceptionally tiny fossil teeth belonging to *Cebuella*, the Pygmy Marmoset, reported in 2016, and a Peruvian-U.S. team found the marmoset-sized *Parvimico*, described in 2019.

FIG. 9.2. Geographic distribution of platyrrhine fossils in South America and the Caribbean.

In the Caribbean, since 2009 teams of scuba divers have opened up a new venue for exploration in the Dominican Republic: submerged freshwater caves, where they have found troves of subfossils that are often miraculously complete because the bones did not suffer from being scattered before fossilization by carnivores or scavengers, deformed by the crushing weight of overlying sediment, or carried away by wind and water after eroding out, to renew the cycle of scattering, damaging, and burying.

Despite material shortcomings, the platyrrhine record establishes historically important details. This is a fortunate consequence of the fact that the modern genera are quite different from one another anatomically. This means there is good potential to recognize their relatives in the fossil record, theoretically up to a point—the time when one line splits from another to form a new lineage. The chances of encountering such specimens, which means finding what is one of the paleontologist's most fervent wishes—the oldest, the earliest, the first—is remote because the splits happen rapidly in a geological sense. And the likelihood of any individual of a species actually turning into a fossil is very small, no less the species that will be found that is direct evidence of one of evolution's most elusive questions: What are the ancestral origins? Yet, because platyrrhine genera are so different from one another and some have evolved for long periods of time without much detectable change, the chances of discovering and connecting ancestors with descendants may be greater than we usually expect in mammal paleontology.

Geographically, the fossil record demonstrates that New World monkeys were more widespread in the past than they are today (fig. 9.2). Miocene epoch fossils (table 9.1) continue to be recovered near the southern tip of the continent that has been prospected off and on for well over 100 years, well south of the current range of any extant species. It was around 20 million years ago, and possibly before that, when an abundance of platyrrhine taxa were living in the southern cone, surviving there for millions of years before it cooled down and began turning into grassland roughly 14 million years ago. At the northern margin of the continent, thanks to a paleontological salvage operation attached to the Panama Canal Expansion Project that began in 2007, a 20-million-year-old New World monkey was found in a small bloc of terrain that once sat within the seaway that connected the Pacific Ocean and the Caribbean Sea before the isthmus arose. It's surely a platyrrhine, but a paleogeographic hair-splitter at the same time. Does the geology qualify *Panamacebus transitus* as South American, Middle American, or North American? What *Panamacebus* definitely reveals is the continentwide distribution of New World monkeys at this time interval, when, as we have long known, monkeys were well established in the southern reaches of Patagonia, about 6 million years earlier than they had been documented to exist in the vicinities of Amazonia, in Colombia, living in a habitat that is effectively a historical part of today's rainforest. Toward the east, bones from the Greater Antilles islands within the Caribbean, including some that had for decades been unidentified and stowed away in museum drawers as miscellany that needed safekeeping for posterity, and then the new underwater finds, proved that platyrrhines established an indigenous splinter radiation that arose millions of years ago and evolved on Cuba, Hispaniola, and Jamaica before meeting with recent extinctions.

All the families and subfamilies represented by extant genera have genetic relatives among the fossils. This provides critical information about the pattern of platyrrhine evolution, indicating when ecophylogenetic clades were present and what some of their key characteristics were like during earlier phases of their history. The evidence concerning body size, dietary preferences, and jaw shape that can be demonstrated or inferred from fossils confirms the outlines of the Long-Lineage Hypothesis and its complement, the Ecophylogenetic Hypothesis. These hypotheses posit that the early evolution of the modern Adaptive Zones within the major phylogenetic groups was followed by the stability of these lifestyles, their anatomical bases, and their ecological relationships.

It is interesting to note that these hypotheses were developed from paleontological information when the fossil record consisted of a collection of bones and teeth which could fit comfortably in a single shoebox, the eight genera Szalay and Delson discussed in 1979. Yet, as hypotheses they hold predictions. And, as the fossil record accumulated and the molecular phylogenetics of platyrrhines developed in technique and taxonomic coverage, the core predictions held true: newly found fossils proved to be closely related to extant genera or to extant low-level taxonomic clades such as subfamilies, and the time stamps of molecular phylogenetic trees have demonstrated independently that all extant Neotropical monkey lines are long-lived.

It is important to emphasize that three of the pre-Pleistocene fossils are, in fact, classified in the same genus as their living counterparts, *Aotus*, *Cebus*, and *Cebuella*, representing the Owl Monkeys, Capuchin Monkeys, and Pygmy Marmosets. The fossil specimens assigned to *Cebus* and *Cebuella*, which are about 11 million years old, have not been given species names by the authors of these finds as of this writing, pending a closer examination of the living species. The fossil species of *Aotus*, which is 12–14 million years old, is *Aotus dindensis*. It is diagnostically different from living owl monkey species.

That these three fossils fall neatly among extant genera means several things. It means they occupied ecological niches and lived lifestyles comparable to the living, congeneric species; that their positions in the distant past make it possible that they are in or near the direct ancestry of living forms; and that there is ecological continuity between the paleo and the modern faunas. Look-

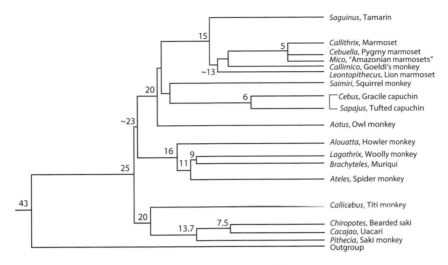

FIG. 9.3. Molecular cladogram and timeline for the evolution of clades and genera recognized by most molecular phylogeneticists, in millions of years. Outgroup refers to catarrhines. Adapted from Schneider and Sampaio (2015).

ing both broadly and narrowly at the 12–14-million-year-old fossils from La Venta, Colombia (table 9.1), one finds nearly every "monkey type" one would expect to find today in the Amazon. There are callitrichines, a squirrel monkey, a Goeldi's monkey, a titi monkey, a saki-uacari relative, an owl monkey, and two species of howlers. What's missing, an ateline like a spider or woolly monkey, will probably be found there some day. This is another prediction based on the Long-Lineage Hypothesis and the Ecophylogenetic Hypothesis.

To summarize, the fossils demonstrate the antiquity of five of the major platyrrhine subfamilies: cebines, callitrichines, homunculines, pitheciines, and alouattines. For the atelines, there is as yet no fossil record that is pre-Pleistocene. The oldest dated fossils pertaining to each of these families are between 12–14 and 20 million years old. The 12–14-million-year-old La Venta site in Colombia includes fossils from the non-ateline clades. The oldest ones are from sites in Argentina. They demonstrate the existence of cebines and homunculines 20 million years ago, and pitheciines nearly 17 million years ago. The molecular clock is consistent with the dates of differentiation estimated from the fossil record. A survey of eight molecular phylogenetics articles published between 2007 and 2016 indicates that the cebids existed at 22 million years; callitrichines began splitting up by 16.5 million years; pitheciid and atelid families existed at 23 and 22 million years, respectively.

A more detailed molecular analysis targeting the genus level is also consistent with the paleontological ages (fig. 9.3). As a whole, molecular studies that compare the origination dates of the major lineages of primates have demonstrated that extant platyrrhine groups are older than any of the Old World monkey and ape clades. South America as an island continent with its own physiographic parameters, many conditioned by the gradual evolution of the Andes as the mountains developed from south to north during the last 20 million years or so, has been a unique place for primate evolution and the New World monkeys have responded in their own enduring way.

Linking a fossil with a living monkey

The Long-Lineage Hypothesis

The skull arrived at the American Museum of Natural History in New York City on loan at my request from the Museo Argentino de Ciencias Naturales "Bernardino Rivadavia," in Buenos Aires, Argentina. I was a graduate student studying the evolution of New World monkeys, and this 20-million-year-old fossil was one of only nine species known to science at that time, in the 1970s, the eight recognized by Szalay and Delson and one that had been more or less forgotten by the mainstream until the 1990s, when spectacular new finds drew attention to the enigmatic *Protopithecus brasiliensis*. This particular fossil skull was about 2.5 inches long. It had never been cleaned properly and was covered with cemented sediment. As soon as its anatomical secrets became visible, a discovery was made from this fossil that led to the important understanding that it was, in fact, a relative of today's Squirrel Monkey genus, which meant that the *Saimiri* lineage has existed for 20 million years. The procedure at the museum was to have an expert preparator clean the specimen, preserve it, and expose as much anatomy as possible. In the museum's Vertebrate Paleontology prep lab, fossils are freed from and cleaned of their adhering matrix, repaired, reconstructed, chemically coated to harden them against physical damage, and, by molding and casting in epoxy, fashioned into perfect high-resolution replicas for exhibition and study.

This type specimen of *Dolichocebus gaimanensis* was first described in 1942, but it had not been given that special kind of treatment until I borrowed it 35 years later. Its teeth were gone, the cranium was squashed a bit from side to side, and the snout was damaged; it did not seem like a promising source of new information. Nevertheless, it needed to be cleaned and prepped properly. The small skull was heavy, despite having obviously thin bones. During fossilization, gray sediment had filled the eye sockets and the braincase, through the large foramen magnum opening on the cranium's underside, and the whole thing had solidified into a single mass of rock (fig. 9.4).

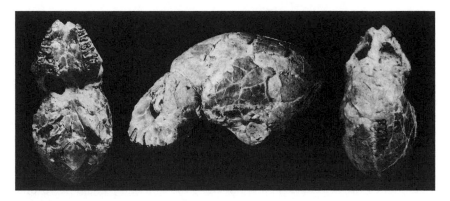

FIG. 9.4. Type specimen of *Dolichocebus gaimanensis*, a relatively complete edentulous (toothless) cranium. Courtesy of John G. Fleagle.

The chief vertebrate paleontology preparator at the American Museum at that time, Otto Simonis, a master, undertook the job of readying *Dolichocebus* for study. He was patient, and endowed with the dexterous skills and sharp eyes of an ultrafine jeweler. With dental picks, a tray of water, sponges, artists' brushes, solvent, and a diluted mixture of glue, he worked on *Dolichocebus* for weeks, often under a microscope, dissolving the hard matrix into fine sand and removing it strategically, almost grain by grain. The aim was to uncover more anatomy, more information, and to produce research-quality casts that would minimize wear and tear on the original, so that it could also be shared with others. Today, lasers and CT scanners have replaced the pick-and-brush tool kit, and the 3-D printer has replaced manually built template molds and hand-poured, liquid resin casting methods. From digital files that map the external surface by laser, or external and internal surfaces by computerized X-ray tomography, we now build three-dimensional virtual specimens *en silico*, study, measure, and reconstruct them on-screen, and print off replicas automatically by stereolithography. But those technologies were decades away at that time.

The fossil's damaged and misshapen orbits were a focus of the operation. In cleaning them out, Simonis frequently worked under black light, which fluoresced the bone and distinguished it from the gray-colored matrix. One day he called me into the lab to show me something unusual that he had discovered. Deep inside the orbits, behind the narrow bony pillar that separates right and left eyes along the bridge of the nose, there was a gap rather than a dividing wall. Anthropoid eye sockets are set up like side-by-side, cone-shaped funnels, with the eyeball in front and nerve endings and circulatory vessels passing out the rear and elsewhere through an array of small, anatomically well-mapped holes.

The paired inner walls of the orbits are typically separated by a space between them, a hollow sinus that is effectively part of the internal, bony nose. But *Dolichocebus* was different. With the matrix removed, it was clear that its orbits were set so close together that no sinus existed. Instead, contiguous right and left medial walls were fused into a single sheet of bone perforated by a large interorbital gap. Middle-deep into the paired orbital cavities, the right and left sockets were confluent through a fenestra, which is Latin for window. Simonis took great pains to be certain the opening was truly anatomical and not an artifact of his own work. We reasoned that part of the space could be a postmortem breach because there was no detectable, unbroken bony margin that outlined the void in that area. But along the roof of the conjoined orbits near their apices he was able to trace the gap's natural arc formed by petrified bone, a bit of the fenestra's perimeter. Wisely, he left a sliver of matrix affixed to that part to protect its smooth bony edge and hold it in place.

Otto knew that no other mammal he had ever worked on had a natural hole between the eyes like this. In fact, the only living mammal with a true interorbital fenestra is *Saimiri*, the Squirrel Monkey. This was a fact reported in the literature, and I was quickly able to confirm that fact by surveying a large series of adult *Saimiri* crania in the collections of the Department of Mammalogy just down the hall from the museum's paleo prep lab. It became clear that the second mammal found to exhibit the interorbital fenestra is the 20-million-year-old fossil *Dolichocebus gaimanensis*.

What did it mean? Paraphrasing Darwin, when an exceptionally rare anatomical novelty is found in two or more species, the null hypothesis—the simplest and most logical one that needs to be accepted until disproven—is that they share a propinquity of descent, they are closely related. This principle led me, first, to explore other possible novelties shared by *Dolichocebus* and *Saimiri*, and there were several that strengthened the null. Second, I again enlisted the services of Otto Simonis to prep another fossil skull from Argentina of approximately the same age, *Tremacebus harringtoni*. These two New World monkey fossils were the best preserved crania known at the time. Another belonged to *Homunculus patagonicus*. It consisted of only a broken face, yet it was potentially very useful for making the necessary anatomical comparisons. *Homunculus* was discovered in the 1890s, the second fossil platyrrhine ever described, and it was also now in the hands of Otto Simonis for proper preparation and reconstruction.

Coincidentally, both *Dolichocebus* and *Tremacebus* were lacking what mammal paleontologists tend to value most as evidence: teeth, and in both specimens the critical features of phylogenetic importance were orbital. In *Dolichocebus* it was the interorbital fenestra. In *Tremacebus* it was the enlarged size of the orbit. The only modern platyrrhine with enlarged eye sockets is the nocturnal Owl Monkey, *Aotus*. With the exception of *Tremacebus*, to date the latter is the only nocturnal anthropoid that ever lived. The only modern New World monkey with an interorbital fenestra is the Squirrel Monkey, *Saimiri*. Except for *Dolichocebus*, it is the only primate known to have ever evolved this trait. Therefore, based on these and other shared, derived features, I concluded that *Dolichocebus gaimanensis* and *Tremacebus harringtoni* are extinct platyrrhines that belong to the Squirrel Monkey and Owl Monkey lineages, respectively. Coming from the same region and the same time period, that was no coincidence.

Dolichocebus and *Tremacebus* were the fossils that inspired the Long-Lineage Hypothesis when I first proposed it in 1979, and they remain cornerstones because of their time depth. It was unknown then that the lineages leading directly to a living primate genus or species could be traced back that far in time using the fossil record. Scientists didn't know how to connect them—or had not developed a widely accepted method to do so—and didn't try. At that time, there was also no credible molecular clock that could be brought to bear to test the idea.

Thinking about the implications of the Long-Lineage Hypothesis from the top down, the modern genera linked to these ancient fossils are themselves living fossils. This idea was a departure from what was then settled thinking about the Miocene-Recent time span of evolution in the Old World, where fossils of catarrhine monkeys and apes, a database vastly larger than the record for platyrrhines, were being studied intensively. During the prior decade, the idea that the lines of living chimpanzees and gorillas and orangutans could be traced back as much as 14 million years was proposed and discarded. At the same time, the Long-Lineage Hypothesis was consistent with another idea presented in a classic, transformative 1972 article by Niles Eldredge and Stephen J. Gould. It introduced the Punctuated Equilibrium Model of evolution, which emphasized that *stasis*, little or no change over long stretches of time, was a fundamental part of the evolutionary process governing species.

To be clear, the long-lineage concept applies to the platyrrhine lineages that are now alive, and that make up a single adaptive radiation; however, there will be older platyrrhine radiations discovered in the fossil record that left no living descendants, other than the single line leading to the modern monkeys. They would have evolved as a side branch, or set of side branches, of the great platyrrhine Tree of Life, separate from the extant forms. These more basal platyrrhines undoubtedly existed and probably flourished because New World monkeys originated about 45 million years ago, or more, which is roughly 20 million years before the age of fossils we know belong to the extant groups.

More lineage offshoots are surely a large part of deep-time platyrrhine history, as there is little doubt that the evolution of early platyrrhines was shaped by the same dynamics experienced by other groups after arriving in a new ecological environment empty of close competitors. They undergo an adaptive radiation.

The La Venta fossils look like modern monkeys

The La Venta fossil site in northwestern Colombia is an area that was part of the Amazonian rainforest when vertebrates were fossilized there 12–14 million years ago. The fossil primates found there include close relatives and likely ancestors of several New World monkeys living today. These fossils are the clearest demonstration of the predominance of long lineages as a major theme in the evolutionary story of the platyrrhines.

The following pages focus on six species that occur exclusively at La Venta as far as we now know. That they represent four different subfamilies, of the six that now exist, and had coexisted within a narrow band of time highlights the point that they represent an ecophylogenetic community that is continuous with today's Amazonian primate community. Looking forward in time, two of these fossil forms, *Neosaimiri fieldsi* and *Aotus dindensis*, are related to extant Squirrel and Owl Monkey genera, respectively. Looking backward, they are also links to 20-million-year-old fossils belonging to the same lineages, specifically *Dolichocebus gaimanensis* and *Tremacebus harringtoni* found in Argentina. Moreover, one of the La Venta monkeys has been reassigned taxonomically, making this connection to the remote past even more definitive. The fossil previously called *Laventiana annectens* was reassigned in 2019, by me, to *Dolichocebus* because their molars are barely distinguishable, and it now carries that genus name.

The idea of a close relationship between Middle Miocene fossils and living platyrrhines was evident when the original La Venta platyrrhines were first published. *Neosaimiri* was immediately recognized as a close relative of *Saimiri* when the first specimen, a mandible, was announced in 1951 (fig. 9.5), hence the genus name, which essentially means "new squirrel monkey." Scholars agree that *Neosaimiri* and *Saimiri* are closely related. Yet, in the prevailing research paradigm of the 1950s, paleontologists were more prone to proclaim ancestors in the fossil record, whereas today's approach involves more rigor and caution. To be regarded as such, the fossil must first meet a more general standard based on shared, derived features, showing that it is at least a sister taxon to a living form. Following that, other criteria must be satisfied. For instance, there can be no novel specializations in the fossil that distinguish it from its putative living relative that go against the possibility of a direct change in morphology over time. Any difference from the living form must

Dolichocebus annectens   Neosaimiri fieldsi

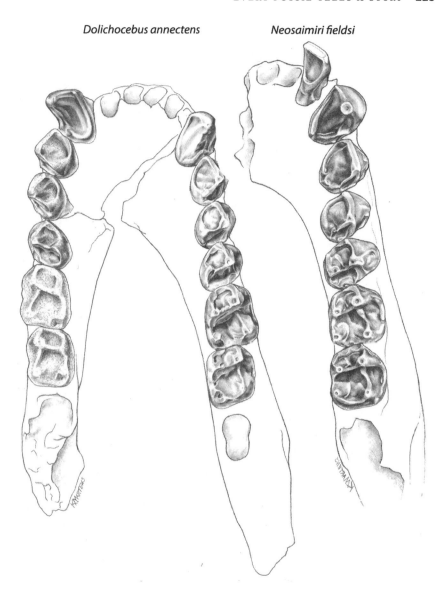

FIG. 9.5. Right side of the lower jaw of the type specimen of *Neosaimiri fieldsi* (right) and the type mandible originally of *Laventiana annectens* (left), now referred to genus *Dolichocebus*.

be more primitive or logically transformable into the extant pattern. And, the fossil must be older, which is normally the first precondition satisfied. With the *Neosaimri-Saimiri* comparison, all these requirements are met, as additional dental and postcranial remains have demonstrated since the initial fossil discovery. Like extant squirrel monkeys, *Neosaimri* was a middle-sized platyrrhine with a frugivorous and insectivorous diet and a quadrupedal style of locomotion that incorporated leaping.

Stirtonia tatacoensis was discovered in the same field season as *Neosaimiri* but the species was originally referred to one of the fossil platyrrhine genera known at the time from Argentina, *Homunculus*. In 1970, the mandible was referred to a new genus and renamed *Stirtonia* after the paleontologist who discovered it, R. A. Stirton. It is widely accepted now that *Stirtonia* and *Alouatta* are also close relatives. More fossils, which now document lower and upper dentitions, have since been found, confirming the connection. *Stirtonia* matches the howlers in sharing a unique dentition, with large molars whose shearing crests are arranged in an instantly recognizable, intricate, zigzag pattern. Two species of *Stirtonia* are now known from La Venta, and they fall within the size range of living *Alouatta*. The smaller one is estimated to weigh 5 kg, 11 lb, and the larger one is as large as the largest living howlers, weighing about 10 kg, 22 lb. Like *Aotus dindensis* and *Neosaimri fieldsi*, *Stirtonia tatacoensis* is an excellent indication of ecophylogenetic continuity between a La Venta platyrrhine and an existing genus. It also meets the stringent criteria mentioned above that are required of a potential ancestor, at the genus level, of living Howler Monkeys.

The fossil Owl Monkey, *Aotus dindensis*, was the first pre-Pleistocene platyrrhine assigned to a living genus. Though that interpretation was questioned when first published in the mid-1980s, discovery of additional dental and postcranial remains since then have erased those doubts for most scientists. This middle-sized species had enlarged, nocturnally adapted orbits, a frugivorous dentition, a uniquely shaped mandible matched only by the living Owl Monkeys among the living New World monkeys, and a quadrupedal talus.

Aotus dindensis is also a link to the 20-million-year-old *Tremacebus*. The edentulous *Tremacebus* cranium (fig. 9.6), remarkably similar to living owl monkeys (fig. 9.7), exhibits two features indicative of its affiliation: expanded orbits and, on the inside of the braincase, a large natural depression where the olfactory lobe is lodged. It shows that the lobe was larger than it is in most platyrrhines, as it is in extant *Aotus*. Together, these features indicate that, like the living Owl Monkey, *Tremacebus* was nocturnal and highly reliant on olfactory communication. A younger Patagonian fossil, *Homunculus patagonicus*, did not have enlarged orbits, but the contour of its mandible is intriguingly similar to those of *Aotus dindensis* and living *Aotus* (fig. 9.8). *Homunculus*, *Aotus*, and *Callicebus* are pitheciids, a family-level clade that has evolved a variety of mandible shapes in connection with hard-fruit husking and a large-gape anatomical setup that enables species to shuck fruits that are relatively large for these middle-sized platyrrhines. The similarities they exhibit in jaw

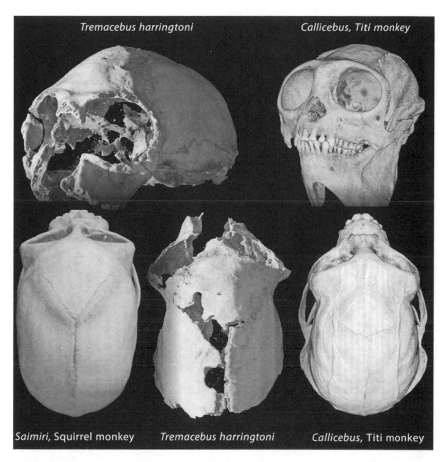

FIG. 9.6. Type specimen of *Tremacebus harringtoni*, an edentulous cranium, compared with same-sized monkeys. Based on three-dimensional virtual specimens developed from CAT scans, courtesy of DigiMorph, http://digimorph.org/.

morphology, roughly dating as far back as 16 million years, may be indicative of an ancestral pattern for the family, all with deep jaws.

When originally discovered, *Aotus dindensis* and another fossil from La Venta, *Mohanamico hershkovitzi*, were sometimes confused and thought to be the same genus. With Takashi Setoguchi and colleagues, a team with whom I had worked exploring the badlands of La Venta for fossil monkeys, I restudied the *Mohanamico* dentition and jaw, the only specimen known, and we identified a unique set of resemblances that the fossil shares with Goeldi's Monkey, *Callimico*, that are not found in *Aotus dindensis* or living *Aotus* species. What we focused on was the canine tooth complex, the combined structure of the lower canine and the premolar adjacent to it. They are designed to accommodate the

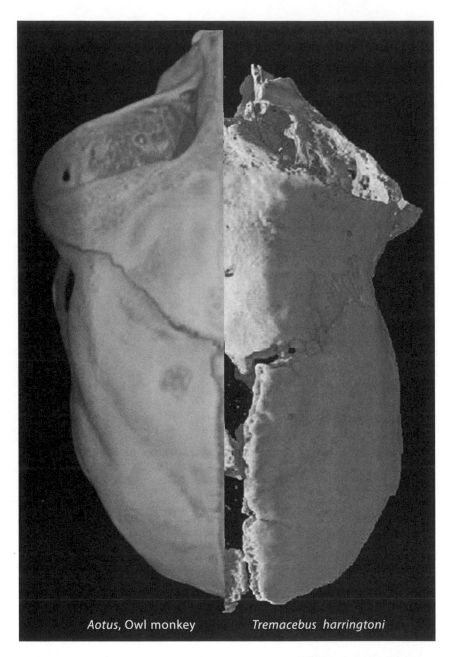

FIG. 9.7. Left and right halves of the crania of an owl monkey (left) and *Tremacebus harringtoni* (right), here digitally fused at the midline and aligned along the cranium's narrowest diameter, a standard landmark. Based on three-dimensional virtual specimens developed from CAT scans, courtesy of DigiMorph, http://digimorph.org/.

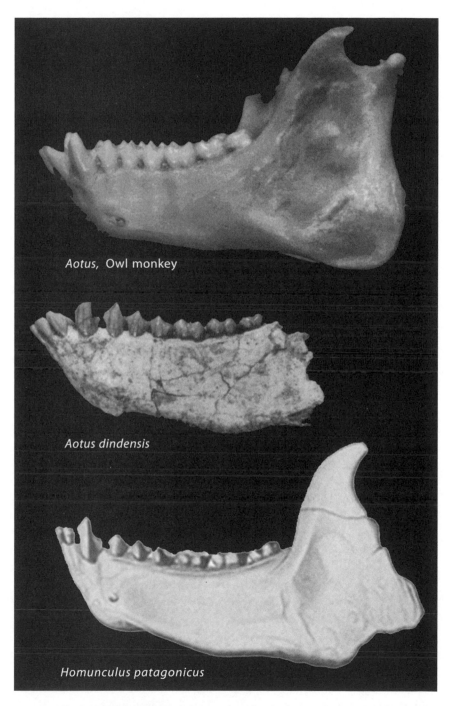

Aotus, Owl monkey

Aotus dindensis

Homunculus patagonicus

FIG. 9.8. Side views of the mandibles of a living owl monkey (top), *Aotus dindensis* (middle), and the reconstructed mandible of *Homunculus patagonicus*. Note the small, stout canines and jutting incisors in both *Aotus* species. Bottom image adapted from Bluntschli (1931).

upper canine that slots between these two mandibular teeth when the teeth meet in occlusion. We showed that *Mohanamico* clearly would have had a large upper canine that complemented the daggerlike lower canine evident in the jaw, while *Aotus dindensis* would have had a lower-crowned upper canine that reciprocated the stubby morphology of the lower canine preserved in its mandible. Matched, tall upper and lower canines are typical of callitrichines, as we have seen, the particular form of canine sexual monomorphism that is found only in that group among the platyrrhines.

We also clarified the contrasting shapes of the jaw's silhouette as seen in side view: essentially flat from front to back in *Mohanamico*, as in the cebids, but curving sinuously and deepening toward the rear in *Aotus dindensis*, as in a modern owl monkey. Furthermore, we demonstrated how similar, yet distinguishable, the cheek teeth of *Aotus* and *Callimico* are, a factor that carries over to the fossils and might thus have contributed to the earlier confusion about taxonomic identification. Both sets of monkeys are about the same size and they have frugivorous-insectivorous dentitions. By eliminating characters from its morphological profile that were mistakenly comingled with *Aotus dindensis*, the affinities of *Mohanamico* as a member of the *Callimico* lineage were made clear. This nullified the only significant argument ever put forward that the species *dindensis* did not belong to *Aotus*, and was not a fossil owl monkey.

Miocallicebus is a fossil represented by one poorly preserved specimen also found at La Venta, a piece that holds together a heavily worn upper second molar flanked by the first and third molars on either side of it that are even more damaged. The crown anatomy of the second molar shows a distinctive pattern that resembles the extant Titi Monkey, *Callicebus*. The dimensions of this tooth are larger than that of *Callicebus*, about 30% more in length and 20% more in breadth. Such a size disparity, even when found in conjunction with dental similarity, is usually an indication of taxonomic distinction at the genus level.

The fossil evidence of *Miocallicebus* is still meager, but in suggesting the presence of the *Callicebus* lineage at 12–14 million years it is no surprise. Titis are a very widely distributed, ecologically versatile genus, and La Venta has yielded many signs consistent with being a proto-Amazonian flora and fauna. The molecular clock places the origins of the genus *Callicebus* in the range of 19–25 million years. Another fossil that may support this idea of long-lineage, Titi Monkey continuity comes from Patagonia, where *Homunculus* existed about 16 million years ago. Its affinities have long been debated for various reasons, but there is a good possibility that this genus is closely related to *Callicebus* as well, though this linkage is also treated as a tenuous connection here in order to err on the conservative side of the argument.

A sixth La Venta primate is mentioned here, and further below, because it is quite clearly related to living forms, though without additional study it cannot be linked with a particular genus—though the possibilities are very narrow, sister genera that are exceedingly similar morphologically. I refer to *Cebupithecia sarmientoi*, which is definitely related to sakis and uacaris. It deserves to be mentioned here because *Chiropotes* and *Cacajao* are quite specialized adaptively as shell crackers and seed eaters, and *Cebupithecia* documents the presence of that niche in the La Venta fauna.

Fossil evidence for longevity with little change

Fossil-to-living pairs like *Dolichocebus-Saimiri*, *Neosaimiri-Saimiri*, *Tremacebus-Aotus*, and *Stirtonia-Alouatta* provide evidence of the processes and patterns of platyrrhine evolution. Only by searching for links like these, and specifically probing the fossil record for evidence of ancestral-descendant relationships—a mode of research that does not easily square with mainstream cladistics as it is currently being practiced—can we begin to learn what was the main phylogenetic framework of the radiation. Was it an abundance of straight-line phyletic evolution, or a pattern of lineage-splitting speciation, which is essentially the type of phylogenetic affinity that cladistics is best designed to find? Both surely occurred, but in what proportions, and can one or the other be associated with a particular geological time period and a particular ecological context? With these examples in mind, it may seem remarkable that a living genus has been around and evolving, while not changing discernibly, over a 20-million-year period—maybe more, because a fossil cannot pinpoint a beginning of a species' existence, only that it was a reality at a certain time.

What we need to know is how many such pairs there are, and what they are ecophylogenetically. How far back in time do these long histories stretch? Are there others with ancient relatives or ancestors that we have missed because their most recent members are extinct and not part of the living ensemble? How and why did the long-lineage pattern come to be a dominant theme during the last 20 million years in South America?

As we have seen, the original evidence for long lineage comes from La Venta, an arid, Middle Miocene landscape that continues to reveal fossil monkeys 12–14 million years old, in the upper Magdalena River valley of northwest Colombia. It is a locale of heavily dissected topography, with interconnected, multihued red and gray hills. Shaped by ages of wind and rain, which erodes the softer rocks and soils, fossils buried within are occasionally revealed. La Venta was known to be an important vertebrate fossil site since the 1920s, but its significance for platyrrhine evolution emerged only near 1950, after primate fossils were first found there. From the late 1970s and into the 1990s, a number of international expeditions that partnered researchers from Kyoto University, Duke University, and local Colombian institutions, scoured La Venta with a focus on finding more fossil monkeys. Eleven fossil platyrrhine genera have been identified at La Venta, and it is likely that more will be uncovered.

The place is a desert, hot, riddled with dry gullies and sparse vegetation; cactus and sagebrush are common. It is a valley nestled between two rows of the northern Andes, the Eastern and Central cordilleras. The dryness results from the rain shadow cast by the mountains, which block the wet equatorial trade winds blowing off the Pacific. But in the past, before local Andean uplift, the Magdalena River valley was continuous with the habitat that would become today's Amazon basin, and the area supported a large fauna of tropical animals.

The first discoveries of La Venta platyrrhines were interesting in that a pair of them closely resembled living species, as mentioned, seemingly unlike the other fossils known at the time that came from Argentina, *Homunculus* and *Dolichocebus—Tremacebus* was not then recognized as a distinct genus— neither of which was well understood during the 1940s. Eric Delson and I reassessed the La Venta primates in a 1984 paper that discussed living fossils, and we suggested that *Neosaimiri* and *Stirtonia* were not only close relatives of their living counterparts; we thought it likely they were potential ancestors. This was particularly exciting because in the Old World, with a superior fossil record, the lineages of modern monkeys and apes then known did not go back that far. The two oldest generic lineages Delson could be confident about in the 1980s belonged to rhesus macaques and orangutans, that were only about 7 million years old.

This reinforced the Long-Lineage Hypothesis, proposed five years earlier, and indicated that the South American long-lived pattern of evolution was different from that seen among Old World anthropoids, in which adaptive radiations were successively replaced over time through faunal turnover. In some cases, turnover occurred when new intercontinental connections between Europe, Asia, and Africa allowed the entry of new primates, as competitors, or when grasslands expanded and encouraged, via selection, the in situ evolution of new radiations that exploited those habitats and pushed other groups out. This never happened in South America. It was an isolated continent for at least 35 million years after monkeys first arrived and, as far as we know now, there were no other anthropoids potentially dispersing from nearby North America that took a foothold and could have usurped them. The issue of South American grasslands as suitable monkey habitat is another matter, but there is no evidence that New World monkeys ever made it their home, so it probably did not serve as an incubator of other, potentially competing platyrrhine radiations.

If a pattern of stasis, genus-level and fauna-level continuity, could be maintained for the 12–14-million-year timespan postdating the La Venta, it is equally likely that southern fossils predating La Venta could be taxonomically continuous as genera with Colombian primates over a period half as long. Actually, this constitutes an indirect prediction of the Long-Lineage Hypothesis informed by fossils and also the molecular evidence, which points to the great longevity of modern generic lineages all across the radiation. Genus-level continuity with earlier forms is possibly the case with the fossil owl monkeys, represented by *Aotus dindensis* in the north and *Tremacebus harringtoni* in the south. But there is no way yet to investigate this further, because one species is best known from teeth and a few postcrania, and the other from only a toothless cranium. So, their standing as separate genera is maintained.

However, for the other early Patagonian form of the same age there is a pertinent part of the anatomy that can be examined comparatively with younger primates from the north. The collection of *Dolichocebus* fossils now includes lower molars and upper and lower canines that are well preserved, in addition to the original cranium that initially prompted the Long-Lineage Hypothesis. The molar crowns turn out to be basically indistinguishable from those belonging to a pristinely preserved mandible from Colombia that was first classified as *Laventiana annectens*. The *Dolichocebus* canines also match the morphology of living squirrel monkeys in size, shape, and degree of sexual dimorphism, an unusual pattern among the moderns that supports the more general idea that this fossil is related to *Saimiri*.

For these reasons, I now consider these two fossils one and the same genus, which must both be called *Dolichocebus* rather than *Laventiana* since it is the earlier of the two genus names published. In addition to fleshing out more of the *Dolichocebus-Saimiri* phylogenetic connection, here we have the first link binding these geographically disparate faunas together, with one being older and generally more primitive morphologically than the other. That this link exists at the community level, rather than being restricted to one or two genera, is corroborated by the presence of Patagonian fossils that are certainly related to Colombian pitheciines, namely *Proteropithecia*, though it does not have a genus-level counterpart at La Venta. Evidently, the southern cone harbored an early-phase outpost of the modern radiation.

A 12–14-million-year-old Owl Monkey fossil

I visited La Venta several times in search of fossil monkeys. My first trip did not yield much at all, but it introduced me to the complex labors of paleontological fieldwork. I was a new assistant professor at the University of

Illinois shortly after receiving my doctoral degree. The college was interested in helping my career advance with a small grant that would support a brief reconnaissance trip into the field by me and one graduate student. Armed with little more than a road map and a destination, Neiva, the largest town in the region and the jumping-off point for the Tatacoa Desert where La Venta is situated, we began our road trip in the cheapest car we could rent. But not until we could maneuver out of the parking garage, where that little French Peugeot introduced me to the mechanics of the manual automobile transmission, how a car would buck and jerk when the clutch is mishandled, and, later, how a cushion of grant money on a credit card would always be necessary in fieldwork to cover the unforeseen—like a gentle fender bender where metal hits concrete wall. Once properly in gear, though, the dense, cacophonous traffic of Bogota proved easier to negotiate than getting that car into reverse. We headed out in late afternoon, not knowing how long the drive would take and not thinking it was unwise for Americans to be traveling in darkness on unlit, one-lane blacktop in a remote territory of political unrest and guerilla activity. Perhaps we had some level of nighttime protection because neither of us could figure out how to turn on the Peugeot's headlamps, and we drove the last leg in darkness until we neared the lights of Neiva. During the following days, we did get automotive help getting to and from the badlands from a generous gentleman with a jeep who was attached to a local oil company. I had previously contacted this company by aerogram to ask for mapping assistance in the Tatacoa Desert, where they had been drilling. As for fossils, nothing but scraps came of that trip but I was not discouraged from trying again, in a smarter way. And as for the badlands of La Venta—30 years on they became a destination for trekking tourists.

A few years later, I joined a team of Japanese paleontologists from Kyoto University who had been working at La Venta with more success, and far more experience. With grant money from the National Science Foundation's U.S.-Japan Cooperative Science Program, which drew federal dollars from both countries to foster collaborations, we mounted another two expeditions and also spent some time at each other's universities, studying the small Kyoto collection of fossil material and collaborating on various platyrrhine projects. The fieldwork was productive in finding the new fossil species that is now transferred to *Dolichocebus*. I hadn't considered such a reassignment would be required until decades later, but the work became quite important after the team visited La Venta at a time when I was not able to go because of teaching commitments. As we agreed, my partners, led by Takashi Setoguchi, would pass through Chicago on the way home from Colombia, and we would have some time together to study whatever interesting fossils that field season had produced. Tak called me when they landed and asked me to meet him at a hotel near O'Hare International Airport. They were in a celebratory mood when I arrived as Tak, smiling broadly, pulled out a small cardboard box from his leather shoulder bag. In it, wrapped carefully in a scroll of toilet paper, the primate paleontologist's most important piece of field matériel, was a small, perfectly preserved, comma-shaped half-mandible with all its teeth. It was smaller than the length of a USB flash drive.

We had some ideas of what it might be—the species was at that time surely new to the La Venta inventory—but without a series of comparative specimens in hand there was no way to be certain, until Tak dug more deeply into that shallow box and pulled out the other bone that belonged with the mandible. It was smaller, about 10 mm long, less than half an inch; broken into an irregular shape but undistorted; a relatively flat piece of slightly pocked bone resembling a cornflake in size and shape. Turning it over in my hand, I saw that one surface, a bit of the palate, held a pristine third molar that fit perfectly atop the third molar of the lower jaw when I occluded them. This exposed the topside surface of the bone, making the specimen clearly identifiable as a fossilized chunk of the rear part of the left lower face. Luckily, it was a fragment that preserved a piece of the orbital floor upon which the eyeball rests and a few subtle, telltale anatomical signs that we could recognize as significant. The whole bit was shallow from top to bottom. The floor was quite flat, not concavely rounded as most orbits are at the bottom where bone interfaces with the spherical eye. Flatness meant the eye was large. The bone's intact rear edge revealed another clue suggesting the same, a remnant of a peculiar shape I knew as the notchlike incisure that accompanies an enlarged eye socket with a relatively large natural opening in the back—not a wall of bone that sealed the orbit from behind as in all but one anthropoid primate. With those clues, and the mandible, the identity of this specimen came sharply into focus. It was a fossil owl monkey.

After completing our study at the Field Museum of Chicago, we named it *Aotus dindensis* in 1985, coupling the living Owl Monkey's genus name with a species name referring to the location in the Colombian badlands where the fossil was found, El Dinde. Here was another La Venta primate with links— possibly ancestral-descendant ties—to a living platyrrhine species. That we could not distinguish it at the genus level from a modern owl monkey—where the shape of the mandible beneath the toothrow is taxonomically diagnostic— added even more credibility to the Long-Lineage Hypothesis.

Additional evidence supporting the Long-Lineage Hypothesis has since come from La Venta, as noted. *Mohanamico hershkovitzi*, also found in the 1980s, may be ancestral to Goeldi's Monkey, *Callimico*, and *Miocallicebus villaviejai,* described by Tak and his colleagues in 2001, may be an ancestor of the living Titi Monkey, *Callicebus.* This means that very close relatives, congeners, or even direct ancestors of five of the sixteen extant Amazonian genera (*Saimiri, Aotus, Alouatta, Callicebus, Callimico*) were living in northwestern South America at that time. The status of a sixth, *Cebupithecia,* is still not clear because its phylogenetics need to be reevaluated, but it shows that the Bearded Saki and Uacari sister group was present by La Venta times as well. It may be linked with one of these genera, to *Chiropotes* or *Cacajao,* or it may be a member of the joint *Chiropotes-Cacajao* clade that lived before those two lines split from their common ancestor. Similar extended lineages are evident among other La Venta mammals, including opossums, armadillos, and bats.

Outside La Venta, other fossils confirm the long-term phylogenetic continuity of Amazonian primates. From a slightly younger age, about 11 million years ago, a deposit in Peru has produced several teeth attributed to living genera, *Cebus,* the Capuchin Monkey, and *Cebuella,* the Pygmy Marmoset. In all, we still have very limited remains of any of these northern fossils, and, as usual, they consist mostly of single specimens of jaws and teeth, or a few isolated teeth, and even fewer postcranial skeletal remains. Only one fossil primate skull has been discovered at La Venta, *Lagonimico conculatus,* and it is crushed, flat as a pancake. Nevertheless, we can discern intriguing, highly detailed resemblances shared by the La Venta monkeys and the modern ones. The dentitions especially are sensitive to the ecological zones and niches occupied by these animals in offering information on two crucial factors, body size and diet. The evidence suggests that little morphological and adaptive evolution has taken place in the identifiable generic lineages over 12–14 million years. And with a reconsideration of the relationship between *Dolichocebus* and the genus formerly known as *Laventiana,* we are beginning to empirically fill in the earlier gaps of a continuity that has extended about 20 million years, stretching down to the older tropical forests of Patagonia.

La Venta was not a mirror image of the Amazonia we know now. Those modern-looking fossil primates interacted ecologically with several other extinct primate genera that cannot be placed in the framework of potential ancestors or sister genera to the living, though none appear to be platyrrhines outside the modern adaptive radiation. Yet even those species appear to be typical platyrrhine primates with no anatomical hints of being adaptive outliers—nothing more gigantic than atelids, nothing more miniscule than a callitrichine, nothing terrestrial, nothing with teeth any more bizarre than a pitheciine's. The La Venta predecessors of today's primate species lived in a landscape that would also undergo vast changes. At about 11 million years ago the region's current geophysical state began to take shape in the form of a transcontinental waterway, the Amazon River, that developed in response to the rising Andes. Before that, for millions of years, large sections of the northern expanse of South America encompassed what was probably the world's largest swamp, lacking the extensive platforms of terra firma and the massive network of tributaries that now connect to the river backbone and set the geography and ecology inhabited by today's living New World monkeys.

Fossils that tell us where they once lived, what they ate, and more

With the fossil platyrrhine record now registering more than 30 genera spanning roughly 36–40 million years, several others deserve special attention. Some are important because of the time when they were discovered, in the late 19th century and even in the 1960s, which gave them undue influence because we did not have much else. Others are spectacular finds visually, revealing a trove of crucial information. Some are keys that unlocked information vital to interpreting aspects of platyrrhine evolution that could not be accessible by studying the extant forms alone. Others because they disclose unexpected findings that shape our thinking about the morphological evolution of New World monkeys. Several have redrawn the distribution map of the platyrrhine radiation. A few have been tangled in controversy that needs to be sorted out in order to advance the field. And several recent discoveries may have special importance for understanding the very early phases of primate evolution in South America.

One controversy involves the cranium of *Dolichocebus gaimanensis*, the type specimen, discovered in the 1940s when there was little interest and less expertise in the morphology of living platyrrhines, and when the fossil record consisted of a couple of specimens discovered in the 1800s, assigned to two disparate genera. Decades later, cleaning and preparing the cranium led to the discovery of the interorbital fenestra, and the recognition that this fossil was part of the Squirrel Monkey lineage. During the 1980s, a small collection of teeth and one ankle bone were recovered that eventually reinforced this view; however, the original cranial evidence behind this idea was not universally accepted.

The differences of opinion concerning *Dolichocebus* centered on the interorbital fenestra, whether or not it was an authentic anatomical feature or merely an artifact of fossilization or poor paleontological preparation. It seemed like a fitting issue for Timothy Smith, Valerie DeLeon, and I to investigate once again, in detail, using new technologies that were not even imagined by morphologists in 1979, when I first published on it. We had been working together since 2010 on the unusual cranial morphology of tarsiers, the tiny Far

East Asian primates that have been a flashpoint in big-picture interpretations of primate evolution. Three-dimensional imaging and microCT technology, one of the research specialties of DeLeon, were now part of our tool kit. Both Smith and DeLeon are experienced in the growth and development of primate heads, and Smith is also a histologist who works on the morphology and microscopic anatomy of soft tissue. We reexamined the 3-D morphology of the *Dolichocebus* cranium using microCT data and extended the study to include living squirrel monkeys to better understand the nature of the interorbital fenestra, visualizing its cranium and examining the anatomy and histology of the soft tissue that actually seals the fenestra in life and provides a functional separation between the eyeballs.

We developed new sets of high-resolution 3-D models of *Dolichocebus* and *Saimiri* and, in addition to poring over these virtual specimens, studied the cross-sectional images, one by one, that are composited in software to produce the models. We traced the outlines of the oval fenestra in histological preparations of a cadaveric specimen of a squirrel monkey to learn how the interorbital membrane stretches across the opening and attaches to its bony rim. We learned how the bony perimeter is formed by the orbital roofing bones of the right and left sides arching downward and fusing together to produce a razor-thin edge where the filament-thin membrane attaches. We looked at the membrane, too. We rediscovered the wisdom of Otto Simonis who, in preparing the fossil for study some 40 years earlier, had decided to strategically leave intact a few millimeters of matrix in the hope that a small piece of rock kept in its original position would protect the observable fraction of the arc that formed the fenestra's natural oval outline, which it did (fig. 9.9).

How much does this contribute to our understanding of *Dolichocebus* and the *Saimiri* lineage? First, the uniqueness of the fenestra as a derived trait demonstrates that the Squirrel Monkey lineage is at least 20 million years old. Second, the continuity of craniodental morphology suggests that the fossilized species was similarly adapted at that time. Third, it implies that the ecology of the environment in Argentina where *Dolichocebus* lived, now grassland far removed from today's wet, heavily forested squirrel monkey environments, was suitable to support a *Saimiri*-like lifestyle.

How much of that lifestyle can we attribute to the fossil? The fortuitous discovery of a fossil that closely resembles and is allied with a morphologically highly specialized extant genus means that a lot can be learned from it about ecology and behavior. The fossil's body size, a core ecological feature, is comparable to a squirrel monkey's medium-sized build. The single ankle bone allocated to the genus is of a basic quadrupedal and leaping design, resembling squirrel monkeys, capuchins, and titi monkeys. The cheek teeth are consis-

Saimiri, Squirrel monkey   *Dolichocebus gaimanensis*

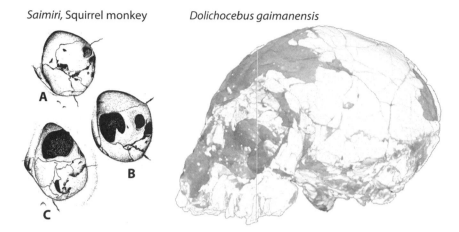

FIG. 9.9. Oblique view of the *Dolichocebus* cranium (right). The vertical marker is placed at the front end of unbroken bone that formed part of the elliptical fenestra. Left panel is a growth series (A–C: juvenile, subadult, adult) depicting the normal process of development as bone is resorbed to form the fenestra in a squirrel monkey. *Dolichocebus* is a three-dimensional virtual specimen developed from CAT scan data, courtesy of Richard F. Kay. Growth series adapted from Maier (1983).

tent with a frugivorous-insectivorous diet. The pronounced degree of sexual dimorphism of the canines suggests *Dolichocebus* lived in social groups where male-male competition was pronounced, and troops were organized as uni-male or multi-male groups with several females. The long, rounded braincase, irrespective of its deformation, indicates that the brain was relatively large. The overall narrowness of the entire cranium, a factor that contributes to the formation of very closely spaced and fenestrated orbits, suggests a squirrel-monkey-like pattern of growth and development which is unique among the platyrrhines. The narrowness of a squirrel monkey head mirrors its narrow physique, which has been explained as an adaptation to ease parturition of a large-brained baby, and that would also be expected to have been the case in an extinct species. It is correlated with the rapid prenatal growth of a large-brained, precocious newborn in a lineage of cebines that evolved by a process of dwarfing.

In these critical aspects, *Dolichocebus* conforms to the highly distinctive biological profile of an extant *Saimiri*. There are morphological differences, to be sure. But the overriding evidence emphasizes a pattern of similarities shared with an emblematic living platyrrhine—or an iconic living fossil—*Saimiri*. *Dolichocebus* fossils, from deposits that are 20 and 12–14 million years old, constitute a remarkable set of specimens documenting a long history, and the persistent ecological role of the Squirrel Monkey lineage in the greater platyrrhine community as it unfolded over time.

Homunculus patagonicus is another noteworthy fossil from Argentina. It put platyrrhine paleontology on the map, partly because it was mistakenly thought to be a one-and-only, the first fossil platyrrhine discovered that had no competition for center stage for decades. Actually, the first one consisted of a few assorted pieces, mostly long bones, collected in a cavern in Brazil, in 1838, *Protopithecus brasiliensis*. But that collection was neglected for a long time because paleontologists thought it was a muriqui and long bones held little interest for most of them.

Then, in 1891, after the Darwinian revolution gave paleontology new meaning, a small collection of fossils from Argentina that included *Homunculus* was described, and attention was turned to those as being scientifically meaningful for the burgeoning study of evolution. They were analyzed and publicized by Florentino Ameghino, the founder of South American paleontology along with his partner, his brother Carlos, the intrepid, skilled collector. *Homunculus* was promoted by Florentino as a fossil central to the origins of humans, which he argued had begun in his homeland, Argentina. His hypothesis was expressed in the genus name he gave the fossils, which refers to an imaginary, fully formed, miniature human body from which a human being grows and develops, according to medieval lore. That was obviously absurd, still the Miocene fossils of *Homunculus* were the only ones pertinent to the evolution of platyrrhine monkeys for many decades, which gave them an outsized importance relative to the quality of the specimens Carlos found.

The original Ameghino material is problematic for various reasons, ranging from incompleteness, poor documentation of its geological age, confusion about where it was actually found, little paleontological preparation, poor scientific illustration for a century, further damage caused by indelicate handling, and taxonomic misidentification. What was to become the archetypical specimen of *Homunculus* had been sketched many times in crude drawings, copied and recopied sometimes, probably, based on deficient plaster replicas.

The fossil actually consisted of broken, misaligned pieces of the cranium, until I contracted Otto Simonis to reconstruct it in the late 1970s. It preserves a remnant of the forehead, the frontal bone, a left orbit, the adjacent pillar of bone that includes the bridging nasal bones, and the left side of the face's nasal opening below the orbit (fig. 9.10). The crowns of its teeth were essentially destroyed. When first describing *Homunculus*, Ameghino had selected a small mandibular fragment with teeth in it as the name-bearing type specimen. Unfortunately, at some time during or after the 1930s, the fragment disappeared and has never been found. In its absence, since a type serves as the true voucher specimen as we learned above, this means there would be no

FIG. 9.10. Reconstructed partial face of *Homunculus patagonicus* and the mandible (top and side views) that was designated a new type specimen (neotype) to provide a formal voucher bearing that name. Adapted from Tejedor and Rosenberger (2008).

way to rigorously and directly determine the identity of any other fossils as *Homunculus patagonicus*.

A remedy which Marcelo Tejedor and I invoked in 2008 as per a provision in the code of zoological nomenclature was to formally designate a substitute type specimen, called a "neotype." We followed Ameghino's lead by selecting another mandibular specimen with a full row of teeth from his original series of fossils, after it was properly cleaned and reconstructed, in spite of the fact that the crowns of the teeth were badly damaged and/or worn down, obliterating their surface morphology. In any event, our efforts would give scientists a nominally proper *Homunculus* specimen for comparison should any new mandibular fossils or lower teeth be found, to determine if they are the same genus and species or something different. In platyrrhines this comparison could be easily made based on jaw shape and tooth proportions. Fossil mammal teeth and partial jaws are, after all, the most abundant types of fossils ever found.

In spite of our best intentions to avoid taxonomic confusion, after a paleontological drought that lasted a hundred years, skulls—not lower jaws—turned up. Several crania were among the next important sets of primate fossils found from the same region in Patagonia, but there was still no direct way to determine whether any of them were *Homunculus*. Such a problem is not unusual in paleontology, and researchers are often faced with making informed decisions about taxonomic identity from less than ideal specimens, but it means that the taxonomic questions can linger until a lucky find makes it possible to better piece the evidence together. Another chance discovery did just that. In 2017, a paleontologist interested in fossils belonging to the extinct relatives of South American armadillos, Laureano Gonzalez, who also was part of Marcelo Tejedor's fossil team, came across Patagonian fossil monkeys in a small, forgotten museum collection in France. He recognized a fairly well-preserved mandible that matched the Ameghino mandible Tejedor and I had offered up as a replacement neotype for *Homunculus*. The mandible confirms the reconstruction done by Otto Simonis and, more importantly, it contains one good molar tooth, which means we are now finally able to conduct detailed comparisons based on the rich, diagnostic evidence afforded by dentition.

The affinities of *Homunculus* are not well established. One idea, that it belongs to a monophyletic group along with *Dolichocebus*, *Tremacebus*, and *Soriacebus*, outside the modern radiation, the so-called Stem Group Hypothesis, is not viable. There is strong evidence that each of these three fossil genera is demonstrably linked either with living genera or with extant clades. Dentally, *Homunculus* compares best with homunculine pitheciids. The *Homunculus* mandible, in lateral profile, is very similar to the Owl Monkey, although the *Homunculus* orbits are not enlarged. Other important similarities are shared with the Titi Monkey, *Callicebus*, including the relatively large molars, the size of the orbit and condition of the orbital floor, and the small canine teeth. The limb proportions also resemble owl and titi monkeys. These and other elements indicate *Homunculus* was a quadruped that also engaged in leaping and climbing, as is typical of *Aotus* and *Callicebus*. Thus, while the phylogenetic and ecological picture of *Homunculus* still relies on limited information, the evidence tilts toward the hypothesis that this genus is related to *Callicebus* and *Aotus*. If so, it is a fossil that records the existence of the homunculine subfamily at 16.4 million years, at which time both the *Callicebus* and the *Aotus* lineages would have existed according to the molecular evidence.

Cartelles coimbrafilhoi was named in recognition of two South American luminaries whose work spans the extinct and extant platyrrhines. Castor Cartelle is a paleontologist who has made major discoveries in Brazilian cave deposits containing thousands of subfossil mammals, including this one. Adelmar Coimbra-Filho was a founding father of the Brazilian conservation movement, and primates were his passion. Their namesake subfossil has important bearing on the evolutionary past and present of a central New World monkey genus, *Alouatta*, the Howler Monkey, specifically as to how and why its two most specialized adaptations evolved—the booming vocalizations and the most folivorous diet of any New World monkey

Cartelles coimbrafilhoi is known from one spectacular, nearly complete skeleton. It was found by cavers of the Grupo Bambuí de Pesquisas Espeleológicas of Belo Horizonte, Minas Gerais in 1992, in one of the longest caves in the world, Toca da Boa Vista, in the north of the state of Bahia. While its age has not been established, it is assumed to be Pleistocene, because the cave is believed to be that old, 11,000–2.6 million years. The species is estimated to have weighed more than 20 kg, more than 44 lb, making it the largest known platyrrhine living or extinct, one-and-a-half to twice the mass of a large spider monkey or a muriqui. That alone makes *Cartelles* exceptional. Yet there are other intriguing facets that are revelatory, that would never have surfaced without discovering this fossil. It exhibits a surprising mosaic of characteristics that, on the one hand, tell us *Cartelles* is phylogenetically related to the Howler Monkey, and, on the other, that the alouattine subfamily was once more diversified dietarily and ecologically than any living howlers, and included monkeys that were not committed folivores. That comes as a surprise.

The scientific history of *Cartelles* began when the expat Danish naturalist Peter Wilhelm Lund explored the Lagoa Santa cave system in the Brazilian state of Minas Gerais in the 1830s. He recovered and named a host of subfossils representing living and extinct platyrrhines. The aforementioned *Protopithecus brasiliensis* was one of them. It consisted of a few miscellaneous postcranial bones, the most important of which was a partial humerus. Apart from one relatively inconspicuous yet valuable 1895 publication, in Danish, by Herluf Winge, *Protopithecus* lingered in near obscurity until 1992, when the Toca de Boa Vista cave, approximately 1,200 km, or 750 miles, to the north of Lagoa Santa yielded two nearly complete skeletons of comparably large size. One of them was immediately allocated to a new genus and species, *Caipora bambuiorum*, and the other was attributed to *Protopithecus brasiliensis*. Lauren Halenar-Price, one of my PhD students, extensively studied these skeletons some 20 years later as part of her dissertation project. She found that the new fossil material attributed to *Protopithecus* was quite distinct from the original Lagoa Santa remains, hence the identification of a new platyrrhine, *Cartelles coimbrafilhoi*, in 2013.

Cartelles is a lesson in mosaic evolution, the phenomenon that results when different parts of the body evolve at different rates or in different directions, such that some characteristics preserve a more primitive morphology while others are more modified, or derived. In this case, broadly speaking, the post-cranial anatomy of *Cartelles* exhibits features we would expect of a primitive member of the atelid family: long limbs, a long tail, large hands and feet, mobile joints, and a skeleton basically designed for quadrupedal climbing and clambering, suspensory and, possibly, tail-assisted locomotion. At the same time, the hip of *Cartelles* shares unusual resemblances to that of Howler Monkeys that are possibly associated with inverted foot-hanging, something that *Alouatta* commonly does. The evidence suggests it was built like the fundamentally quadrupedal howler rather than the highly acrobatic Spider Monkey or Muriqui.

The cranium is even more convincingly howler-like, with exaggerated crests indicating a significant development of the neck muscles and mechanical leverage necessary to balance the head like *Alouatta*, attached to the vertebral column at the back so it projects from it, doglike (fig. 9.11). In the related atelines, the spine attaches to the head at the bottom of the cranium, closer to the middle. The shape of the *Cartelles* cranium as a whole is also howler-like in that the rostrum is large and upturned, not hafted low relative to the braincase. This design is probably related to the evolutionary beginnings of a similar hyoid-laryngeal complex, that is, the vocal resonating mechanism for which *Alouatta* is most famous (fig. 8.2). There are other resemblances as well. But what is most striking about this anatomical context is that the teeth are decidedly unlike a howler monkey's dentition. The incisors are large and broad, and the cheek teeth have rounded, blunt cusps. The combination is typically frugivorous and seems well suited to masticate a diet that included tough seeds. As discussed previously, a dentition and gut designed for seed eating may be a preadaptation to the evolution of committed folivory.

What does the mosaic mean? The hypothesis presented by Halenar-Price and me suggests that howler monkeys are derived from an animal like *Cartelles* that existed long before the Pleistocene and had already begun to evolve the mechanism supporting its roaring vocalization. The transition also involved an ecological shift, from a fruit-based diet to a leaf-eating diet, which would have been facilitated, or mandated, at the start by a large body size, with a large gut. If so, this might have involved selection for a significant reduction

Alouatta macconnelli

Cartelles coimbrafilhoi

Caipora bambuiorum

Brachyteles arachnoides

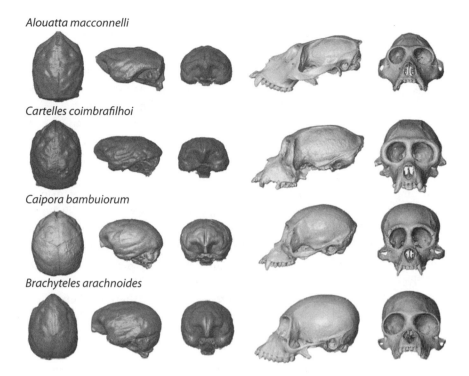

FIG. 9.11. Three-dimensional models of endocasts and crania based on CAT scans of the fossils *Cartelles coimbrafilhoi* and *Caipora bambuiorum* (middle rows), compared with a howler monkey (top) and muriqui (bottom). Endocasts are oriented (left to right) in top, side, and anterior bottom views. Adapted from Aristide et al. (2019).

in body size in the *Alouatta* lineage from a weight comparable to *Cartelles*, which seems like an enormous mass to maintain with low quality food, not to mention the demands it would place on arboreal locomotion. Since folivory is well complemented by an energy-minimizing ecological stratagem, it might have been advantageous to evolve a smaller body mass and a relatively small brain, thus lessening the need to engage in long-distance daily travels in search of high-energy fruit to sustain a massive body and fuel a big brain.

It may seem counterintuitive that this model sees the howler's origins involving body size reduction, when most primate and mammal folivores or herbivores tend to be larger in size than their near relatives with different diets. Even so, two observations suggest it is credible in this case. First, living howler species exhibit significant size disparities, with a range of roughly 5–10 kg, 11–22 lb. This suggests selection for body size has gone both up and down the weight scale during the history of the genus, or at least that both relatively large- and small-sized folivorous howlers are ecologically copacetic. Second, one can see the merits of selectively downsizing a monkey as large as *Cartelles* living in an arboreal milieu for simple physical reasons concerning locomotion. There are only several arboreal Old World monkeys that are heavier, and the living apes, the largest of all extant primates, are arbo-terrestrial. Of course, without more evidence it remains impossible to exclude the hypothesis that *Cartelles* is simply a gigantean genus that evolved from the last common ancestor it shared with *Alouatta*, which was smaller.

Halenar-Price suggests *Cartelles* also gets us closer to understanding what seems like an anomalous pairing of adaptations in *Alouatta*, long-distance booming calls in a leaf-eating monkey. Long calls are an effective form of communication when protecting limited resources, a useful adaptation when foods are widely or patchily distributed, like fruits, rather than being ubiquitous, like leaves. Such calls may also be effective in dissuading males of nearby groups from attempting to displace a resident male and take over his females, by advertising the alpha male's vigor. So, why is howling so fundamental to *Alouatta*? Perhaps it is a carryover from a time and situation when a pre-folivorous alouattine was more reliant on scattered fruits and/or seeds that were protected in territories, with boundaries enforced by vocal behaviors. The calling may have found a new selective benefit if howlers became more efficient folivores, limited their daily roaming, became more sedentary as a digestive adaptation, and consequently began living in higher local population densities, thus exacerbating the challenge of an alpha male to maintain social control and repel potential competitors to protect his most precious resource, his mated females.

The great body mass of *Cartelles* has another interesting implication, speculative though it may be. It redefines the body size bracket associated with this family of New World monkeys, and that has ecological consequences. *Cartelles* was about the size of an African baboon, an Old World monkey genus that is highly terrestrial while also able to climb and utilize trees with ease. There is nothing in the anatomy of *Cartelles* that points to terrestriality, but there is nothing that precludes it either. The large spider monkeys and muriquis, as well as howlers, do come to the ground at times with no evident locomotor handicaps. The same would likely have been true of *Cartelles*. One reason this is an intriguing notion is that a related genus of the *Alouatta* lineage even further along toward a howler in a number of cranial features, the Cuban *Paralouatta*, has finger bones and elbows that look like they are adapted to ground-based locomotion. This may be a sign that near the start of the evolutionary line leading to *Alouatta*, the skeleton was sufficiently flexible in design to serve as a preadaptation to terrestrial quadrupedalism, moving in a direction away from the acrobatics of its sister clade, the atelines. That might be another reason these monkeys were predisposed to evolving their unique style of cautious quadrupedalism. Parenthetically, in spite of the cranial resemblances shared with living howlers, *Paralouatta*, like *Cartelles*, lacks dental indications of advanced folivory, another example of the diverse adaptive nature of this clade that, without fossils, we would have believed had originated as a leaf-eating group.

Mosaicism also reveals clues about the origins of the most committed platyrrhine seed eaters, the pitheciines, including *Pithecia*, the Saki Monkey, *Chiropotes*, the Bearded Saki, and *Cacajao*, the Uacari. Like *Dolichocebus*, *Soriacebus* is one of several Patagonian platyrrhines constituting a fauna that predates the proto-Amazonian community at La Venta, Colombia. *Soriacebus* is a pitheciid, although its affinities were not immediately apparent because its morphology exhibited characteristics that, upon first inspection, seemed to resemble some callitrichines (fig. 9.12). For example, the anterior teeth in the mandible are not aligned in a transverse arc of incisors and canines. They are staggered, as in the Marmoset or Pygmy Marmoset genera, with the first incisors set forward of the second incisors, followed by the canines. This gives the jaws and its dental arcade a V-shaped aspect when looking down on the mandible. Also like callitrichines, the posterior upper molars have a triangular crown outline and the lower last molars are relatively small; however, the latter's V-shaped jaws and staggered anterior dentition are now well accepted as highly specialized adaptations that support intense gumivory in the smallest forms, *Callithrix* and *Cebuella*. These traits are thought to be derived modifications from a more U-shaped jaw, more quadrate upper molar crowns, and final molars that are not conspicuously small.

Analysis of the *Soriacebus* morphology and its stated resemblances to the gum eaters in a functional and adaptive context has overturned that interpretation and suggested another one. Its dental features are primitive versions of the highly derived masticatory complex shared by the shell-cracking, seed-eating pitheciines. *Soriacebus* specifies the earliest paleontological date marking the existence of this subfamily clade and its ecological niche at 17 million years. The fossil also indicates the sequence by which the functional elements that make up the masticatory complex of pitheciines were put together via selection, over time. The anterior teeth and deep jaws, which relate to ingesting heavily protected fruits and seeds, and muscle attachments and force production favoring large gapes, were among the first component parts to evolve. The incisors were already tall and narrow, the canines massively robust, and the anterior premolar shaped like a nut-cracking wedge. At that stage, the cheek teeth did not look modern—although there is no reason to believe that the modern saki-uacari

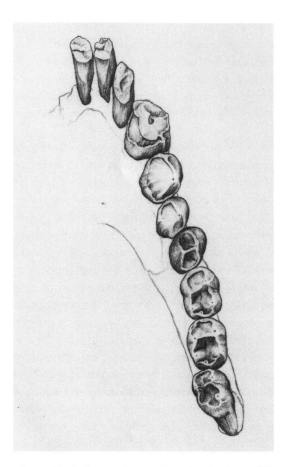

FIG. 9.12. Type specimen of *Soriacebus ameghinorum*, the right side of a mandible, preserving all the cheek teeth and the broken crowns of three incisors (one from the left side) and a canine. Courtesy of John G. Fleagle.

design is the only pattern that is optimized for seed eating—and the dental arcade was still V-shaped. In the younger Colombian form, *Cebupithecia*, as in the living pitheciine genera, these features had already been transformed. The jaws are U-shaped, the canines are more massively built and tilted outward as in the Bearded Saki and Uacari genera. Ankle bones attributed to *Soriacebus* are also comparable to the saki, and they suggest a quadrupedal locomotor mode that involved climbing, leaping, and foot-suspension, an adaptation that expedites bi-manual food handling, an advantage when positioning hard shells between the teeth so they can be cracked open or removing a fruit's seeds with the lower incisors.

While these behavioral and adaptive inferences are widely accepted, there is still debate about the precise phylogenetic position of *Soriacebus* that has bearing on the overall pattern of platyrrhine evolution. The unlikely alternative view has been advocated, that *Soriacebus* belongs to a phylogenetically exclusive early Miocene group along with *Dolichocebus* and *Tremacebus*, rather than being part of the modern radiation. That view also maintains that the fossil's hard-fruit and seed-eating adaptations are parallelisms to the pitheciine pattern. These ideas are rooted in the Stem Group Hypothesis. They are extremely unlikely, because a deeper analysis of *Soriacebus* suggests it is a pitheciine, and the notion of a monophyletic "stem group" is itself a very challenging idea. Concerning the latter point, as we have seen, the phylogenetic evidence is stronger that *Dolichocebus* is not a stem platyrrhine that arose before the modern radiation differentiated. Rather, *Dolichocebus* is part of the *Saimiri* lineage. Likewise, *Tremacebus* is part of the *Aotus* lineage. Second, to be viable, the Stem Group Hypothesis requires rampant parallelisms in functional morphology and ecology—of shell cracking and seed eating; of the interorbital fenestra and the Squirrel Monkey-like cranium; and, of a second large-eyed nocturnal monkey with an augmented olfactory apparatus. All of this is theoretically possible. Yet what are its chances of actually occurring among New World monkeys?

In addition to *Dolichocebus* and *Tremacebus*, *Soriacebus* would be the third of three independent cases where lifestyles and their underlying morphologies would have evolved by coincidence among the living and extinct genera of Neotropical monkeys that are, according to the preponderance of evidence, actually closely related, thus negating the invocation of parallelism as a way to explain resemblances. The existence of these lineages, and the discovery of fossils that may be sister taxa or ancestors of these living forms is, in fact, predicted by the molecules. These fossils fall within the expected time periods fleshed out by the molecular clock.

In short, the Stem Group Hypothesis requires that we interpret the Argentinean monkeys as an ecological community composed of three or more clades that mirror the same ecophylogenetic units that exist today and have existed at least since La Venta. It is far more reasonable to view these 17–20-million-year-old early Miocene fossils as part of an older, regional disposition of the evolving New World monkey radiation, which makes a robust appearance at La Venta in the middle Miocene, 12–14 million years ago, after a 5–8-million-year gap in our knowledge of the fossil record.

The whole concept of the stem group is based on one extensive, complex study designed to arrange living platyrrhines and various New World monkey fossils into a coherent cladistic picture in spite of major holes in its database.

The bigger picture of platyrrhine cladistics that grew out of the exercise was shaped by molecular evidence of the affinities of the moderns and a sweeping data set of cranial and dental information. However, the vast majority of the traits in the data set were lacking in the fossils attributed to the stem group. Its most suspect conclusions, identifying all but one fossil genus older than the La Venta forms as an outlier assemblage, and regarding the Colombian fossils as the oldest—except for *Proteropithecia*—possible relatives of extant genera or clades, is inconsistent with the molecular evidence pointing to their greater longevity, across the board. It does not accord with the morphological evidence of adaptation in the fossils, which points to the enduring ecology of platyrrhines and their community structure, or to the functional-adaptive examination of the most distinctive morphological features of the individual genera that are pertinent to their phylogenetic histories.

The mystery of fossils found on Caribbean islands

The understanding that native platyrrhines had once inhabited Caribbean islands evolved haltingly once it became accepted that *Xenothrix mcgregori* was a primate, however unusual its morphology appeared to be, and *Antillothrix bernensis* was not an African vervet or mona monkey, a leftover scrap of many that had been introduced as exotics over the centuries, perhaps starting with slave traders in the 16th century. In his comprehensive historical review of worldwide primate relocations, Benjamin Beck reported that thousands of these monkeys, who were aboard ships as "living food" or as pets and barter, may have been released over time beginning as early as 1560, on Barbados, Anguilla, Granada, St. Kitts, and Nevis.

Xenothrix, which has been discovered only on Jamaica, has an interesting scientific backstory. It had lain fallow, apparently unstudied for more than 30 years, before scholars realized that its remains belonged to a New World monkey. Not only was the type specimen confusing anatomically. Its recognition was also delayed because there are no indigenous extant Neotropical monkeys in the Caribbean and so it was not expected that any would have lived there in the past. As North American mammalogists began to systematically explore the biodiversity of the Greater Antilles early in the 1920s, they were keenly interested in surveying rock shelters and caves where remains of extant and recently extinct mammals could sometimes be found. Yet it took another 50 years for it to become abundantly clear that a splinter radiation of extinct platyrrhine monkeys constituted a significant natural part of a vanished Caribbean mammal fauna, and that lone specimens of *Xenothrix*, then *Antillothrix*, were not flukes—individually marooned monkeys that departed the mainland or were transported by people.

The first to describe *Xenothrix*, Ernest E. Williams and Karl F. Koopman, in 1952, were unable to classify the original specimen among the other platyr-rhines because of a mixed set of resemblances. They noted that "The Jamaican jaw combines the dental formula of a marmoset, the expanded angle of *Callicebus*, and a dental pattern approximating that of *Cebus*." Indeed, the genus name is a token to the odd combination of characteristics. It means "strange monkey," and the peculiarities must have stymied the scientist responsible for its initial discovery decades earlier, for he made no mention of it at all in his publications. The specimen had been discovered in 1920 by American Museum of Natural History mammalogist H. E. Anthony when excavating a collapsed cave. Williams and Koopman, who were not primate specialists, dug it out decades later from a museum drawer that held bone boxes containing Anthony's unidentified materials, and they described it. Twenty years later, Philip Hershkovitz, the leading, mid-20th-century expert on extant platyr-rhines, offered minor revisions to the Williams and Koopman morphological analysis. Yet at the same time he authoritatively announced a major taxonomic decision, saying, "The remarkable distinctiveness of *Xenothrix* is here recog-nized by designation of the genus as type of the new family Xenothricidae." Such a move, erecting a taxonomic family of mammals solely for the reception of a single species, or a single specimen in this case, happens occasionally but it is not highly recommended. It may serve well to communicate distinction, but it fabricates an unattached branch of the phylogenetic tree, thus consigning the species to a confounding limbo.

The interlocking mixture of *Xenothrix* characters posed challenging prob-lems at least partly because Williams, Koopman, and Hershkovitz were op-erating under the gradistic paradigm, where tooth count, jaw shape, and molar crown pattern each pointed in different directions, toward different platyrrhines that were taken as archetypes of different evolutionary grades. By applying the method of character analysis that is the basis of morphological cladistic analysis, I learned to approach the problem differently. As an exercise, and with the intent of writing a term paper for a graduate course on primate evolution taught by Eric Delson and Fred Szalay at the City University of New York, I periodically withdrew the specimen for study, by permission of the curatorial staff of the Department of Mammals at the American Museum of Natural History. It is kept under lock and key with the other type specimens in the collection.

At the urging of Eric and Fred, the term paper turned into an article, titled "*Xenothrix* and ceboid phylogeny," published in 1977. It was the first time the cladistic approach was applied to New World monkeys, when we still knew next to nothing about the interrelationships of any platyrrhines, living or ex-tinct. As we carried out the *Xenothrix* project, anatomical keys came to mind that could potentially reveal previously unseen links forming the foundations of a workable phylogenetic family tree for the platyrrhines, and that prospect developed into my PhD dissertation during the next few years. With this preliminary analysis, Capuchin and Squirrel Monkeys could be seen not only as sister genera, but as close relatives of the callitrichines as well, which meant *Cebus* and *Saimiri* had been misclassified for more than a century. The other platyrrhines seemed to be held together as a monophyletic group by jaw shape and molar morphology. These insights turned the gradistic interpretation of platyrrhines on its head.

Addressing the two-molar dental formula was the first challenge. Was it signal, evidence of a phylogenetic connection with callitrichines, or noise, a product of parallel evolution with no real phylogenetic significance? If the resemblance was homologous, it meant *Xenothrix* was a callitrichine. If it was analogous, then loss of the third molar happened coincidentally and other features would potentially demonstrate a link with a different group of platyrrhines. This led me to thinking more carefully about tooth proportions within the jaw, how upper and lower molars fit together functionally in occlusion, and how the dentitions of cebines and various callitrichines evolved stepwise via selection, first reducing third molars, and then eliminating them entirely, later reducing the second molars when they became the last postcanine tooth in the dental arcade. A key here was the functional morphology of the molars. They were large relative to jaw size, thickly covered in enamel and with a non-callitrichine cusp pattern. In *Xenothrix*, it appeared that first and second molars were enlarged evolutionarily to enhance their crushing and grinding capacity, and fitting them in the space of the jaw was accomplished by eliminating the third molar, quite unlike the patterns and purpose seen among callitrichines.

The deep jaw shape that *Xenothrix* shared with *Callicebus*, noticed by Williams, Koopman, and Hershkovitz, turned out to be an important phylogenetic marker once the 2-1-3-2 dental formula was understood to be analogous to the callitrichine condition. The evidence now pointed to a phylogenetic relationship with *Callicebus*, with the deepest jaws of all New World monkeys. Today, the precise phylogenetic position of *Xenothrix* is a little clearer, but the scope of possibilities remains as it was in 1977: *Xenothrix* is part of the clade whose living members are *Callicebus* and *Aotus*. Adaptively, with relatively large incisors, inferred by tooth roots and position, and large puffy molar crowns, coupled with unusually stocky limb bones and an estimated body mass of 2–4 kg, 4.4–8.8 lb, the middle-sized *Xenothrix* was probably a slow, climbing quadruped that relied heavily on fruit or other plant parts that were tough, consistent with its phylogenetic position as a pitheciid.

Mammalogist R. D. E. MacPhee, a leading authority on primate morphology and evolution, led several expeditions to Jamaica in the 1990s and recovered more *Xenothrix* subfossils. He and his colleagues have suggested on the basis of morphology, as I did at first, that *Xenothrix* is most closely related to Titi Monkeys, *Callicebus*. I now think it is more closely related to Owl Monkeys, *Aotus*, though both ideas seem plausible. Newer specimens that preserve the lower part of the face suggest that the orbit is relatively large, in my view, and the spacing of the sockets for the upper incisors indicate a wide incisal battery. *Aotus* has exceptionally wide central upper incisors and very large eyes. But MacPhee and his team have added another important datum by extracting and analyzing ancient DNA from a *Xenothrix* specimen. It indicates a close affinity with *Callicebus*. Though this is a strong piece of evidence, it does not come without questions, because in my view the molecules have not been able to resolve the affinities between *Callicebus* and *Aotus*. Molecular phylogenetics consistently places owl monkeys among cebids rather than pitheciids, closest to *Callicebus*. Is this another case of phylogenetic misalliance?

The DNA evidence also suggests that *Xenothrix* split from *Callicebus* about 11 million years ago, which gives us something of a baseline when considering when the *Xenothrix* lineage might have dispersed into the Caribbean. If the Jamaican monkey is more closely related to *Aotus*, or to *Callicebus*, 20 million years would be the oldest possible date for its divergence as a distinct lineage. Either way, it is likely that *Xenothrix* inhabited Jamaica for millions of years. One of its close subfossil relatives in the Caribbean, *Antillothrix*, lived in nearby Hispaniola about 1.3 million years ago.

The extinction of *Xenothrix* was probably hastened once people began to colonize the Greater Antilles, estimated at 6,000 years ago, though the archaeological evidence is scant for Jamaica. They probably died out gradually as the result of ecological pressures and/or disease rather than being overhunted, as there is little direct archaeological evidence from bone damage that hunting of mammals was an important part of the food economy of the indigenous populations. Benjamin Beck notes that disease, like the yellow fever virus that humans host and has been devastating to primate populations in Brazil, could have been ruinous to small island populations of New World monkeys inhabiting the Caribbean. In 2017, a team led by Siobhán Cooke used radiocarbon dating to determine that at least one *Xenothrix* monkey was alive about 900 years ago.

Antillothrix bernensis, the first extinct Caribbean monkey brought to light, has a similar scientific history. It was not properly identified at first either. It was mistaken for a transplanted Old World monkey rather than seen as a platyrrhine, so it sparked little interest when published. On one level, the confusion was understandable because the original material provided very little information. It was the lower end of a monkey tibia with a broken shaft only a few inches in length. Although the bone also preserved a diagnostically important part, the joint surface that articulates with the upper ankle bone, the talus, that anatomy would have been something of a black hole to any mammalogist at the time: postcrania were of little interest then. The second reason is that the scientist who discovered the bone had no reason to believe that platyrrhines had ever existed in the Caribbean, so he reasoned it belonged to an African monkey.

Gerrit Miller, when serving as a curator of mammals at the Smithsonian's National Museum of Natural History, dug the fragment out of a pre-Columbian kitchen midden in 1929, only a few years after H. E. Anthony noted his unpublicized discovery of the *Xenothrix* in his field notes at the American Museum. It was one of several discoveries written up by Miller in the same year, in the creatively titled article, "Mammals eaten by Indians, owls, and Spaniards in the coast region of the Dominican Republic." Scientists were aware that several of the Antilles islands had become home to populations of the African Green Monkey that descended from animals that had traveled on slave-trading vessels centuries ago. Thinking the fragment came from one of those, or perhaps from another Old World species, the Mona Monkey, from the stock released long ago on Grenada, Miller attributed his tibial fragment to a subfamily of Old World monkey, with a question mark: "?cercopithecine." Unnamed and almost forgotten, this specimen would reemerge in the 1970s, when it was recognized as a member of a new genus and species of what would become the best documented Caribbean subfossil platyrrhine, *Antillothrix bernensis*.

At about that time, the Dominican mammalogist Renato Rimoli joined a team of archaeologists doing a trial excavation in a dry cave that was yielding pre-Columbian artifacts and bone, Cuevo de Berna, around 50 miles from Miller's discovery site. Rimoli identified one of the specimens as part of an upper jaw of a platyrrhine monkey with three cheek teeth in place, and he classified it as a new extinct species of the Squirrel Monkey, calling it *Saimiri bernensis*. The publication of Rimoli's paper in 1977 occurred during the same year my *Xenothrix* paper appeared. I had been working on it while on a predoctoral fellowship at the Smithsonian when Rimoli visited in order to compare his specimen with monkeys maintained in the mammal collection, but somehow we missed one another. In reading his paper, I could see the points of Rimoli's argument, but since I had been poring over skull after skull in the collections for months, I doubted from the illustrations he published that this new species was a Squirrel Monkey. I thought it was a new genus, as

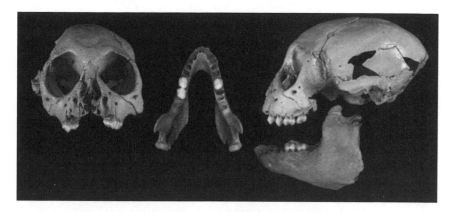

FIG. 9.13. Cranium and mandible of *Antillothrix bernensis*. Adapted from Cooke et al. (2016).

I suggested in a brief follow-up note published in the same Dominican journal that announced Rimoli's find.

Twenty years later, MacPhee and colleagues gave Rimoli's dental specimen a proper new genus name, *Antillothrix*. Then, prompted by the discovery of this important specimen, things came full circle with respect to the midden mammals that Miller believed were eaten when his tibial specimen was re-examined by me during my dissertation work and by Susan Ford, an expert on platyrrhine postcrania who also worked on her PhD dissertation at the Smithsonian soon after me. We both concluded that it was, in fact, a platyrrhine, not an Old World monkey. Decades later, backed by a stock of *Antillothrix* postcranial specimens that had been collected, detailed statistical studies by Siobhán Cooke and her colleagues would prove that *Antillothrix* was the proper taxonomic designation of Miller's tibia as well.

Antillothrix is now known by three remarkably well-preserved and rather complete crania (fig. 9.13), several lower jaws, and postcranial elements. This new material has all been found since 2009, in incredible fashion. It comes from two submerged freshwater caves situated roughly 30 miles from the archaeological sites from which Miller and Rimoli collected. After the first accidental discovery by recreational divers, all the specimens were recovered by teams of scuba divers recruited by Rimoli and me specifically to collect vertebrate fossils, led by Phillip Lehman and the Dominican Republic Speleological Society. One of the bones from the Padre Nuestro cave, until now the one with the densest accumulations of vertebrates, was found partially encased in the same type of rock that forms stalagmites and stalactites, flowstone, a deposit that forms when minerals are dissolved and carried away by water before solidifying. The geochronologists Robyn Pickering and Helen Green

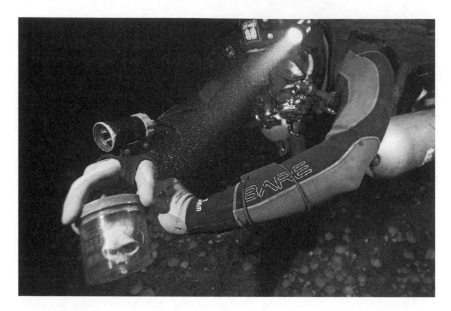

FIG. 9.14. Scuba diver Oleg Schevchuk brings a monkey skull to the surface from a submerged cave in the Dominican Republic. Courtesy of Phillip Lehman.

proved that the rock was datable, giving us the first direct, specific geological age of a fossil Hispaniolan platyrrhine. The rock is roughly 1.3 million years old, so we assume the tibia is just as old. A new bone-bearing cave in the Dominican Republic is being explored by Lehman as of the time of this writing, and it shows great promise: already one monkey has been discovered among its rich fossil remains (fig. 9.14).

Interpreting the phylogenetic interrelationships of *Antillothrix* is complex, as it is for *Xenothrix*, but it too has interesting implications for the origins and evolution of native Caribbean monkeys. The teeth of *Antillothrix* resemble some older fossils from Argentina, and its lower jaw, which is a very informative anatomical element, bears a close resemblance to the lower jaws of the Titi Monkey, *Callicebus*, and the Owl Monkey, *Aotus*, the two extant New World monkeys that have been linked to *Xenothrix*. Our three-dimensional shape analyses of mandibles proved that samples of titi and owl monkeys overlap with one another, and *Antillothrix* falls squarely within what we call the "shape space" they jointly occupy. This high-end method of assessing similarity does not simply translate into a phylogenetic conclusion, but it is a welcome complement to the character analysis that is designed to do that precisely.

There is a third Caribbean fossil genus, *Insulacebus*, from Haiti, on the opposite side of Hispaniola from where *Antillothrix* was found. It is about the same size and represented by a pristine dentition. Anatomically, it, too, recalls Argentinean Miocene platyrrhines in some features. Other traits resemble both *Antillothrix* and *Xenothrix*, and some resemblances link it to *Xenothrix* and *Aotus*. A fourth Caribbean genus is known exclusively from another island, Cuba. It is represented by a very good cranium, scattered dental remains, and postcranial bones. Cranially, it shares an uncanny resemblance with the Howler Monkey, *Alouatta*, as mentioned. Dentally, however, it exhibits a more primitive, non-folivorous cast. Postcranially, *Paralouatta* is intriguing. It appears to be the only platyrrhine, extant or extinct, which has features of the elbow and hands that suggest a terrestrial lifestyle.

What does all this mean with regard to the big picture? Within the Caribbean, the affinities among *Xenothrix*, *Insulacebus*, and *Antillothrix* provide evidence of past inter-island faunal connections between Hispaniola and Jamaica, suggesting that *Xenothrix*, the anatomical outlier, may be an offshoot of a Hispaniolan clade that became isolated on nearby Jamaica. *Paralouatta* is another matter, with no evident phylogenetic links to the other three genera. A second point offers deep-time clues that involve mainland South America. Dental resemblances shared with early Miocene fossils from the southern cone hint that the entry of platyrrhines into the Caribbean involved an early branch, or branches, of the modern radiation. Keep in mind that the Argentinean forms range from about 16 to 20 million years old, and several of them are more primitive anatomically than the Colombian fossils. The oldest monkey found in the north is *Panamacebus*, also at 20 million years, and one of the primate bones allocated to *Paralouatta*—by default, because it is Cuban and extremely difficult to diagnose on its own—is 16–17 million years old.

This context tells us that platyrrhines were widely distributed on the mainland at about 20 million years and that the radiation of Caribbean primates may have begun before the Amazonian primate fauna was modernized, as shown by the ensemble of fossils at La Venta, Colombia. Two lineages are represented on the islands, which suggests either two separate entries or a one-time colonization event that brought both in simultaneously, possibly as a small fauna. This can be explained zoogeographically in the following ways, after rejecting as too far-fetched the idea of a natural vegetation raft floating with two species aboard and depositing its cargo on Cuba or Hispaniola, or that a pair of waterlogged monkey species that had been cast into the sea managed to swim or float to one of the islands.

If only one original dispersal was involved it most likely happened overland as a range expansion, perhaps via a land bridge or island stepping stones in the east that intermittently linked Venezuela and the Greater Antilles when parts of the latter were geologically conjoined or very close to one another. A likely pathway followed what is now the Aves Ridge, a geological feature that under-lies the Lesser Antilles islands. If two independent dispersals were involved, without better dating it is impossible to infer which of two scenarios is more likely in either case. Was it a dry crossing of several hundred miles via a land bridge extending from the coast of South America that is now underwater? Was it an overland route from Central America via proto-Cuba that emerged during a sea level fall, or one that eventually allowed their progression by intermittent tectonic and climatic events? Or, was it a wet-water dispersal—possibly combined with a land bridge of stepping stones? Accidental rafting on a floating tangle of vegetation has often been proposed as a mode of transport to explain seemingly impossible across-water dispersals. This mechanism is highly implausible for long-distance, oceanic dispersal, as discussed in detail in chapter 10, but it may be viable across short distances. Its chances are also better if the monkeys are able to swim a bit; howler monkeys can and do swim.

Fossils prior to 20 million years ago

More questions than answers

Since its discovery in 1969, *Branisella boliviana*, from the Bolivian Andes, though long represented by a single broken cheekbone containing only three worn upper teeth, was considered one of the most important platyrrhine fossils because of its geological age, approximately 26 million years. Until the discovery and description of *Perupithecus ucayaliensis* was announced in 2015, estimated to be 36–40 million years old, *Branisella* was the oldest known fossil New World monkey and thought to provide clues of platyrrhine origins. The keen interest in origins naturally led to lots of attention on the fossil, its adaptations, use of habitat, and environmental setting. Some of this was overinterpretation. After all, *Branisella* lived perhaps 20 million years after monkeys arrived in South America, and the collection of *Branisella* fos-sils that has come to light since the initial discovery amounts to little more than a dozen specimens consisting mostly of worn and battered teeth, one broken piece of the lower face, a few partial jaws, and no limb bones at all. Without postcranial evidence, it seems rash to infer that *Branisella* may have been semiterrestrial, as has been claimed. And without paying attention to the fact that the only platyrrhines that are obligatory gum eaters are the tiny Marmoset and Pygmy Marmoset, genera that supplement their diet with insects and have cheek teeth of utterly different design and with no similar wear patterns, it seems incongruous to assert, as some have, that *Branisella* may have had the same type of diet.

The affinities of *Branisella* and another genus from the same site that is easily confused with it, *Szalatavus attricuspis*, are still uncertain, but it is highly
doubtful that these fossils offer much with regard to the origins of Neotropical
monkeys. The critical question has always been whether or not these genera
belong to the modern radiation, like all the younger fossils, or if one or both are
outside it and occupy a more basal position on the cladogram. There were undoubtedly platyrrhine monkeys that existed before the branch of the phylogenetic tree leading to the moderns arose, and the age of these fossils is close to
the oldest ages of the living New World monkeys estimated by the molecular
clock. Therefore, the older the fossils are, the more likely it becomes that one
may be part of that archaic set. For *Branisella* and *Szalatavus*, the uncertainty
about their phylogenetics is reflected in a vacillating taxonomy. They have
bounced around from being allocated to one or another of the platyrrhine
families, or placed formally in a Linnaean category meaning of "uncertain
placement," *Incertae sedis*.

One reasonable interpretation of *Branisella* dental morphology is that it
aligns with cebids, the clade that includes living *Cebus*, *Saimiri*, and the clawed
callitrichines. This hypothesis is consistent with the fossil's four-cusped upper
molars, the large final upper premolar, the relatively large canines, and the
shallow lateral profile of the mandible. As to *Szalatavus*, its upper molars exhibit a more triangular crown outline, and a fourth cusp that is very small
if legitimately identifiable at all. The presence of an authentic, small fourth
cusp, as opposed to being categorized as a slight bump on a crest or ridge that
is much smaller than the three others, is sometimes a judgment call. Such a
crown shape is a generalized resemblance to the callitrichines, but it is difficult
to say whether or not the similarity is phylogenetically significant. It is likely
that these Salla primates were frugivore-insectivores based on their teeth and
small-to-medium body size.

One of the details noted by many scholars concerning this collection of
fossils is that the teeth tend to be very heavily worn, which suggests the animals chewed a grit-laden diet. Airborne grit settles on leaves and fruit and is
known to be responsible for wearing down a monkey's molar teeth at a faster
rate in seasonally dry sites than in wetter habitats. It is possible that their excessively worn teeth reflect the fact that the Salla locality and its mammals were
exposed to heightened levels of coarse particulates swirling in the atmosphere
and falling onto trees and grasses, because they lived during a period when
active Andean uplift and volcanism, including ash falls, were known to have
had a local impact.

The discovery of *Perupithecus ucayaliensis*, older than the Bolivian fossil
monkeys, has the potential to have a more lasting influence on the reconstruction

of primate evolution in South America. It consists of a small number of fossils, tiny isolated teeth that were discovered by sifting several tons of sediment dug out of a fossiliferous riverbank at a site in Peru, called Santa Rosa, following wet- and dry-screening methods. While this is an excellent way to separate tiny individual fossils on the basis of their size, corresponding to the square millimeter density of the filtering screen mesh, it cannot provide contextual information about any of them. Thus, there is no way to know by association which teeth belong together once they are found. When separate premolar and molar teeth are found in isolation it may even be difficult to determine on anatomical grounds exactly what tooth locus one is dealing with: Is it the first or second molar; is it the final or penultimate premolar?

One other difficulty pertaining to *Perupithecus* is that neither the matrix from which the fossils were taken, nor the site itself, using other rocks, has been dated using geochemical methods. We do not know exactly how old the Santa Rosa fauna is. Estimates based on the geology and assemblage of mammals, which have been compared with other, dated sites, suggest it is of the Late Eocene epoch, between 36 and 40 million years ago; some believe it may be a few million years younger.

In any event, the Santa Rosa site is extremely important for understanding South American mammal evolution because it is midcontinent geographically and it has yielded many unique genera and species of mammals not known from the Patagonian sites of similar age that have taught us much about mammal evolution on the continent. Of particular interest are Santa Rosa's rodents that resemble equally old or older fossil African rodents. These have figured prominently in the argument that platyrrhine primates are paired historically with rodents that are also thought to have dispersed from Africa to South America.

The type specimen, the only fossil of the small collection formally attributed to *Perupithecus*, is a single upper molar. It has a triangular crown shape quite similar to several living callitrichine genera. The similarity is so impressive that I was first concerned the specimen may be an intrusive, material of a younger age, or even a modern monkey, that somehow became mixed in with much older sediments and fossils. But the paleontological context seems secure and the type specimen is also similar to an African Middle Eocene primate, *Talahpithecus parvus*. That led to the claim that *Perupithecus* is evidence supporting the highly unlikely scenario of the transoceanic dispersal of ancestral platyrrhines from Africa via rafting, though the authors of *Perupithecus* acknowledge the possibility that the resemblances shared with *Talahpithecus* may be parallelisms and not indicative of an African ancestry for New World monkeys. Even if there is a possibility of an African ancestry for *Perupithecus*, that does not speak to the mechanism of dispersal from Africa to South America. A comprehensive discussion of how platyrrhine ancestors reached South America is found in the next chapter.

A lower first or second molar that has been described in connection with the *Perupithecus* series is even more difficult to understand. It does not seem to occlude functionally with the type specimen, which indicates it belongs to another taxon. It has been identified as a primate of uncertain affinities, but there is room for other interpretations: it may belong to another mammal group. If this lower tooth is a primate, it is one that is extremely primitive, sharing resemblances with archaic non-anthropoids from the Eocene epoch in the northern continents and others claimed to be anthropoids from North Africa—a stretch, to my way of thinking.

These problems of identification, and the fossil's age, pose significant interpretive issues, not least of which stems from the fact that we are dealing with less than a handful of specimens without anatomical connections. Though some of the Santa Rosa specimens present clear-cut anatomical links to certifiable platyrrhines, there is no reason to assume that all the primates from there are New World monkeys, or even their closest relatives. Geography is relevant to our understanding, but it is not valid evidence of phylogeny. It is possible that platyrrhines or pre-platyrrhines were not the only primates ever to have lived in South America. And it is not necessary to assume that South America was colonized in a single event by a few individuals belonging to a single primate species. Other continents harbored multiple clades of primates during the Eocene that dispersed from elsewhere. That epoch was a period of significant intercontinental exchange in a warm, humid world that was replete with rainforest, involving many mammal groups that bypassed what would appear to be significant intercontinental water barriers. It is conceivable that the Peruvian fossils include clues that other primates, non-platyrrhines, reached South America and became extinct without producing living descendants.

Of course, these are early days in the paleontology of Santa Rosa primates, and the discovery of *Perupithecus* and the other specimens may be seen someday as a watershed moment toward a better understanding of primate evolution in the Americas. However, as was the case with *Branisella*, being the oldest sample of a South American primate known to us at any one point in time does not guarantee that it offers specific insight into platyrrhine origins. If the molecules are right, the platyrrhine clade existed roughly 45–50 million years ago—an even earlier origin is also quite possible—and Santa Rosa is only questionably dated to 36–40 million years. Some believe it may be a few millions of years younger. Either way, as with *Branisella* and *Szalatavus*, this is a case where, at the moment, geological age is less pertinent to the analysis than morphology, and the morphology is difficult to sort out.

Despite the popularized notion that *Perupithecus* is evidence that platyr-rhine primates came to South America from Arica by rafting across the Atlantic Ocean, it actually "proves" nothing of the sort. Working out the phyloge-netics of *Perupithecus* has bearing on the biological *history* of dispersal but not on the *mechanism* of dispersal. Phylogenetics may point to the identity of the primate clade that shared an immediate common ancestor with *Perupithecus* or other platyrrhines—and that clade may or may not be an African endemic—but it tells us nothing about *how* any animals actually moved through space.

Ucayalipithecus

Another fascinating discovery from Santa Rosa published in April, 2020 (Sei-ffert, E.R. et al. 2020. A parapithecid stem anthropoid of African origin in the Paleogene of South America. *Science* 368, 194–197) has profound signifi-cance for the history of primate evolution in South America, proving it is far more complex than the long-standing model that held it was colonized by a single dispersing species giving rise to the platyrrhine adaptive radiation. It means several taxonomic groups of primates, possibly three, somehow got into South America during the Eocene, but only one survived to the pres-ent. It described four small teeth with a highly distinctive morphology that matches that of primates previously known only from Egypt's Fayum deposits, the Parapithecidae.

The specimens leave little doubt about a link between Eocene primates of South America and Africa. However, because phylogenetic and morphologic evidence consistently indicate New World monkeys must have descended from a different kind of primate, not from any closely related to *Ucayalipithe-cus perdita*, it also demonstrates with high confidence that South America was colonized by more than one primate clade, as I suggested above based on the 2015 collection of Santa Rosa fossils. Whereas in the past scientists were hard pressed to explain how exchange between the hemispheres happened one time and have resorted to accepting it was a singularly unusual *chance event*, i.e., transoceanic rafting, explaining how this occurred two or three times means explaining a *recurring pattern* that can hardly be produced by chance.

These and other factors discussed in the next chapter come into play and need to be objectively considered as we try to understand and reconstruct how monkeys managed to arrive on what was then the great continental island of South America, a scientific and, to the monkeys, a physical challenge that easily attracts boundless speculation.

CHAPTER 10

SOUTH AMERICA WAS ONCE AN ISLAND

How Did Platyrrhine Ancestors Get There?

We may be able to reconstruct, as working hypotheses, where, when, and from what clade New World monkeys branched from the anthropoid family tree, but figuring out how they got into South America, the mechanics, is the stuff of scenarios, beyond the reach of our evidence. This primate puzzle is one of the broader geographic questions about the roots of South American mammals, and it has played a historical role in evolutionary biology. It makes for inspired discussions, but the questions associated with the primates' arrival on the continent do not generate hypotheses that are falsifiable, amenable to rejection by empirical evidence, which is the litmus test of scientific proposals and interpretations.

There is no fossil record or analysis that definitively demonstrates from what taxon the pre-platyrrhines arose, who their nearest relatives are, where they lived before arriving in South America, and how they made the journey. Unlike today, South America was for millions of years an island continent surrounded by a mega-ocean before any primate arrived. This was proven during the 1960s by the new science of plate tectonics and continental drift. It demonstrated that continents and other landmasses are perpetually moving about the surface of the globe and reshuffling their connections. South America was not attached to any other continents at the time that we think the ancestor New World monkeys split off from early Old World anthropoids to become a distinct group. Most estimates are that this happened 45–50 million years ago, during the Early Eocene, a date based on the molecular clock. But no fossil primates have been discovered in South America during that or earlier time periods, although an abundance of other mammals had already been living on a landscape also suitable for primates for many millions of years. While primates were evolving elsewhere, the flora and fauna of South America were evolving in splendid isolation, a phrase coined in 1980 by George Gaylord Simpson, the leading American paleomammalogist—and a premier expert on South America mammal evolution—for much of the 20th century. This concept is still the guiding postulate behind the biogeographic history of South America during the Age of Mammals, the last 66 million years, strongly supported even by today's paleontologists who have access to a much richer fossil record by far than what Simpson had at his disposal.

How does this background elucidate the enigma of New World monkey origins? There are several pertinent facts and uncertainties that cloud the picture. To address the questions of where, when, and from what, the first thing to consider is phylogenetics, how the pertinent taxonomic groups are related to one another. The second is time and geography. The third is deployment, how a group might have shifted its distribution, moving in time and through space. The last poses the most treacherous questions because in the end this matter relies mostly on nonbiological data. Knowledge that pertains to long-distance deployment is typically not the kind that paleontologists can access or reconstruct, and that leads to speculation.

There are two possible scenarios to explain the dispersal pathway behind platyrrhine origins. I call these the Americas Scenario and the Transatlantic Scenario (table 10.1). The term scenario is used as opposed to hypothesis because there is an important distinction. A hypothesis is a testable inference or interpretation based on observable evidence aimed at establishing facts, patterns, and principles. A scenario is a sketch or plotline of possible events that, although also based on information we might consider relevant, is a story that cannot be tested. In science, we need scenarios, historical narratives, to make sense of phenomena we observe, but those accounts may be one or more steps removed from the underlying data, which means they are speculative rather than inferential.

The Transatlantic Scenario, which has been the most prevalent one in recent years, describes platyrrhine ancestors dispersing from Africa to South America by crossing the South Atlantic Ocean. How did anthropoids cross the ocean at that time, when the shortest distance between shorelines may have been 600–900 miles? This scenario posits that several individuals were swept away into the ocean on a raft of vegetation that naturally broke away from a riverbank and was discharged from a large river that emptied into the Atlantic, and that the castaway monkeys survived the long journey at sea. To highlight this type of unlikely transfer between lands separated by significant water barriers, Simpson called the process "waif" dispersal by means of a "sweepstakes" route. He noted that such routes are low-probability occurrences, arguing they are not only rare but also unpredictable because they are

TABLE 10.1. Americas Scenario and Transatlantic Scenario compared

	AMERICAS SCENARIO	TRANSATLANTIC SCENARIO
ITINERARY		
Hemispheric crossing	Overland/Thulean route	South Atlantic Ocean
Mode of travel	Locomotion/range expansion	Floating raft of vegetation
Arrival time	~45 million years agp	~45 million years agp
Minimal duration of dispersal sequence	~25,000 years	7–11 days or more
Travel distance	<10,000 miles	~600–900 miles
Founders	Pioneering population	Two or more survivors
	HIGHLY LIKELY	HIGHLY UNLIKELY

not related to any other travel avenues, such as the location and timing of a downed land bridge that had once spanned a barrier and could do so again. Transport, however, was not the only challenge facing them. Upon arrival, as waifs the travelers would have consisted of a small number of individuals so their chances of becoming successfully established as a long-term breeding population would have been small. We know that for genetic reasons, small populations, like those of many endangered wild species living today and those held in captivity, are at a high risk of extinction in generational time.

The Americas Scenario posits that a normal breeding population of ancestral monkeys came through North America and then the Caribbean Basin, first by crossing into the New World via either Europe or Asia in an ordinary pattern of geographic range expansion, possibly from an ancestry in Africa or Asia. Like present-day pioneering species, those belonging to a community that are the ones earliest to spread into a newly available habitat, they would have followed the geophysical, climatic, and environmental changes that enabled the regional spread of flora and fauna. Following the spreading forest, the animals' range would have shifted along more or less overland routes at a natural pace, perhaps over a period of 25,000 years or more, to reach South America and begin to evolve there as platyrrhines. Shifting tectonic or other physiographic events may have temporarily removed water barriers to allow the final passage into South America, or a modest waterway may have been crossed over time by intermittent island-hopping.

Even during the early 1500s as the Age of Discovery gained momentum, Europeans recognized that the New World mammals were oddly distinctive. By the middle 1700s, naturalists such as Georges-Louis Leclerc, the Comte de

Buffon, were speculating about the significance of the New World–Old World biotic similarities and differences, and they realized there had been a major geographic separation that needed explaining. By the middle to late 1800s, Charles Darwin and his contemporary, Alfred Russel Wallace, the preeminent biogeographers of their time, talked about the disjunct geography of modern New World platyrrhines and their Old World catarrhine counterparts in an evolutionary perspective, though they lacked fossil evidence. Then, during the last decade of the 19th century and the entire 20th century, while fossils of many early South American mammals were being discovered from the Paleocene and Eocene epochs, a period spanning 34–66 million years, there was still a total absence of early platyrrhines among them. This logically led to the idea that platyrrhine ancestors arrived from elsewhere, abruptly, unlike the non-primate native mammals whose evolution over time could be traced in the rock record.

Catching an event like that in geological time, a basically instantaneous occurrence, is like being able to see the second hand of a clock tick by on a nanosecond scale. No one truly expects to find the ancestral monkey species where and when it departed, or the monkeys that actually arrived in South America, or a remnant of the raft if that's the way they came, or a footprint trace or other mark in North America to indicate that was the itinerary. The problem of origins is always a difficult one, and in this case it intersects biology and geography. Though it is a question naturally addressed by paleontology, it is one of the most difficult ones to resolve, especially without a highly detailed fossil record. Fossils found in situ, in a tight stratigraphic sequence that factually demonstrates morphology, time, and space, can provide layer-cake-like, vertical evidence of evolutionary change, and signs that a species that occurs on top of the geological column possibly evolved from one found at the base. However, with such detailed sets of records, as we know from the well-sampled, dense record of fossil primates from the Eocene of North America, animals dispersing from nearby areas can intrude into the column and obscure ancestry and descent.

Working horizontally across continents, using information from different time periods, is even more prone to error. It would rarely provide definitive evidence of the direction of evolutionary change, and it cannot show that the putative ancestor was geographically restricted to the one place its fossils were discovered. That renders essentially invisible the dynamic time-and-place evolutionary process that produces new species and links ancestors with their descendants. Adding to this difficulty is that the most common method we use to infer evolutionary relationships is cladistics, as discussed throughout this book, an approach designed to link the closely related branches in a phylogeny, but not ancestors and their descendants. In practice, it is analogous to a genealogist's quest to find siblings, which is one reason we call cladistic affinities sister group relationships, rather than searching for parents, that is, the ancestors of offspring. The discussion of platyrrhine origins is thus likely to be wrapped in informed speculation. The zoogeographic upshot is this: finding a set of disjunctly distributed sister taxa, say, one in Africa and the other in South America, is not the same as identifying one as the ancestor of the other. Nor does that information rightly form the basis of an explanation of how such an ancestor physically crossed between the hemispheres.

These concerns have not been widely shared by the experts, as I was stunned to learn during a conference in Frankfurt, Germany, in 2011. I was among an international group of paleontologists who were examining the paleobiology of the Eocene world 34–56 million years ago. At a roundtable session I asked, "Who in this room actually thinks New World monkeys came to South America from Africa by rafting across the Atlantic Ocean?" More hands shot up than I could ever have expected. These people were scholars, students, and interested nonprofessionals, and I had to assume some of them simply were not moved to respond either way, leading me to think there were even more believers in the rafting scenario in the room.

The Transatlantic Scenario is still widely accepted, despite the many difficult assumptions it presupposes. Proponents have rarely openly acknowledged that ocean rafting is an extremely unlikely method because of the particularly extreme conditions that would have been encountered while making the crossing over many hundreds of miles in open waters. Intuitively, one can rationalize that the chances of safely getting to the other side are something like the chances of winning a lottery jackpot, very small though not quite zero. But being able to conceive that transoceanic rafting is a possibility does not make it probable, which means it does not pass the threshold required of a scientific explanation.

The Americas Scenario

The important points that establish the Americas Scenario as a preferred narrative that can explain how ancestral New World monkeys arrived in South America are these: (1) It is consistent with the normal everyday life of primates; (2) It is consistent with the long-term histories expected of primate populations and species; (3) It fits with a geographic and temporal pattern of long-distance, intercontinental dispersal that has already been demonstrated for a wide variety of other mammals; (4) Postdispersal, it is consistent with genetic factors that enhance the potential success of a founding population;

(5) It requires none of the far-fetched assumptions that shore up the alternative, the Transatlantic Scenario.

The Americas Scenario is derived from a geographic worldview of evolution and continental geography in which North America supplied the ancestral stocks of several groups of South American mammals. Regarding the platyrrhines, it was made explicit with the first discoveries of primate teeth during the late 1800s in Rocky Mountain sediments, rich in fossils of Eocene age. Some of those teeth resembled those of living howler monkeys; others resembled extant callitrichines. Paleontologists thought at that time that these fossils were connected to the origins of South American primates living on a continent which they understood was not yet linked to the north via Panama, and they tended to believe that dispersing animals would have crossed over water gaps in stepping-stone fashion as dry land appeared, or that they could have rafted on natural floats of vegetation.

We know much more today. We do not presume that the platyrrhine forerunners must have originated in North America and, in additional to paleontology, we look to new evidence of plate tectonics, continental drift, sea levels, and climate to reconstruct the possible routes and timing that may have benefited dispersing monkeys. There is also a strong consensus that places the origins of anthropoids somewhere in the Old World, most likely in Africa. In this scenario, the land link to the Americas that enabled passage from the Old World occurred in the northern circumpolar region, across a multicontinent landmass, Laurasia (fig. 10.1). It could have happened in the west, via the Bering Land Bridge that gave humans the foothold to migrate from Asia into North America some 50 million years later, but was also in existence at times during the Eocene. Or, it may have happened in the east, via a passage reconstructed by continental drift and known as the Thulean route. The last is the most plausible, if only because it was the more commonly used deployment avenue for a larger range of mammalian taxa.

To understand and reconstruct how such a dispersal could have occurred, we need to consider what the capabilities and behaviors of the animals themselves might have been. The global geographic distribution of primates that we plot on the world's map today is a testament to their adaptability, which can be attributed to their flexible feeding habits, locomotor dexterity, curiosity, intelligence, and sociality. Primates are adroit at moving about the environment as the environment changes. They track the size and shape of ever-changing forests as they ebb and flow through space and time on their own evolutionary journeys. There are many arboreal and terrestrial primates that have adjusted evolutionarily to many different environments—swamp, mangrove, jungle, grassland, snowy highlands, and rocky mountains—and many species exist on continental and archipelagic islands.

FIG. 10.1. Polar projection of the northern megacontinent of Laurasia during the early Eocene showing the disposition of landmasses that facilitated intercontinental dispersal routes for terrestrial mammals, including the Thulean route. Drawing by Mark Hubbe.

Unlike migratory mammals such as gnu, caribou, bison, and whales that seasonally move long distances to and from alternative feeding grounds or birthing sanctuaries, primate groups are attached to a local habitat. As that habitat shifts, the primate community flows. On a regional or continental scale, some species in a community are more inclined than others to become pioneers as climate, environment, and habitat are altered, and they always live in a state of potential change. Those vanguard species tend to become very widely distributed. They are often composed of smaller taxa, like marmosets, squirrel monkeys, and capuchin monkeys, that take advantage of edge habitats, the forest fringes where food is plentiful and accessible, where new growth is most intense and insect life most prolific. Or, they may be large monkeys like the folivorous howlers, able to sustain themselves on leaves, a nutritionally poor but abundant food source. Either way, once the ancestral platyrrhines were established as one or more breeding populations, which probably took generations, they would have benefited from ecological release, a phenomenon that refers to the burst of phylogenetic and adaptive success that a species experiences when introduced to a new favorable environment that lacks the constraints of its former situation, such as native predators, competitors, and diseases. The platyrrhine adaptive radiation then ensued.

This means that the simplest reasonable idea of a dispersing ancestral platyrrhine is a medium-sized monkey whose lifestyle resembled a squirrel monkey's, relying on a mixed diet of fruits and insects rather than a specialized staple. The species would be advantaged by having good, agile locomotor abilities, and it would have been able to exploit edge habitats for foraging and movement. These are descriptions that fit well with the anatomical features evident among the majority of early Egyptian anthropoid fossils. Sociality is harder to reconstruct, but living in social groups of a modest size, rather than in small, pair-bonded units, which are in any case exceptionally rare among primates, would have been conducive to maintaining genetic diversity and buffering inbreeding over generational time, which would have been a critical advantage following their arrival in South America.

Considering that the entry of platyrrhines into South America was a singular event that involved no more than two groups of mammals, that is, primates and rodents, if indeed they did come together, we know that the route of the Americas Scenario did not involve a continuous long-distance corridor as a last leg that broadly afforded direct land connections and widespread faunal interchange. Otherwise, many groups of mammals would have entered at the same time. The primates must have been constrained geophysically by filtering effects. Undoubtedly, there were water barriers that temporarily checked range expansion and the animals would have been blocked until those were passable, which could easily have happened several times at different places. This would have weeded out less flexible species of mammals as the fauna was on the move, tracking the spreading habitat. The same factors would have applied to rodents, if they arrived simultaneously or a few million years earlier, which is beginning to seem likely based on recent discoveries of South American rodent fossils.

How and where did the primates actually cross between hemispheres? The Thulean route, evidently used by a host of mammals transferring between Europe and North America during the early Eocene, is the most likely causeway. Although it has long been known from the fossil record that the mammal communities of the Paleocene and Eocene epochs resembled one another, suggesting an enduring, though intermittent capacity to cross over, only recently have plate tectonics, paleontology, and paleobiogeography provided the means to reconstruct how this took place, its timing, and the details of paleoclimate and habitat.

The Thulean route connected the western margin of what is now continental Europe with the landmasses of the British Isles, Iceland, Greenland, and Ellesmere Island in the northeastern Canadian Arctic Archipelago. Reconstructions of the condition at about 56–57 million years ago, when sea level was at a lowstand, indicate that an above-water Thulean bridge existed. It is the younger of two connecting routes in the region and probably the one that explains the Eocene Euro-American faunal continuity. The older one existed about 10 million years prior. However, it is important to be cognizant of both because they show that proximity was long-lasting during the first epochs of the Cenozoic when mammals were moving between the northern continents. It is also important to be mindful that there are limits to our powers of resolution: while the dating of sea level rise and fall in this one place is likely to be imprecise, as a time-and-place bracket it is broadly indicative of patterns that existed. Passages along the Thulean route probably acted like waterway locks that allow ships of certain configurations to move through canals. They were passable for some land mammals at some times and places when sea level was low, the ground was above water, and it had been subaerial long enough to sustain the growth and expansion of a terrestrial flora and fauna.

Two paleontology sites of the period are very important for understanding the Euro-American connection that is central to the Americas Scenario. One is on Ellesmere Island, a relatively large landmass, roughly the size of Washington State in the United States, and we know a great deal about its environment and biological history. It was a transit point along the Thulean route. The fossil site is Early-Middle Eocene age, 53–38 million years ago. Now it lies within the Arctic Circle as it did then. Yet during the Eocene, Ellesmere supported plants that are typical of lush, wet, subtropical forests, including palms and other fruit trees. It has produced a rich record of animals, including alligators, lizards, snakes, giant tortoises, and early, archaic mammals called plesiadapiforms that are either primates or very close kin. More than a dozen orders of mammals are represented. They lived in an ecology resembling today's Borneo, which is inhabited by more than a dozen living primates, including lorises, tarsiers, monkeys, and apes. Ellesmere was not as warm then as Borneo is now; it averaged about 20°F cooler. It was a seasonal environment, with the coldest month just below freezing, the warmest about 67°F, and the average annual temperature 47°F. Today on Ellesmere the warmest month is 43°F, the coldest is −36°F or more, and the average annual temperature is −2°F. It is also interesting that the island's longitudinal position in the far north means that it experienced about 100 days of total darkness, comparable to one of the northernmost cities inhabited by people in the world today, Longyearbyen, Norway, which is dark for about four months of the year. Perhaps that means diurnal monkeys would have been motivated to be on the move.

Another critical site for understanding intercontinental dispersal is the Middle Eocene Messel Pit near Frankfurt, Germany, which is about 47 million years old. It is one of the richest Eocene fossil sites in the world. About half of the 34 genera of Messel mammals, including primates, are closely related to fossils from the Rocky Mountain region of North America and from Asia. Several genera co-occur in the Eocene of Wyoming in the United States and at Messel. One of the plants found at Messel is the cashew nut genus *Anacardium*, known in South America today. This tree has also been identified in Eocene fossils from Texas, indicating a massive intercontinental distribution and a circumpolar connection during the Eocene.

If an overland dispersal journey of pre-platyrrhines started in Africa, how long would it have taken to get to South America via the Thulean route? There is one detailed, pertinent study that attempts to trace an Early Eocene long-distance, multicontinental dispersal of a primate across Laurasia. It involves the fossil *Teilhardina*, known from a number of species represented by many excellent dentitions found at several localities in Asia, Europe, and North America, and it is but one element of a coherent primate fauna that existed at the time. *Teilhardina* is a small, early, tarsier-like primate with ecological requirements that would have matched those of any early anthropoid. Morphological studies suggest that the direction of anatomical change and the phylogenetic branching pattern among the species of this genus go from east to west, from Asia (China) to Western Europe (Belgium) to North America (Wyoming). This sequence coincides with the stratigraphic positions and ages of the first occurrence of each of several species at their respective fossil sites. The data indicate that dispersal from east to west along a northern route through the continents, presumably taking the Thulean route in the west, took approximately 25,000 years. If an ancestral platyrrhine were to follow the second half of the trail taken by *Teilhardina*, leaving from Africa, for argument's sake, and moving through Europe, one might expect a similar rate of travel over about half that distance. Afterward, getting from North to South America would depend largely on when sea level and the geospatial alignment between those continents made the last leg possible. Thus 25,000 years can be thought of as a reasonable figure covering the entire dispersal sequence from start to finish.

Other fossil discoveries in North America indicate that *Teilhardina* had a widespread distribution. Their southernmost occurrence is in Louisiana, which is at the edge of the Caribbean Sea. This is a significant finding because it demonstrates the extensive presence in the region of habitat suitable for primates, particularly in an area that could have been near a launching point for dispersion into South America.

The other important point implied by the *Teilhardina* Trail is that along the way platyrrhine ancestors would have lived a normal existence. Their dispersal would have come about as a natural range expansion within a supportive ecology. The monkeys would have lived in their own social group, amidst a community of other monkeys and mammals. In this scenario, it is possible that more than one primate population could have gotten into South America; however, only one species would have survived to give rise later to the last common ancestor of the living platyrrhines.

The final step of the journey to South America may have involved traversing temporary landmasses to cross the Caribbean Sea. The rationale for this hypothesis comes from the fact that the Caribbean Basin was a tectonically active zone in which there would have been a good probability that uplift and sea level falls periodically exposed blocks of landmass there, such as today's shallow Nicaragua Rise platform that spreads from eastern Mexico to the vicinity of Jamaica but is now submerged. In other words, dry land extensions into the Caribbean may have provided an exit from North America, for at least part of the way. We know that at least one Eocene terrestrial mammal found in North America lived in Jamaica during that time. It was *Hyrachyus*, a relative of the rhinoceros clade that has also been found in Eocene deposits in Europe and Asia. It is the oldest fossil mammal known from the Caribbean thus far and, as a good-sized hooved mammal, it probably was not disposed to wet-water travel. Other mammals may have gotten there even earlier, like the Antillean insectivores that are related to North American forms. Again, with our limited powers of resolution, it is patterns that are important, and the fossil record has established that early in the Age of Mammals South America received dispersers from the north. It is also noteworthy that platyrrhine primates actually did manage to enter the Caribbean and establish themselves there, as discussed in the previous chapter. This demonstrates the viability of inter-island, trans-Caribbean dispersal even though in the recent cases the mammals came from the south.

The main objection to the Americas Scenario is that no anthropoid fossils that are potentially ancestral to platyrrhines have been discovered in North America, nor have any been discovered along an earlier phase of the route in Europe. While this is a legitimate concern, the absence of evidence is not evidence of absence. We just may not have found any yet. There are large geographic and temporal gaps in the fossil record, as witnessed by the fact that major discoveries of new taxa and new primate faunas continue to emerge even today in the Eocene of Laurasia, as they do in South America. The Santa Rosa fauna and *Perupithecus* come to mind, found in 2015.

There are also numerous fossils that we do not understand well that may change several major working hypotheses about the course of primate evolution. For example, in the late Eocene in Texas, there are primate fossils with clear affinities to European forms, which buttresses the Americas Scenario, and at least one other that is intriguingly enigmatic. That is *Rooneyia viejaensis*, which is about 37 million years old. Its status has been debated since the fossil, an essentially complete cranium, was discovered in the 1960s. Oddly, with so much information available, scientists have not been able to resolve confidently the phylogenetic position of *Rooneyia*. The majority of researchers have believed it belongs to one or the other of the two groups that were quite successful in North America: to the adapids that are related to today's strepsirhines, or to the fossil tarsiiforms that are related to today's tarsiers. It is quite possible that the disagreement is influenced by what we have assumed rather than demonstrated. *Rooneyia* may seem to be in the wrong place, at the wrong time, and endowed with the wrong anatomy, but that is based on suppositions about what to expect.

I have proposed another view. I suggest that *Rooneyia* belongs to a clade that is the sister group of the anthropoids, based on shared resemblances found in the orbit and ear region. Few fossils classified as non-anthropoids present even hints of anthropoid-like morphology but, in my view, *Rooneyia* does. If *Rooneyia* is, in fact, what we might call a proto-anthropoid, it means there is a North American element to the anthropoid origins story that we have missed. It means that there is likely a diverse clade of fossil primates whose remains we have missed as well; *Rooneyia* cannot be the only one. This would not make deciphering the geographic origins of platyrrhines any simpler. But it reminds us that all our evolutionary and taxonomic hypotheses are *working hypotheses*, to be tested again and again against new evidence and ideas.

The Transatlantic Scenario

The idea of animals rafting across large bodies of water was vigorously promoted in the late 1800s and early 1900s to explain particular instances of mammal dispersal in different parts of the world. It was an antidote to the fanciful myths of midocean sunken continents, like Atlantis, that were imaginary way stations said to facilitate an ocean crossing. Still, overwater dispersal was considered a likely success only under the rarest circumstances. A major early biogeographer was W. D. Matthew of the American Museum of Natural History, who wrote the classic *Climate and Evolution* in 1915. His discussion of how terrestrial mammals might have colonized continental islands such as Madagascar, those located in relatively shallow water on continental shelves usually not more than a few hundred miles from the mainland, inspired the model of long-distance, intercontinental oceanic dispersion. Matthew thought that such islands were settled by mammals that rafted there on floating islands, chunks of land that were pushed out to sea by powerful, disgorging rivers. He noted that there was a one in a million chance that this process could have brought a pregnant female to shore on an island not more than a couple of hundred miles from the mainland, and thus start a new population.

During the 1970s, a new understanding of the mechanisms of plate tectonics provided the means for paleogeographers to reposition the continents on a global map of deep time, and this reinvigorated the Transatlantic Scenario. A smaller Atlantic Ocean, say, 40–50 million years ago, meant a measurably shorter water gap for monkeys to traverse, but it may nevertheless have been 600–900 miles between nearest points of land, according to various reports summarized in an influential 1999 paper by Canadian biologist Alain Houle. Among the first paleontologists to propose an updated version of the Transatlantic Scenario were the French paleontologists Robert J. Hoffstetter, who had extensive experience in South America, and René Lavocat, who made important discoveries of early mammals in Africa, including fossils belonging to the clade of rodents believed to also have made the crossing. At the same time, important anthropoid fossils roughly 30–35 million years old were being found by the great American paleoprimatologist Elwyn Simons and his teams working in the Sahara Desert 50 miles south of Cairo, in the Fayum Depression. This helped solidify the idea that Africa was the geographic source of ancestral platyrrhines.

In the Transatlantic Scenario, the extrinsic challenges are daunting and the intrinsic, biological challenges severe, getting across an ocean many hundreds of miles wide compounded by a spate of life-threatening trials that the founders would have had to overcome. From a monkey's perspective, what would it have been like for castaways to live on a raft for the time it would take to drift across at least 600–900 miles of the Atlantic Ocean without any of their natural needs—food, water, and shelter—exacerbated by a world of new, incomprehensible phenomena?

Simons laid out physiological and behavioral reasons why a voyage on a floating raft of vegetation had an infinitely small chance of successfully crossing the Atlantic, if the monkeys had any chance at all. He reasoned that a small primate would have become comatose after 4–6 days at sea, suffering from dehydration accelerated by heat, sun, wind, and no refuge. He noted that the plants would similarly have withered due to salt spray, meaning the monkeys' only food and water resources, if they had any to begin with, would quickly be gone. Heat stress and sun stroke would quickly be debilitating or fatal. Simons doubted that monkeys would in the first place allow themselves to be trapped on a raft that detached from a riverbank and began to be swept away by a rapidly flowing river. He also stressed that reproduction in the tiny founding group would have been impeded by the innate drive to avoid incest. He surmised that in a weakened physical condition, after such a harrowing journey, it would be difficult to cope with a new environment where the animals quickly had to find new foods and combat new parasites, predators, and diseases. They would have already experienced motion sickness, perhaps, hazardous storms at sea, and been thrown into a state of unrelenting sheer terror. Even if all that was overcome, inbreeding and small initial group size would severely impact genetic diversity and the long-term chances to survive.

Saying it another way, the Transatlantic Scenario requires many leaps of faith to believe that unimaginably good luck was how the limitations of animal biology and the chaotic forces of nature on the high seas were defeated. It requires believing that luck kept them alive, sent them straight-as-an-arrow across the ocean and steered them safely to shore. Moreover, this scenario requires a belief that the waif monkeys were exceedingly lucky over and over again, at every point of their journey.

To put this into perspective, it is useful to look at a factual, well-documented modern-day oceanic voyage, also of unusual character. In 2011, Anthony Smith, an 85-year-old Briton, set out with a crew of three to cross the Atlantic Ocean on a large, 40×18 foot, sail-powered raft. Smith was a best-selling author and science journalist, an experienced sailor and lifelong adventurer. He was inspired to undertake this trip by the famed Pacific Ocean voyage of Thor Heyerdahl and his crew in 1947, on his raft the *Kon-Tiki*, and by an account of two sailors during World War II who survived a similar transatlantic crossing in a lifeboat after their merchant ship was sunk by the German navy. Smith's vessel, *Antiki*, was called a raft but it was designed and constructed to be oceanworthy and livable and navigable. It was outfitted with a cabin and a mast with a sail, and onboard there was a camping stove, supplies of water, fresh and packaged foods. He also had a computer and a radio.

The *Antiki* traveled a farther distance, 2,900 miles, than the castaway monkeys would have. It took 66 days to navigate from the Canary Islands off the coast of Morocco to St. Maarten, in the Caribbean, which makes Smith's rate of travel about 2.9 miles per hour. Waif monkeys traveling at the same rate with the same directional accuracy would have taken 9–13 days to make the trip. Another calculation about the travel time of monkeys making such a voyage on a mat of vegetation was presented by Houle. He estimated the transatlantic travel time as 7–11 days at 40 million years ago and 5–7 days at 50 million years ago, propelled by current and wind across distances of about 900 and 600 miles, respectively. It is well to recall at this point that, unlike Mr. Smith and his crew, the monkeys would have not benefited from a rudder to keep them on a nearest-points course, in spite of favorable currents and wind.

There are many negative factors and phenomena that one can consider in evaluating the credibility of the transatlantic rafting scenario, and there is a scientific literature that describes some of that. The starting point is one detail. During the Early Eocene, the Congo River, which has been identified as a likely launching site because of its massive ocean-directed flux, may have flowed to the east, not into the Atlantic. Getting through typical nearshore ocean currents, which tend to flow in north-south loops, would also have been a major obstacle to reaching the main east-west Atlantic Gyre that heads toward South America. Yet, even if the monkeys did, currents close to the equator, which is near the required, closest straight-line heading, also involve major west-to-east flows that could have captured a powerless craft and set it on a different course. The ocean surface is also full of spiraling eddies, some as large as 6 miles in diameter that can last for weeks and months. Any one of them would have trapped the raft in a maelstrom that prolonged the passage, at the very least. This is precisely what happened to another intrepid person in 2019, the Frenchman Jean-Jacques Savin, who attempted a South Atlantic crossing in a large, seaworthy, barrel-shaped vessel that was not pilotable but relied on the ocean itself to carry him to the New World. He was caught in currents that stalled his voyage for more than 19 days in midocean. But if the monkeys did manage to get close to shore, how would they have disembarked if the raft did not come to a safe landing on its own? Studies of the colonizing potential of stowaway rats whose host ships landed in the Pacific's Marshall Islands showed that the animals were not able to reach shore unless the ships were moored to a pier, which provided a chance to exit via ropes and gangways. Did the monkeys swim ashore through the breakers?

In a very basic way, efforts to calculate the minimal amount of time for monkeys to drift across a massive, powerful ocean on a raft that are based on today's navigation charts of ocean current, wind directions, and wind speed, are unrealistic. As a motive force, the midocean environment is far more complicated than is implied by the arrow-straight map markers placed on standard charts to provide a macro view for ship navigators today. Oceanographers have demonstrated this complexity in studies that used telemetry to follow the travel paths of experimental buoys. It is highly unrealistic to imagine that

Eocene monkeys adrift on a raft could have accomplished anything comparable to what the adventurers Anthony Smith and Jean-Jacques Savin did with their high-tech, well-provisioned vessels.

Calculating the likelihood of the Transatlantic Scenario

The Transatlantic Scenario may have taken hold as a model of platyrrhine origins, but it still needs to be evaluated in an objective fashion, treating it as if it were a testable hypothesis rather than an untestable scenario. Biological anthropologist Mark Hubbe and I attempted to do this by looking at the question as an exercise in calculable probabilities relating to specific parameters pertaining to such a scenario. Hubbe's main research interests, how humans disperse across continents, dovetail perfectly with mine in this regard. We asked: What are the odds that the Transatlantic Scenario could have been successful? To do this, we listed a series of 20 steps, hurdles which the animals would have had to negotiate successfully, each and every time. Failure to get beyond any one stumbling block in the series would spell disaster, the end of the voyage, death in most cases (table 10.2).

An analogy to our exercise is trying to toss a coin 20 consecutive times in an effort to win by drawing 20 heads in a row. But our problem is more complicated than that, because coming up heads vs. tails is a 50/50 chance. Many of the individual challenges of the Transatlantic Scenario are much more likely to disfavor the monkeys than others. For example, the probability that the castaway group was not composed entirely of males but included at least one female seems likely. That would be favorable to the monkeys. But how do we rate the chance that no hurricane smashed the rafts to bits in the middle of the ocean? Hubbe and I thus developed a four-part measurement scale of likelihoods, and we assigned corresponding probability values of each category: 0.5 (likely), 0.1 (unlikely), 0.01 very unlikely, 0.001 (extremely unlikely). The numerical part of this system is arbitrary, but it is mathematically balanced and progressive because it applies orders of magnitude differences from one level of difficulty to another. The system is also arbitrary because many of our determinations about degree of difficulty were intuitive, not fact-based. They could not be otherwise. There is no relevant data. Nevertheless, our approach is a way of computing the odds of success or failure given the real-life situations and conditions implied by the Transatlantic Scenario, which is usually presented in the simplest terms I have used in describing it above: a bunch of monkeys got caught on a chunk of land that broke off from a riverbank in a storm, floated out to sea, and were carried across the ocean to land in South America and begin reproducing. End of story.

TABLE 10.2. The Transatlantic Scenario in sequential steps: one chance in 12.8 octillion (billion, billion, billions) of succeeding

Step	Events	Relative Likelihood
1	In a weather-related event like a storm or flood, near the mouth of a river in western Africa, a piece of land bearing natural vegetation, including one or more trees, breaks away from the riverbank, forming a substantial raft that floats toward the ocean.	Likely
2	Monkeys were in the tree(s) at the time and did not abandon the moving raft while it began to drift away.	Unlikely
3	They consisted of a small number of individuals, fewer than 10.	Likely
4	The original population of castaways was a mixed group including females and males.	Likely
5	The trees were of the right species, i.e., providing physically edible and nutritionally appropriate foods, essentially fruit, for that particular species of monkey.	Likely
6	The raft was discharged with enough force to be swept into the near-ocean.	Very Unlikely
7	Shoreline currents that normally run in the north-south axis did not capture the raft but sent it out to sea.	Unlikely
8	Currents sent the raft westward toward South America, with a direction and rate of travel that was essentially constant, i.e., they did not drift aimlessly during the voyage and were not delayed or re-routed by being caught in eddies that form in the midocean (and near shore).	Extremely Unlikely
9	At 40 million years ago, the raft would travel roughly 900 miles propelled constantly by a combination of currents and wind, taking 7–11 days to complete the crossing, based on Houle's analysis.	Extremely Unlikely
10	Weather conditions were favorable throughout the voyage, so the raft held together. There were no high winds, choppy seas, or powerful tall waves to damage the integrity of the raft or capsize it, no severe rainstorms to erode it, no hurricanes to destroy it.	Extremely Unlikely
11	Enough fruit remained edible for the duration of the voyage to provide adequate nutrition rather than desiccating under intense sun, being doused by saltwater spray, and becoming inedible because the tree roots were being saturated by saltwater.	Very Unlikely
12	Insects and other actual or potential food sources (leaves, flowers, bark, pith, etc.) associated with the vegetation were additional edible provisions if fruit became inedible or depleted.	Likely
13	The fruit provided adequate water to hydrate the animals while they were physiologically and psychologically challenged by the extreme physical conditions, compounded by exposure to sun, heat, and saltwater.	Very Unlikely
14	Despite being subject to prolonged, intense exposure to the elements, at least one female and one male survived, avoiding heatstroke, dehydration, starvation, injury, and the sheer terror of being adrift at sea. If there was only one survivor, it was a pregnant female who did not spontaneously abort due to the hazardous journey.	Extremely Unlikely

(continued)

TABLE 10.2. (continued)

Step	Events	Relative Likelihood
15	The raft traveled through deflecting shoreline currents that follow a north-south axis and any beachfront waves to approach a landing site	Unlikely
16	The landing site was at an environmentally favorable spot, not a barren coastline, located at the opposite end of the shortest linear distance between Africa and South America.	Unlikely
17	The monkeys safely disembarked the raft by swimming ashore.	Very Unlikely
18	The party survived the immediate challenges of finding food in a new habitat, warding off new diseases, parasites, and avoiding new predators.	Likely
19	Enough individuals survived long enough on land, or recovered from their hazardous journey, to eventually become reproductively active and form a viable population.	Likely
20	Over the first few generations the group did not succumb to inbreeding depression due to inherent genetic factors and small population size, despite being prone to extinction as an isolated, insular population.	Very Unlikely

Our thought experiment shows that the odds against a successful crossing of the Atlantic by rafting monkeys are astronomically high, in the ultra-high billions using our calculations, well beyond the numbers that most of us ever deal with. An exact figure is 1 in 12.8 octillions—a billion billion billions—against success. But since our intention was to rigorously explore the theoretical likelihood of this scenario in a realistic way and not to determine what they truly are in a definitive sense, which is impossible, we then looked at the problem as devil's advocates, to validate the essence of our conclusion by challenging our original determination. In an effort to favor the Transatlantic Scenario as much as possible under the conditions we outline, we relaxed the requirements to make success more probable by reducing the venture to the final 11 steps, accepting that the raft formed, the monkeys were on it, the craft made it to the open ocean intact, etc. Now, the odds against success were less, of course, but still unreasonably high: 1 in 106 billion. We ran other types of simulations, giving the monkeys as many as 1 million tries to get aboard and complete the voyage during a 5-million-year interval, a period we chose based on timing suggested by the fossil record, the molecular clock, and paleogeography. Under these conditions, the odds for our conservative scenario are 1 in 2.5 sextillion, that is, 1 in 2.5 million million million million million millions. To improve the monkey's odds, we would have to further manipulate all our assumptions. In the final analysis, it does not seem fathomable that the Transatlantic Scenario merits any serious consideration at all, unless one cooks the books radically to make it seem plausible.

Our point is that when it is broken down sensibly, the Transatlantic Scenario is an extraordinarily unlikely mechanism of dispersal to explain how pre-platyrrhine monkeys made their way to South America from Africa. Even if we toss out concerns about monkey health, weather, the ocean environment, etc., and set up the exercise as if no such constraints existed, the probabilities against success are staggering. If we go back to the coin toss model and rate each of the 20 steps as a 50/50 chance, the odds of passing all of the 20 hurdles in one try are nearly 1 in a million. Another way of saying this is that to be successful only once if everything went their way, nature would have had to launch 1 million oceanworthy, monkey-laden rafts so that a single one would get through and yield the founders.

Is there any empirical evidence that such a thing ever happens, that natural rafts cross the ocean all on their own? The answer seems to be no. Modern, comprehensive studies have examined thousands of records in the literature pertaining to the phenomenon of floating islands, historically and in all their varieties, when they occur in lakes, ponds, rivers, and at sea, with and without animals stuck on them. There are no known examples, credible observations, of any floating islands occurring far enough offshore to give a full-ocean voyage credence. There is, however, a long written history revealing the evolution of this idea—both the raft and the model of dispersal—from myth to traveler's tale to tabloid-like media reports, to unsourced validation and repetition in the scientific literature of the 19th and 20th centuries, and today.

The Transatlantic Scenario is a far-fetched idea built upon an extensive, intricate set of flimsy, implausible assumptions. As with any scenario, because there is no way to test it with hard data, the only way to evaluate its importance for platyrrhine origins is to judge its reasonableness, based on its own merits and with respect to a competing scenario. In this light, it seems much less valuable than the Americas Scenario, which does not introduce a galaxy of unnatural life-threatening circumstances, and unnaturally good luck, along the way to South America.

CHAPTER 11

AFTER 20 MILLION YEARS OF EXISTENCE, NEW WORLD MONKEYS FACE EXTINCTION

Many New World monkeys living today have existed in largely the same way for more than 10 million years, some for 20 million. Yet, of the 16 genera of extant Neotropical monkeys at the heart of this book, 14 include species that are now threatened, classified by the International Union for Conservation of Nature (IUCN) on its Red List as Critically Endangered, Endangered, or Vulnerable. Two platyrrhine conservation success stories exemplify both the threats and solutions. They are the stories of the Muriqui, *Brachyteles*, and the Lion Marmoset, also known as the Lion Tamarin, *Leontopithecus*. If these two genera were to be lost, it would literally obliterate two entire lineages of 10–15 million years duration in nature. The efforts to save them are discussed in this chapter.

The IUCN Red List, which is updated twice each year, provides authoritative assessments of the risk of extinction to living species worldwide. Extinction risk is based on population size, decline, and health (the effective population size measured as the known or estimated number of reproductive individuals), the extent of their geographic distribution, the fragmentation, degradation and loss of their habitats, and the threats they are under.

Every two years the IUCN Species Survival Commission's Primate Specialist Group publishes *Primates in Peril*, a list of the 25 most vulnerable species. There were six species belonging to four New World monkey genera listed in the 2018–2020 report, one each among Tamarins, *Saguinus*, Marmosets, *Callithrix*, Titi Monkeys, *Callicebus*, Capuchin Monkeys, *Cebus*, Howler Monkeys, *Alouatta*, and Spider Monkeys, *Ateles*. The purpose of *Primates in Peril* is "to highlight those primates most at risk, to attract the attention of the public, to stimulate national governments to do more, and especially to find the resources to implement desperately needed conservation measures," said Russell A. Mittermeier in the 2008–2010 report. The Chair of the Primate Specialist Group, Mittermeier is the driving force behind the list and the central figure behind the primate conservation movement worldwide. He is Chief Conservation Officer of the Texas-based NGO (nongovernmental organization) Global Wildlife Conservation and the author and coauthor of more than 35 books and 700 papers and articles.

Not only species, but entire evolutionary streams are in peril

In preceding chapters we have looked at the evolutionary radiation of modern New World monkeys in detail, seen how they exist in their niches, discussed their varied adaptations in terms of ecological and anatomical solutions to feeding, locomotion, and social arrangements, and have traced their lineages through the fossil record where possible. A 2015 review article by leaders in the field of platyrrhine molecular phylogeny, Horacio Schneider and Iracilda Sampaio, indicates that 11 of the 16 genera examined in this volume have long lineages between approximately 9–10 million and 20 million years: *Leontopithecus, Saguinus, Cebus, Saimiri, Aotus, Callicebus, Pithecia, Alouatta, Lagothrix, Brachyteles, Ateles*. There are unique fossil genera linked closely with five of these: *Dolichocebus* and *Neosaimiri* with *Saimiri*; *Tremacebus* with *Aotus*; *Miocallicebus* with *Callicebus*; *Stirtonia* with *Alouatta*. The fossil *Mohanamico* is a tentative link to the extant *Callimico*, a lineage with a molecular age greater than 9–10 million years. There are other fossils as old and older that are congeneric with living genera, distinguished only at the species level: *Cebus, Cebuella*, and *Aotus*. This is an impressive list of longevity that is demonstrated by two independently evolving systems, anatomical and molecular, and it is likely that paleontology will reveal more evidence of genus-level continuity. The molecules and the fossils testify that the radiation of the extant New World monkeys is characterized by an evolutionary pattern defined by a preponderance of long-lived, ecophylogenetic lineages.

A lineage, as has been explained, is an evolutionary stream carried within DNA that is embodied in a species and manifested as a lifestyle. A lineage comes into existence as populations split from a common ancestor to form two branches of evolving species. Over time, each evolves its own distinctive set of ecological characteristics that distinguish both lines as genera, taxonomically distinct from others. What produces an adaptive radiation is the successive repetition of this process, the splitting and accumulating of features, in multiple lines of evolution. This is how the three main branches of platyrrhine families came into existence. A distinctive pattern of the platyrrhine radiation not evident among other major primate groups like the Old World monkeys and apes which have also radiated adaptively, is the multimillion-year longevity, and little change, of genus-level lineages, as is clear from the fossil record.

Taking this macroscopic view of the lineage concept and applying it to the genus level means that the danger of extinction leads to consequences beyond its impact on species, because it is at the genus level that the crucial, enduring ecological adaptations of a radiation unfold and are realized. That is why the odyssey of New World monkeys can be told as the evolution of platyrrhine genera. The ecological significance of the genus as a principal structural component that gives shape to platyrrhine communities is demonstrated by a simple fact: It is rare for a community of primates to include congeneric sympatric species, two species of a genus inhabiting the same area. It is rare for any habitat to be able to support two monkeys of "the same kind," with the lifestyles of congeners being distinguished only at the margins because species of the same genus are fundamentally the same. When it comes to reconstructing evolutionary history, it is also rare that fossils exist or will be found to enable us to trace the actual, particular line of descent for each species as opposed to each genus. Even if this were not the case, when fossils have the potential of being identified as ancestors of living species, there are methodological issues that make it difficult to recognize them. What we can do is recognize whether a fossil belongs to a living genus or genus-level lineage based on its sharing of critical lifestyle characteristics.

Paleontology can distinguish lineages at the genus level if we are fortunate enough to find the right fossils. It can tell us how old a genus is and what its ecological situation has been. Conservation literature and laws, internationally, are predicated on the species level, but this often does not give a realistic representation of the potential loss of an entire lineage. The risk status of only two platyrrhine genera, the Pygmy Marmoset, *Cebuella*, and Goeldi's Monkey, *Callimico*, are fully covered by these species-focused laws. Nearly all scientists familiar with the animals consider each as consisting of only a single species, both now in a perilous state and formally listed as Vulnerable. Such is also the case if a genus contains more than one species and all of them are threatened, as with *Leontopithecus* and *Brachyteles* according to some taxonomic interpretations. These two genera, both endemic to the Atlantic Forest, are at risk of having their lineages completely eradicated. And, it must be emphasized, it is not due to natural causes, but rather, because of human interference.

Therefore, the threats to extinction that platyrrhines face are not only to the loss of species that exist in the here and now, but also to entire lineages that have existed for millions of years. This adds historical and evolutionary dimensions to the threat of extinction and makes the prospect even more dire. Looking at the New World monkeys in this way, as a collection of lineages as well as species, significantly alters the effects of extinction. It means the extirpation of evolutionary lines that have successfully existed in the same harmony of nature for eons without interruption.

What is happening currently that has the potential to wipe out entire lineages of platyrrhines is a consequence of the destruction of the habitats in which they have lived. This is more than an impression: we have data showing why is this happening. A major 2017 article published in the journal *Science Advances*, titled "Impending extinction crisis of the world's primates: Why primates matter," analyzed years of facts and figures collected worldwide and predicted future trends. The authors, Alejandro Estrada and a large international cadre of colleagues, summarized the situation this way: "Unsustainable human activities are now the major force driving primate species to extinction. . . . [It is] the result of escalating anthropogenic pressures on primates and their habitats—mainly global and local market demands, leading to extensive habitat loss through the expansion of industrial agriculture, large-scale cattle ranching, logging, oil and gas drilling, mining, dam building, and the construction of new road networks in primate range regions. Other important drivers are increased bushmeat hunting and the illegal trade of primates as pets and body parts." The authors go on to assert that "Despite the impending extinction facing many of the world's primates, we remain adamant that primate conservation is not yet a lost cause, and we are optimistic that the environmental and anthropogenic pressures leading to population declines can still be reversed."

Evolution is the process that situates species in nature and keeps them alive. The pressure on all species, not only primates, to adjust to the rapid alteration and eradication of their preferred habitats, is extraordinary, and it has led to enormous global reductions in population sizes, especially since the mid-20th century. The speed with which vast tracts of the landscape that New World monkeys inhabit has been modified, destroyed, and eliminated has made it impossible for the natural processes of evolution and adaptation that produced the radiation to continue as it has for tens of millions of years. The food and shelter they need are in the trees, which continue to be cut down.

The Atlantic Forest, a biodiversity hotspot, is being decimated

Charles Darwin, during his history-making voyage around the globe, recorded in his diary on February 29, 1832, his first awed feelings of the grandeur of the Atlantic Forest: "To a person fond of natural history such a day as this brings with it pleasure more acute than he ever may again experience." Today, Brazil's

Atlantic Forest is called a biodiversity hotspot. It is one of the most prolific regions on earth in measures of biodiversity, containing one of the world's densest collections of endemic land-based plants and animals, found nowhere else on the planet, and is one of the most endangered tropical and subtropical forests in the world, having lost almost all its territory to development.

Scholars generally agree on the original size of the Atlantic Forest and its ecological status, as described by Warren Dean in his comprehensive history of its destruction, published in 1995:

> On the eastern margin of South America there once stretched an immense forest, or more accurately, a complex of forest types, generally broad-leaved, rain loving, and tropical to subtropical stretching from about 8 degrees to about 28 degrees south latitude and extending inland from the coast about 100 kilometers in the north, widening to more than 500 kilometers in the south. Altogether the forest covered about a million square kilometers. This complex has been referred to as the Brazilian Atlantic Forest, related to the much larger Amazon Forest but distinct from it. Together, these two great forests formed a life zone distinct from and richer in species than those of other tropics of our globe.

Estimates vary about how very little of the Atlantic Forest remains today— from about 8% to 15%. The remnants of the forest are in the form of fragments separated from one another. While the pace of its destruction has been accelerating rapidly for at least a half-century, it actually began in prehistoric times. The successive waves of human invaders of the Atlantic Forest started with the indigenous hunters and hunter-gatherers of 13,000 to 11,000 years ago. They were followed by the farmers of 3,900 years ago, and then by the arrival of the Portuguese mariners more than 500 years ago, in 1500. The Europeans colonized the territory in order to exploit its natural resources, as one main goal. Regarding what ensued, Dean stated that "No restraint was observed during this half-millennium of gluttony, even though, almost from the beginning, solemn interdictions were intoned intermittently and, in latter days, continually and frantically."

The Atlantic Forest's primate inhabitants include 6 of the 16 genera profiled in this book. One can grasp the tenuous circumstances facing them in the account given by primatologist and conservationist Anthony Rylands. He has published widely on New World primates and is, as of this writing, Primate Conservation Director at Global Wildlife Conservation. Rylands and colleagues, in 2002, said of the Atlantic Forest, "Endemism is especially marked in the primates: 21 of the 24 species and subspecies occur nowhere else. Eighteen of them are threatened (9 Critically Endangered and a further three Endangered)." Six of these 18 threatened species constitute all the populations of *Leontopithecus* and *Brachyteles*. This means the entire lineage of each of these genera is on the verge of extinction.

Conservation efforts

Golden Lion Tamarin Project and Muriqui Project of Caratinga

The populations of Golden Lion Tamarins (GLTs), the species *Leontopithecus rosalia*, and Northern Muriquis, the species now often classified as *Brachyteles hypoxanthus*, that existed in the mid-20th century were already very small. Not only was the future viability of the remaining animals in question, but the future of the remnants of the last patches of Atlantic Forest where they live was also a grave concern. The decades-long efforts to rescue these monkeys from extinction serve as conservation success stories today. Their lineages have continued to survive. Once classified among the 25 most endangered primates in the world, they do not appear in the *Primates in Peril* list right now.

Brazil's pioneering primatologist and conservationist was Adelmar F. Coimbra-Filho (1924–2016), to whom generations of primatologists in the United States and other countries around the globe are deeply indebted for his leadership, vision, advocacy, support, and lifetime of commitment to the cause. In 1972, he and his young protégé, Harvard student Russ Mittermeier, raised the alarm about GLTs at an international conference. *Saving the Lion Marmoset*, a compilation of papers given at the meeting, was published that year. The conservation ethos among scientists interested in primates began to coalesce and gain momentum. Coimbra-Filho and Mittermeier soon began to spearhead projects that sought to map out and verify the distribution of New World primates at that time where remnants of Atlantic Forest remained. They recruited researchers to study some of the animals, about which almost nothing was then known, with a view, also, to conservation. The two priority targets were the Golden Lion Tamarins and the Muriquis. In 1972, only 168 GLTs existed in zoos and a few hundred were estimated to live in the wild. They were almost gone, and a commitment was made to save them. As more forest fragments were lost, the numbers of wild muriquis were also dwindling during the 1960s and 1970s when very rough estimates put their total number at about 1,000–3,000. Surveys conducted in isolated forest fragments in 1982 concluded there were fewer than 500 muriquis. There were no breeding groups in captivity.

Conservation projects to preserve the GLT and Muriqui were started in earnest at about the same time in the early 1980s, and they continue as I write,

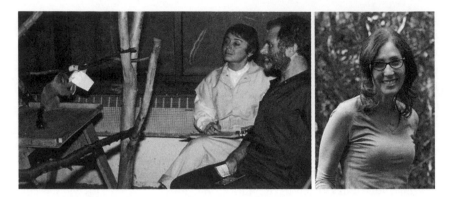

FIG. 11.1. Left, Drs. Devra Kleiman and Benjamin Beck during a Gold Lion Tamarin pre-release training session at the National Zoological Park, Smithsonian Institution, and, right, Dr. Karen Strier at Fazenda Montes Claros. Images courtesy of Benjamin Beck (Smithsonian's National Zoological Park Archives) and Karen Strier.

in 2019. They are formally known as the Associação Mico-Leão-Dourado and the Muriqui Project of Caratinga (Projeto Muriqui de Caratinga). The GLT project, known initially as the GLTCP (Golden Lion Tamarin Conservation Program), was created by Devra G. Kleiman (1942–2010), a research scientist at the Smithsonian's National Zoological Park in Washington, DC. The Muriqui project originated with the first long-term field study of the genus begun in 1983 by Harvard graduate student Karen Strier at the urging of Coimbra-Filho and Mittermeier. Because of the major natural differences between these two genera in their biology, lifestyle, habitat requirements, and ecological niche, the approaches to these conservation efforts differed from the outset. It was believed that the way to conserve the GLT was to put more animals into the forest from the populations that were finally being successfully bred in zoos around the world thanks to the insight of Kleiman, who perceptively removed the unintended, institutionalized social barriers that had inhibited reproduction in captive groups. As for as the extremely rare Muriqui, which very few scientists had even sighted at that time, the strategy was to save the animals by protecting their habitat, remnants of the Atlantic Forest, beginning with Fazenda Montes Claros in the state of Minas Gerais.

Two principals at each of the projects, Benjamin Beck, who managed the reintroduction of GLTs, and Karen Strier, who has continued to lead Muriqui research and conservation, related their perspectives and experiences of 35 years and shared in personal communications in 2018, as reflected here (fig. 11.1). In 1982, Beck moved to the National Zoological Park, and he helped found the GLTCP in the following year with Kleiman and associates. Now, the program is managed by the Associação Mico-Leão-Dourado, a Brazilian NGO created in 1996, currently directed by Luís Paulo Ferraz, and supported in part by Save the GLT, a United States NGO. Beck is the author of more than 80 scientific papers, has published several books, and is a leading authority on animal cognition, reintroduction as a conservation tool, and biodiversity conservation.

Strier, the Vilas Research Professor of Anthropology at University of Wisconsin–Madison, has written and coauthored four books and five edited volumes and has published more than 120 peer-reviewed articles and chapters. She directs the Muriqui Project of Caratinga in collaboration with Sérgio L. Mendes, a leading Brazilian conservationist and currently the director of the Brazilian government's National Atlantic Forest Institute (INMA). Mendes, in 2002, initiated a parallel program to study and protect the highly fragmented and very small populations of Muriqui in the state of Espírito Santo. Strier and Mendes have known each other since 1983. Their joint project at Fazenda Montes Claros is noted for training Brazilian students; more than 60 students and postdocs have participated since its inception. The project maintains strong collaborations with other Brazilian scientists and conservation NGOs, including Preserve Muriqui, the NGO that administers the reserve in Caratinga.

To understand the different challenges, long and short term, faced by these projects, it is important to remember that the GLTs reproduce rapidly. Each social group has litters of twins every year, and females start reproducing at about 3.5 years old. The muriquis, on the other hand, are slow breeders and give birth to singletons. Females produce their first infants at 9–10 years, and reproduce at 3-year intervals throughout their unusually long life spans. The point is that to increase population size as a means of quickly averting extinction, the GLTs will do that on their own, given a controlled environment, a wildlife sanctuary where they are protected. The first protected area established was at the Poço das Antas Biological Reserve, selected by Coimbra-Filho not far from Rio de Janeiro, where there would be ample space for the population to grow. Of course, the animals slated for reintroduction, which had been born and raised in captivity, had to be trained to survive in the wild. The training began before they were to be transported to the forest from which their ancestors had been taken. The hope was that groups of zoo-born animals could adjust to the wild and would eventually merge with the small population of native GLTs on the reserve to increase and invigorate their numbers. Adding genetic diversity to the wild population, and guarding against inbreeding effects that may have arisen in zoos, was also an objective of the GLTPC. The project began to manage the breeding of captive GLTs among the dozens of cooperating institutions, worldwide, that became involved and volunteered to help restock the reserve with their monkeys. The animals selected to form the reintroduced breeding groups were carefully chosen to maximize genetic heterogeneity.

A major monitoring operation was set up. Researchers knew that the best way to recognize individual GLTs and collect data on behavior and ecology in the wild, including the survival of previously caged monkeys, was to anesthetize them and either fit individuals with radio collars or dye their tails with distinguishing markings. Both methods were used in the project. For this and other reasons, the GLT project was inherently hands-on. It fundamentally involved transporting animals internationally and a program to train them to adjust to the wild by teaching those to be reintroduced to forage for food and avoid predators, then following all the Poço das Antas animals throughout their lives to see how they fared.

The aim of the GLT project was to establish a wild population of at least 2,000 animals living in at least 25,000 hectares, about 100 square miles, of protected and connected Atlantic Forest habitat, often forest fragments, and to have a healthy back-up population living in curated collections in zoos. How did the GLTCP seek to achieve this goal? Initially, according to Beck, in addition to documenting the behavioral ecology of the existing population of wild GLTs, by reintroduction of zoo-born GLTs and translocation of jeopardized wild GLTs, public education programs were established and enhanced enforcement of existing Brazilian wildlife protection laws were promoted. Then, more recently, by surveys of wild GLT populations over a wider geographic range, monitoring the wild GLT population, habitat restoration, use of satellite imagery to detect and repair critical gaps in forest connectivity, international fundraising to acquire critical habitat, landowner partnerships, comprehensive, results-based strategic planning, and increased involvement and support of the international zoo community.

Between 1984 and 2000, Beck and his colleagues reintroduced 146 captive-born GLTs from 43 zoos and laboratories, from eight countries on three continents, making this the most collaborative and complex conservation program in history. In a critically important finding, Beck and his all-Brazilian field teams discovered that the reintroductions of GLTs that were trained before their release were unsuccessful; however, they found that supporting and protecting the tamarins after their release, keeping them alive until they could reproduce in the wild, was the key to success. Short-terms gains were not the answer. Successful conservation means playing the long game, in generational time. Close daily monitoring of the monkeys showed that the survivorship of infants born in the wild to captive-born, reintroduced parents was indistinguishable from those born to wild tamarins. The descendants of reintroduced parents now make up about a third of the total wild GLT population.

When Beck and his colleagues started working in Brazil it was estimated there were 150 native GLTs surviving in the wild. Now, there are approximately 3,000. Most significantly, the species has been downgraded from Critically Endangered to Endangered. There is increased participation of private landowners and increased public awareness, which is important since only 21% of the range inhabited by GLTs is legally protected. There is also an increased emphasis on reforestation since much of the range comprises forest islands. Efforts are underway to connect islands where possible so that the GLTs can move more freely, which would advantage genetic diversity. One such example is a bridge planted with trees that is being completed as this book goes to press, built over a highway to connect divided forests and GLT populations.

Despite the successes of the GLTCP and the Associação Mico-Leão-Dourado, the battle to save *Leontopithecus rosalia* will never end in a conclusive victory, Beck acknowledged. Still, one of the broader lessons learned is how the GLT effort can be replicated, its generality, said Maria Cecília Martins Kierulff, one of the leading researchers and conservationists of the GLT project, the first scientist to survey the entire range of the species and to translocate threatened groups to safe forests. "The experience acquired and technology used to save the GLT from extinction can be applied to other species that need similar efforts. Every day the list of threatened species around the world increases, and for the majority of these species, the causes of population decline are similar to those of the GLT: fragmentation and small population size because of habitat destruction, degradation, and overhunting." Kierulff's message about the importance of technology for the future of the GLT holds an ironic twist, because in the absence of technology the genus *Leontopithecus* has been surviving for nearly 13 million years.

The Muriqui Project of Caratinga, dedicated to non-invasive research and conservation of the Muriqui and its habitat, was, from the beginning, decidedly hands-off in contrast to the GLTCP. Strier said in the 2017 book *Muriqui: Kings of the Forest*, "I knew that their distinct facial markings would facilitate my ability to recognize them as individuals, and thus to gain understanding of each of their unique personalities and behavioral peculiarities. Individual recognition is essential to the long-term monitoring of life histories and extended kinship relationships by observational methods alone. I was committed from the outset to study the muriquis in this observational, non-invasive way. This meant that I would never capture them to mark them in order to track them or to keep their identities straight. Instead, I would eventually give each one of them a name." It is because she could recognize them this way and has always had students in the field directing this study that Strier has come to know these monkeys for their entire lifetimes, some for generations. As she said, "Some of the original individuals present in the Matão group at the outset of the study are now in their 40s. There are grandmothers and great grandmothers still living." Though a boon to researchers, being individually recognizable comes with great cost to the species. One of the reasons this is possible is due to their large size. They are the biggest arboreal mammals in the Atlantic Forest, which has made muriquis a favored and easy target for hunters.

According to surveys carried out by Brazilian teams in 2018 in refugial fragments of the decimated Atlantic Forest, the Northern Muriqui remains one of the most critically endangered primates on earth, with fewer than 1,000 individuals known to survive in 15 localities. The Southern Muriqui consists of only 1,200 individuals found in 20 localities. The fragments where they now live are estimated to comprise only 8%–13% of their historical range. According to Strier, both northern and southern Muriqui are a major flagship for conservation efforts on behalf of the threatened ecosystem of the Atlantic Forest.

One of the last remaining strongholds for muriquis is the privately owned, federally protected reserve, known as the RPPN (Reserva Particular do Patrimônio Natural) Feliciano Miguel Abdala, the forest of Fazenda Montes Claros located near the city of Caratinga in the state of Minas Gerais. It is Strier's belief that the success of this population is essential to the survival of the northern species. Although only about 1,000 hectares in size, about 4 square miles, this forest had supported some 350 individuals until the droughts in 2014–2015 and the yellow fever epidemic of 2016–2017 took their tolls. Based on her experience, and the evidence she and her colleagues have collected over 35 years, Strier says, "I am very optimistic about population resilience. The muriqui population had been increasing steadily, so that even with losses from drought and yellow fever, it is still large enough to be able to recover." And, it should be added, the survival of the Northern and Southern Muriqui will prevent the disappearance of a lineage that is about 9 million years old.

Strier described the Muriqui as among the most peaceful primates known. They live in uniquely egalitarian societies in which males remain in their natal groups and females move between groups around the onset of puberty. Since the start of the project, the population at Caratinga has increased greatly in size, and ongoing research is documenting the influence of fluctuating demographic factors, such as infant sex-ratios and survivorship, on their behavioral and ecological adaptations. Strier knew at the start of the project that these monkeys were Critically Endangered. What she also knew, she has said, was that the same kinds of data that would permit her to answer scientific questions would also be useful for developing informed conservation and management plans. The prospect of pursuing scientific research that would have direct applications to conservation was a major motivating factor for her from the start and has remained a sustaining force. In her 1992 book, *Faces in the Forest*, Strier says, "We have learned more about the muriqui in the last 10 years than we had since their scientific discovery in 1806, and there is still a great deal more that they can teach us."

All that is being lost can never be recovered

As of this writing the GLT and Muriqui projects are model conservation success stories. The increases in population sizes that have been mentioned attest to that. Another consequence of these projects has been a phenomenal benefit to science due to the proliferation of research and studies conducted in response. We knew almost nothing about them in 1972 when the clarion call to conserve Brazilian primates was first truly heard by the international community, and now we have a mature, robust body of work that sustains them and contributes to scientific understanding of these and other platyrrhines, and their place in nature.

The research and conservation efforts at this point have allowed for the continuing evolutionary journey of the Atlantic Forest's endangered, endemic primates. The extinction of *Leontopithecus* and *Brachyteles* would be not only the loss of two beautiful, majestic animals, it would also be, in each case, the eradication of a long line of evolution, a lineage, that can never be reconstituted. The conservation projects featured in this chapter address two populations of *Leontopithecus* and *Brachyteles*. There are several others belonging to these genera, but not many. Science is still not able—and it may never be—to confidently discern whether these populations represent real species, or what taxonomists have always called subspecies. Yet, it really does not matter. In the end, the conservation message is about the genera *Leontopithecus* and *Brachyteles*, because the threats to the one forest biome in which these monkeys live have not abated in 500 years and they continue to accelerate rapidly. Only a few, tiny fragments of forest remain that are suitable to give the genera of Lion Tamarins and Muriquis a viable lifeline.

It is important to remember that it is always about the primate community, not just the species. If, in fact, 2 of the 16 genera that make up the radiation of modern platyrrhines are ever eliminated, one of the great tragedies will be that this loss will forever diminish our future capacity for the scientific understanding of the evolutionary odyssey of New World monkeys and the ecological context that shapes it. Compounding the profoundly disturbing thought of losing New World monkeys is the knowledge that many have been living their varied lives harmoniously in their niches in South American forests, basically in the same ways, for 20 or more millions of years. And today as I write in the summer of 2019, the Amazon rainforest is ablaze in fires intentionally set without regard to nature, on a scale not seen for decades, an ecological firestorm of local and global consequences that will forever eliminate species of all kinds, many that have been there for the ages and many that had yet to be discovered.

ACKNOWLEDGMENTS

A profound sense of gratitude is reserved for Anthony B. Rylands. Anthony read every line of this book with his sharp eyes, keen intellect, and sweeping knowledge of the platyrrhines, and made many corrections—the remaining errors are mine—of scientific substance while not drawing attention to our differences of opinion. His buoyant enthusiasm for this project was an important source of sustenance during rewrites.

Ben Beck, Karen Strier, and Anthony Rylands provided significant assistance in developing the final chapter on conservation, with its focus on the model projects dedicated to saving the Golden Lion Marmosets (Tamarins to some) and the Muriquis of the Atlantic Forest in Brazil. I thank them for allowing me to paraphrase their written answers to my questions about their work, and for their suggested revisions. Similarly, Mark Hubbe was instrumental in building the chapter on platyrrhine origins and biogeography, which was based on a research paper that we produced as I was writing it.

While I was a graduate student, three individuals each posed a single question or comment to me that became turning points that set the direction of my research career: Eric Delson, Fredrick Szalay, and Richard Thorington Jr. I thank them for those remarks which, I am certain, were fleeting thoughts at the time, yet profoundly significant to me eventually. Before them, I was fortunate to have Warren Kinzey guide me toward a life in academia and an interest in New World monkeys. To my stalwart colleagues and friends, both platyrrhinophiles, Marilyn Norconk on the ecology and behavior side of things and John Fleagle on the morphology and paleontology sides, I owe decades of thanks for illuminating discussion, encouragement, and camaraderie. My long time coworker and friend, Marcelo Tejedor, warrants special thanks of still another order.

Without the American Museum of Natural History being what it is, one of the finest research institutions in the world and a crucial seat of learning for me, none of this would have been possible. More recently, for the generosity and collaboration in opening doors for me and my students, and for a project that revolutionized fossil collecting in the Caribbean by proving the underwater salvage method, I thank Renato Rimoli and Phillip Lehman. Mark Bowler, Marilyn Norconk, Andrea Martins, Carla B. Possamai, and Natasha Bartolotta/

Owl Monkey Project, Formosa–Argentina, allowed me to use their stunning photographs of monkeys for the plates. Luciano Candisani provided his gorgeous muriqui image for the jacket. Thank you one and all. And Stephen Nash contributed his brilliant artistry to the cladogram portraits in chapter 2. Thank you, Stephen.

I greatly appreciate the support and generosity of Alison Kalett and excellent work of the team at Princeton University Press, including Abigail Johnson, Ellen Foos, Dimitri Karetnikov, and Lucinda Treadwell.

More than anyone else known to the field, Eric Delson, a dear friend for nearly 50 years, has supported me intellectually, professionally, and personally. Writing this book is the best token of thanks I can think of to offer him.

GLOSSARY OF TERMS

adaptation—a process or state of being resulting from evolution, involving a feature considered beneficial to the organisms' lives in nature

adaptive radiation—the array of lineages that descended from a common origin into a variety of adaptive patterns

allometry—how the proportions of features change as body size increases or decreases

anthropoid—the group of primates including monkeys, apes, and humans

catarrhines—Old World monkeys, apes, and humans

character analysis—the analytical approach by which homologous characteristics are distinguished from analogous characteristics, and primitive features are distinguished from derived features

character displacement—the evolutionary exaggeration of traits distinguishing closely related species that occurs when their geographic distributions overlap

clade, cladistics, cladogram—referring to phylogenetic relationships as branches on the metaphorical Tree of Life, analysis of sister group relationships as an element of phylogeny, and the diagram that depicts sister group relationships

derived character—a trait that evolved or changed over time from its original state in that group

dental formula—the system that counts the mammals' four tooth groups, incisors, canines, premolars, and molars

dichromatism—color differences in the coats (pelage) of individuals in a population

differential reproductive success—natural differences in the reproductive output of individuals in a population

ecological release—the burst of phylogenetic and adaptive success that a species experiences when introduced to a new favorable environment

ecomorphology—the relationships between a trait's morphology and its biological role with regard to ecological circumstances

Ecophylogenetic Hypothesis—the proposition that the evolution of platyrrhine diversity has been structured by the evolutionary interaction of ecology and phylogeny within and between taxonomic groups

endemic—living exclusively in one region

faunivore, faunivorous—referring to a species that eats animal prey

folivore, folivorous—referring to a species that eats leaves

frugivore, frugivorous—referring to a species that eats fruit

genus, pl. **genera**—the first taxonomic category above the species level in the Linnaean hierarchy

grade, gradistic—the idea from the Middle Ages that life can be arranged from lower to higher levels of development or evolution

haplorhine—the group of primates including tarsiers, monkeys, apes, and humans

homologous—correspondence in genetic, anatomical, or behavioral traits that occur because the characteristics were inherited from a shared ancestor

insectivore, insectivorous—referring to a species that eats insects

lineage—a genus-level line of phylogenetic descent

Long-Lineage Hypothesis—the proposition that a preponderance of extant platyrrhine genera and clades have been in existence for many millions of years with little change

monomorphic, monomorphism—adult males and females having the same size and shape, in reference to appearance and morphology

monotypic—a taxon consisting of one species

monophyly—descent from a unique common ancestor

morphocline—features found among taxa that are arranged into a series that may reflect the sequence by which the traits evolved

morphology—the study of form in all its aspects

opposable thumb, opposable hallux—referring to the ability to press the thumb and index finger together, or the large toe relative to the sole of the foot, to grasp an object

parallelism—similarities that evolved in relatively closely related forms after they split from their last common ancestor

pelage—the hair or fur covering the body of an animal

phenology—the study of plant cycles with reference to seasonal changes, especially concerning periodicity of leaf and fruit patterns

phylogeny—the genealogical evolutionary history of organisms

platyrrhines—New World monkeys of Central and South America

preadaptation—a feature whose morphology and function set the stage for a later evolutionary, adaptive change

prehensile tail—a tail that is capable of grasping

primitive character—a feature that was present in the last common ancestor of a taxon and remained unchanged

prosimian—an informal term covering living and extinct strepsirhines and tarsiers

sexual dimorphism—difference in the structure and appearance of the sexes

species—a group of interbreeding populations that are reproductively isolated from other such groups in nature

strepsirhine—the group of primates including lemurs, lorises, and galagos

subspecies—the taxonomic category immediately below species in the Linnaean hierarchy

sympatric—living together in the same habitat

tail-twining—twisting tails together in a spiral-like fashion

taxon, pl. **taxa**—a group that has been given a formal name in classification

taxonomy—the arrangement of taxa in a classification

RECOMMENDED READING

Campbell, C. J., Fuentes, A., MacKinnon, K. C., Bearder, S. K., Stumpf, R. M. (Eds.). 2011. *Primates in Perspective*. Oxford: Oxford University Press.

Ciochon, R. L., Chiarelli, A. B. (Eds.). *Evolutionary Biology of the New World Monkeys and Continental Drift*. Boston: Springer.

Fleagle, J. G. 2013. *Primate Adaptation and Evolution*. New York: Academic Press.

Fragaszy, D. M., Visalberghi, E., Fedigan, L. M. 2004. *The Complete Capuchin: The Biology of the Genus Cebus*. Cambridge: Cambridge University Press.

Garber, P. A., Estrada, A., Bicca-Marques, J. C., Heymann, E. W., Strier, K. B. (Eds.). 2009. *South American Primates: Comparative Perspectives in the Study of Behavior, Ecology, and Conservation*. New York: Springer.

Hershkovitz, P. 1977. *Living New World Monkeys (Platyrrhini): With an Introduction to Primates*. Chicago: University of Chicago Press.

Kleiman, D., Rylands, A. B. (Eds.). 2002. *Lion Tamarins: Biology and Conservation*. Washington, DC: Smithsonian Institution Press.

Mittermeier, R. A., Rylands, A. B. & Wilson, D. E. (Eds.). 2013. *Handbook of the Mammals of the World. Volume 3. Primates*. Barcelona: Lynx Edicions.

Norconk, M.A., Rosenberger, A.L., Garber, P.A (Eds.). 1996. *Adaptive Radiations of Neotropical Primates*. New York: Plenum.

Rosenberger, A.L., Tejedor, M.F., Cooke, S.B., Halenar, L.B., Pekkar, S. 2009. Platyrrhine ecophylogenetics, past and present. In Garber, P.A., Estrada, A., Bicca-Marques, J.C., Heymann, E.W., Strier, K.B. (Eds.), 2009. *South American Primates: Comparative Perspectives in the Study of Behavior, Ecology, and Conservation*. New York: Springer, pp. 69–112.

Rylands, A.B. (Ed.). 1993. *Marmosets and Tamarins: Systematics, Behaviour, and Ecology*. Oxford: Oxford University Press.

Strier, K.B., 1999. *Faces in the Forest: The Endangered Muriqui Monkeys of Brazil*. Cambridge, MA: Harvard University Press.

Szalay, F.S., Delson, E. 1979. *Evolutionary History of the Primates*. New York: Academic Press.

Terborgh, J. 1983. *Five New World Primates*. Princeton: Princeton University Press.

Veiga, L. M., Barnett, A. A., Ferrari, S. F., Norconk M. A. (Eds.). 2013. *Evolutionary Biology and Conservation of Titis, Sakis and Uacaris*. Cambridge: Cambridge University Press.

REFERENCES

PREFACE

Blainville, H.-M. D. 1839–1864. *Ostéographie, ou, Description iconographique comparée du squelette et du système dentaire des Mammifères récents et fossiles: pour servir de base à la zoologie et à la géologie. Atlas—Tome Premier. Primates.* Paris: J.B. Baillière et fils.

Clark, W. E. Le Gros. 1959. *The Antecedents of Man: An Introduction to the Evolution of the Primates.* Edinburgh: University Press.

Darwin, C. R. 1859. *On the Origin of Species by Means of Natural Selection, or the Preservation of Favoured Races in the Struggle for Life.* London: John Murray.

Linné, Carl von. 1758. *Systema Naturae per Regna Tria Naturae: Secundum Classes, Ordines, Genera, Species, Cum Characteribus, Differentiis, Synonymis, Locis.* Vol. 1. Holmiae: Impensis Direct. Laurentii Salvii.

Rosenberger, A. L. 1979. *Phylogeny, Evolution and Classification of New World Monkeys (Platyrrhini, Primates).* Ann Arbor: University Microfilms.

Szalay, F. S., Delson, E. 1979. *Evolutionary History of the Primates.* New York: Academic Press.

CHAPTER 1. What Is a New World Monkey?

Ameghino, F. 1894. Enumération synoptique des espères de mammiféres fossiles des formations èocénes de Patagonie. *Boletin de la Academia Nacional de Ciencias en Córdoba* 13: 259–445.

Anonymous. 2019. What is CITES? https://www.cites.org/eng/disc/what.php.

Byrne, H., Rylands, A. B., Carneiro, J. C., Lynch Alfaro, J. W., Bertuol, F., da Silva, M. N. F., Messias, M., Groves, C. P., Mittermeier, R. A., Farias, I., Hrbek, T., Schneider, H., Sampaio, I., Boubli, J. P. 2016. Phylogenetic relationships of the New World Titi Monkeys (*Callicebus*): first appraisal of taxonomy based on molecular evidence. *Frontiers in Zoology* 13: 10. DOI 10.1186/s12983-016-0142-4.

Geoffroy Saint-Hilare, E. 1812. *Tableau des quadrumanes*, 1. Ord. *Quadrumanes. Annales du Muséum d'histoire naturelle Paris* 19: 85–122.

Gregory, W. K. 1920. On the structure and relations of *Notharctus*, an American Eocene primate. *Memoirs of the American Museum Natural History* 3: 49–243.

Hershkovitz, P. 1990. Titis, New World monkeys of the genus *Callicebus* (Cebidae, Platyrrhini): a preliminary taxonomic review. *Fieldiana, Zoology, new series* 55: 1–109.

Huxley, T. H. 1863. *Evidence as to Man's Place in Nature.* New York: D. Appleton & Company.

Mayr, E. 1942. *Systematics and the Origin of Species from the Viewpoint of a Zoologist.* New York: Columbia University Press.

Mittermeier, R. A., Rylands, A. B., Wilson, D. E. (Eds.). 2013. *Handbook of the Mammals of the World. 3. Primates.* Barcelona: Lynx Edicions.

Napier, J. R., Napier, P. H. 1967. *A Handbook of Living Primates.* New York: Academic Press.

Napier, P. H. 1976. *Catalogue of Primates in the British Museum (Natural History) and Elsewhere in the British Isles. Part I. Families Callitrichidae and Cebidae.* London: British Museum (Natural History).

Pocock, R. I. 1918. On the external characters of the lemurs and of *Tarsius. Proceedings of the Zoological Society of London* 1918: 19–53.

Rosenberger, A. L. 2014. Species: beasts of burden. *Evolutionary Anthropology: Issues, News, and Reviews* 23: 27–29.

Schultz, A. H. 1969. *The Life of Primates.* London: Weidenfeld & Nicolson.

Simons, E. L. 1972. *Primate Evolution. An Introduction to Man's Place in Nature.* New York: Macmillan.

Simons, E. L. 1967. The earliest apes. *Scientific American* 217: 28–35.

Smith, T. D., Garrett, E. C., Bhatnagar, K. P., Bonar, C. J., Bruening, A. E., Dennis, J. C., Kinznger, J. H., Johnson, E. W., Morrison, E. E. 2011. The vomeronasal organ of New World monkeys (Platyrrhini). *Anatomical Record* 294: 2158–2178.

Wilson, D. E., Reeder, D. M. (Eds.). 1993. *Mammal Species of the World: A Taxonomic and Geographic Reference.* Washington, DC: Smithsonian Institution Press.

CHAPTER 2. Diverse Lifestyles

Anonymous. 2019. The Code Online. International Commission on Zoological Nomenclature. http://www.iczn.org/iczn/index.jsp.

Ayres, J. M., Prance, G.T. 2013. On the distribution of pitheciine monkeys and Lecythidaceae trees in Amazonia. In Veiga, L. M., Barnett, A. A., Ferrari, S. F., Norconk, M. A. (Eds.), *Evolutionary Biology and Conservation of Titis, Sakis and Uacaris.* Cambridge: Cambridge University Press, pp. 127–139.

Barnett, A. A., Bowler, M., Bezerra, B., Defler, T. R. 2013. Ecology and behavior of uacaris (genus *Cacajao*). In Veiga, L. M., Barnett, A. A., Ferrari, S. F., Norconk M. A. (Eds.), *Evolutionary Biology and Conservation of Titis, Sakis and Uacaris.* Cambridge: Cambridge University Press, pp. 151–172.

Biegert, J. 1963. The evaluation of characters of the skull, hands and feet for primate taxonomy. In Washburn, S. L. (Ed.), *Classification and Human Evolution.* Chicago: Aldine, pp. 116–143.

Blainville, H.-M. D. 1839–1864. *Ostéographie, ou, Description iconographique comparée du squelette et du système dentaire des Mammifères récents et fossiles: pour servir de base à la zoologie et à la géologie. Atlas—Tome Premier. Primates.* Paris: J.B. Baillière et fils.

Boinski, S. 1988. Use of a club by a wild white-faced capuchin (*Cebus capucinus*) to attack a venomous snake (*Bothrops asper*). *American Journal of Primatology* 14(2): 177–179.

Campbell, C. J. (Ed.). 2008. *Spider Monkeys. Behavior, Ecology and Evolution of the Genus Ateles.* Cambridge: Cambridge University Press.

Campbell, C. J., Gibson, K. N. 2013. Spider monkey reproduction and sexual behavior. In Campbell, C. J. (Ed.), *Spider Monkeys. Behavior, Ecology and Evolution of the Genus Ateles.* Cambridge: Cambridge University Press, pp. 266–287.

Digby, L. J., Ferrari, S. F., Saltzman, W. 2011. Callitrichines: the role of competition in cooperatively breeding species. In Campbell, C. J., Fuentes, A., MacKinnon, K. C., Bearder, S. K., Stumpf, R. M. (Eds.), *Primates in Perspective.* New York: Oxford University Press, pp. 91–107.

DuMond, F. 1968. The squirrel monkey in a semi-natural environment. In Rosenblum, L. A., Cooper, R. W. (Eds.), *The Squirrel Monkey.* New York: Academic Press, pp. 87–145.

Fedigan, L. M., Rosenberger, A. L., Boinski, S., Norconk, M. A., Garber, P. A. 1996. Critical issues in cebine evolution and behavior. In Norconk, M. A., Rosenberger, A. L., Garber, P. A. (Eds.), *Adaptive Radiations of Neotropical Primates.* New York: Plenum, pp. 219–228.

Fernandez-Duque, E., Di Fiore, A., Huck, M. 2012. The behavior, ecology, and social evolution of New World monkeys. In Call, J., Kappeler, P., Palombit, R., Silk, J., Mitani, J. (Eds), *Primate Societies.* Chicago: University of Chicago Press, pp. 43–54.

Fernandez-Duque, E. 2011. Aotinae: social monogamy in the only nocturnal anthropoid. In Campbell, C. J., Fuentes, A., MacKinnon, K. C., Bearder, S. K., Stumpf, R. M. (Eds.), *Primates in Perspective.* New York: Oxford University Press, pp. 140–154.

Ferrari, S. F., Rylands, A. B. 1994. Activity budgets and differential visibility in field studies of three marmosets (*Callithrix* spp.). *Folia Primatologica* 63(2): 78–83.

Fleagle, J. G. 2013. *Primate Adaptation and Evolution.* New York: Academic Press.

Ford, S. M. 1980. Callitrichids as phyletic dwarfs, and the place of the Callitrichidae in Platyrrhini. *Primates* 21: 31–43.

Ford, S. M. 1986. Systematics of the New World monkeys. In Swindler, D. R., Erwin, J. (Eds.), *Comparative Primate Biology*. New York: A.R. Liss, pp. 73–135.

Ford, S. M., Davis, L. C. 1992. Systematics and body size: implications for feeding adaptations in New World monkeys. *American Journal of Physical Anthropology* 88: 415–468.

Garber, P. A. 2011. Primate locomotor and positional behavior. In Campbell, C. J., Fuentes, A., MacKinnon, K. C., Bearder, S. K., Stumpf, R. M. (Eds.), *Primates in Perspective*. New York: Oxford University Press, pp. 548–563.

Garber, P. A., Porter, L. M. 2010. The ecology of exudate production and exudate feeding in *Saguinus* and *Callimico*. In Burrows, A. M., Nash, L. T. (Eds.), *The Evolution of Exudativory in Primates*. New York: Springer, pp. 89–108.

Garber, P. A., Porter, L. M., Spross, J., Di Fiore, A. 2016. Tamarins: insights into monogamous and non-monogamous single female social and breeding systems. *American Journal of Primatology* 78(3): 298–314.

Harris, R. A., Tardif, S. D., Vinar, T., Wildman, D. E., Rutherford, J. N., Rogers, J., Worley, K. C., Aagaard, K. M. 2014. Evolutionary genetics and implications of small size and twinning in callitrichine primates. *Proceedings of the National Academy of Sciences USA* 111(4): 1467–1472.

Hartwig, W. C. 1996. Perinatal life history traits in New World monkeys. *American Journal of Primatology* 40: 99–130.

Jack, K. M. 2011. The cebines. In Campbell, C. J., Fuentes, A., MacKinnon, K. C., Bearder, S. K., Stumpf, R. M. (Eds.), *Primates in Perspective*. New York: Oxford University Press, pp. 108–122.

Janson, C. H., Boinski, S. 1992. Morphological and behavioral adaptations for foraging in generalist primates: the case of the cebines. *American Journal of Physical Anthropology* 88: 483–498.

Kinzey, W. G. 1992. Dietary and dental adaptations in the Pitheciinae. *American Journal of Physical Anthropology* 88(4): 499–514.

Kinzey, W. G., Rosenberger, A. L., Heilser, P. S., Prowse, D. L., Trilling, J. S. 1977. A preliminary field investigation of the Yellow-handed Titi Monkey, *Callicebus torquatus* in Northern Peru. *Primates* 8: 159–181. https://doi.org/10.1007/BF02382957

Kinzey, W. G., Rosenberger, A. L., Ramirez, M. 1975. Vertical clinging and leaping in a Neotropical anthropoid. *Nature* 225: 327–328.

Kowalewski, M. M., Garber, P. A., Cortez-Ortiz, L., Urbani, B., Youlatis, D. (Eds.). 2015. *Howler Monkeys. Behavior, Ecology, and Conservation*. New York: Springer.

Maier, W. 1983. Morphology of the interorbital region of *Saimiri sciureus*. *Folia Primatologica* 41(3–4):277–303.

Maninger, N., Mendoza, S. P., Williams, D. R., Mason, W. A., Cherry, S. R., Rowland, D. J., Schaefer, T., Bales, K. L. 2017. Imaging, behavior and endocrine analysis of "jealousy" in a monogamous primate. *Frontiers in Ecology and Evolution*. https://doi.org/10.3389/fevo.2017.00119.

Matthews, L. J., Rosenberger, A. L. 2008. Taxonomic status of the Yellow-tailed Woolly Monkey, *Lagothrix flavicauda*—an object lesson in the implementation of PAUP*. *American Journal of Physical Anthropology* 137: 245–255.

Mayr, E. 1942. *Systematics and the Origin of Species from the Viewpoint of a Zoologist*. New York: Columbia University Press.

Norconk, M. A., 2011. Sakis, Uakaris, and Titi monkeys: behavioral diversity in a radiation of primate seed predators. In Campbell, C. J., Fuentes, A., MacKinnon, K. C., Bearder, S. K., Stumpf, R. M. (Eds.), *Primates in Perspective*. Oxford: Oxford University Press, pp. 122–139.

Norconk, M. A., Grafton, B. W., McGraw, W. S. 2013. Morphological and ecological adaptations to seed predation—a primate-wide perspective. In Veiga, L. M., Barnett, A. A., Ferrari, S. F., Norconk M. A. (Eds.), *Evolutionary Biology and Conservation of Titis, Sakis and Uacaris*. Cambridge: Cambridge University Press, pp. 55–71.

Norconk, M. A., Rosenberger, A. L., Garber, P. A. 1996. *Adaptive Radiations of Neotropical Primates*. New York: Plenum.

Norconk, M. A., Setz, E. Z. 2013. Ecology and behavior of saki monkeys (genus *Pithecia*). In Veiga, L. M., Barnett, A. A., Ferrari, S. F., Norconk M. A. (Eds.), *Evolutionary Biology and Conservation of Titis, Sakis and Uacaris*. Cambridge: Cambridge University Press, pp. 262–271.

Perez, S. I., Klaczko, J., dos Reis, S. F. 2012. Species tree estimation for a deep phylogenetic divergence in the New World monkeys (Primates: Platyrrhini). *Molecular Phylogenetics and Evolution* 65: 621–630.

Robinson, J. G. 1979. Vocal regulation of use of space by groups of titi monkeys *Callicebus moloch*. *Behavioral Ecology and Sociobiology* 5: 1–15.

Rosenberger, A. L. 1992. The evolution of feeding niches in New World monkeys. *American Journal of Physical Anthropology* 88: 525–562.

Rosenberger, A. L. 2002. Platyrrhine paleontology and systematics: the paradigm shifts. In Hartwig, W. C. (Ed.), *The Primate Fossil Record*. Cambridge: Cambridge University Press, pp. 151–160.

Rosenberger, A. L., 2011. Evolutionary morphology, platyrrhine evolution, and systematics. *Anatomical Record* 294 (12): 1955–1974.

Rosenberger, A. L., Halenar, L. B., Cooke, S. B., Hartwig, W. C. 2013. Morphology and evolution of the spider monkeys. In Campbell, C. J. (Ed.), *Spider Monkeys. Behavior, Ecology and Evolution of the Genus Ateles*. Cambridge: Cambridge University Press, pp. 19–49.

Rosenberger A. L., Klukkert, S. Z., Cooke, S. B., Rimoli, R. 2013. Rethinking *Antillothrix*: the mandible and its implications. *American Journal of Primatology* 75(8): 825–836.

Rosenberger, A. L., Tejedor, M. F. 2013. The misbegotten: long lineages, long branches and the interrelationships of *Aotus*, *Callicebus* and the saki–uacaris*. In Veiga, L. M., Barnett, A. A., Ferrari, S. F., Norconk, M. A. (Eds.), *Evolutionary Biology and Conservation of Titis, Sakis and Uacaris*. Cambridge: Cambridge University Press, pp. 13–22.

Ruiz-García, M., Luengas-Villamil, K., Leguizamon, N., de Thoisy, B., Gálvez, H. 2015. Molecular phylogenetics and phylogeography of all the *Saimiri* taxa (Cebidae, Primates) inferred from mt COI and COII gene sequences. *Primates* 56: 145–161.

Rylands, A. B. (Ed.). 1993. *Marmosets and Tamarins*. Oxford: Oxford University Press.

Rylands, A. B. 1993. The ecology of the Lion Tamarins, *Leontopithecus*: some intrageneric differences and comparisons with other callitrichids. In Rylands, A. B. (Ed.), *Marmosets and Tamarins: Systematics, Behaviour, and Ecology*. Oxford: Oxford University Press, pp. 296–313.

Schultz, A. H. 1956. Postembryonic age changes. *Primatologia* 1: 887–964.

Stafford, B. J., Rosenberger, A. L., Beck, B. B. 1994. Locomotion of free ranging Golden Lion Tamarins (*Leontopithecus rosalia*) at the National Zoological Park. *Zoo Biology* 13: 333–344.

Wright, P. C. 2013. *Callicebus* in Manu National Park: territory, resources, scent marking and vocalizations. In Veiga, L. M., Barnett, A. A., Ferrari, S. F., Norconk M. A. (Eds.), *Evolutionary Biology and Conservation of Titis, Sakis and Uacaris*. Cambridge: Cambridge University Press, pp. 232–239.

Wright, P. C. 2013. *High Moon over the Amazon: My Quest to Understand the Monkeys of the Night*. New York: Lantern Books.

Youlatos, D., Meldrum, J. 2011. Locomotor diversification in New World monkeys: running, climbing, or clawing along evolutionary branches. *Anatomical Record* 294(12): 1991–2012.

CHAPTER 3. What's in a Name?

Anonymous. 2019. Darwin Online. http://darwin-online.org.uk/

Elliot, D. G. 1913. *A Review of the Primates*. New York: American Museum of Natural History.

Erxleben, J. C. P. 1777. *Systema regni animalis per classes, ordines, genera, species, varietates cvm synonymia et historia animalivm. Classis I. Mammalia*. Weigand: Lipsiae.

Groves, C. 2008. *Extended Family: Long Lost Cousins*. Arlington, VA: Conservation International.

Hershkovitz, P. 1977. *Living New World Monkeys (Platyrrhini): With an Introduction to Primates*. Chicago: University of Chicago Press.

Lynch Alfaro, J. W., Silva, J. S., Jr., Rylands, A. B. 2012. How different are robust and gracile Capuchin Monkeys? An argument for the use of *Sapajus* and *Cebus*. *American Journal of Primatology* 74: 273–286.

Mouquet, N., Devictor, V., Meynard, C. N., Munoz, F., Bersier, L.-F., Chave, J., Couteron, P., Dalecky, A., Fontaine, C., Gravel, D., Hardy, O. J., Jabot, F., Lavergne, S., Leibold, M., Mouillot, D., Münkemüller, T., Pavoine, S., Prinzing, A., Rodrigues, A. S. L., Rohr, R. P., Thébault, E., Thuiller, W. 2012. Ecophylogenetics: advances and perspectives. *Biological Reviews* 87: 769–785.

Pocock, R. I. 1918. On the external characters of the lemurs and of *Tarsius*. *Proceedings of the Zoological Society of London* 1918: 19–53.

Scott, W. B. 1904. John Bell Hatcher. *Science* 20: 139–142.

Thomas, O. 1903. Notes on South-American monkeys, bats, carnivores, and rodents, with descriptions of new species. *Annals and Magazine of Natural History* 12: 455–464.

CHAPTER 4. Evolutionary Models

Anonymous. 2019. Allpahuayo Mishana National Reserve http://www.dawnontheamazon.com/allpah uayomishananationalreserve.html.

Aristide, L., Rosenberger, A. L., Tejedor, M. F., Perez, S. I. 2015. Modeling lineage and phenotypic diversification in the New World monkey (Platyrrhini, Primates) radiation. *Molecular Phylogenetics and Evolution* 82: 375–385.

Fleagle, J. G., Reed, K. 1996. Comparing primate communities: a multivariate approach. *Journal of Human Evolution* 30: 489–510.

Ford, S. M. 1980. Callitrichids as phyletic dwarfs, and the place of the Callitrichidae in Platyrrhini. *Primates* 21: 31–43.

Gadotte, M. W. 2009. Ecophylogenetics. In Duffy, J. E. (Ed.), *Encyclopedia of the Earth*. https://www .nceas.ucsb.edu/~cadotte/NR_pubs.html.

Hennig, W. 1966. *Phylogenetic Systematics*. Champaign: University of Illinois Press.

Hennig, W. 1950. *Grundzüge einer Theorie der phylogenetischen Systematik*. Berlin: Deutscher Zentralverlag.

Hershkovitz, P. 1977. *Living New World Monkeys (Platyrrhini): With an Introduction to Primates*. Chicago: University of Chicago Press.

Kinzey, W. G., Rosenberger, A. L., Ramirez, M. 1975. Vertical clinging and leaping in a Neotropical anthropoid. *Nature* 255: 327–328.

Luckett, W., Szalay, F. S. (Eds.). 1975. *Phylogeny of the Primates: A Multidisciplinary Approach*. New York: Springer.

Maier, W. 1984. Tooth morphology and dietary specialization. In Chivers, D. J., Wood, B. A., Bilsbororugh, A. (Eds.), *Food Acquisition and Processing in Primates*. New York: Plenum, pp. 303–340.

Martin, R. D. 1992. Goeldi and the dwarfs: the evolutionary biology of the small New World monkeys. *Journal of Human Evolution* 22 (4–5): 367–393.

Norconk, M. A., Rosenberger, A. L., Garber, P. A (Eds.). 1996. *Adaptive Radiations of Neotropical Primates*. New York: Plenum.

Rosenberger, A. L. 1976. Functional patterns of molar occlusion in platyrrhine primates. *American Journal of Physical Anthropology*.

Rosenberger, A. L. 1980. Gradistic views and adaptive radiation of platyrrhine primates. *Zeitschrift für Morphologie und Anthropologie* 71: 157–163.

Rosenberger, A. L. 1981 Systematics: the higher taxa. In Coimbra-Filho, A. F., Mittermeier, R. A. (Eds.), *Ecology and Behavior of Neotropical Primates*. Rio de Janeiro: Academia Brasilia de Ciencias, pp. 9–27.

Rosenberger, A. L. 1983. Aspects of the systematics and evolution of marmosets. In de Mello, M. T. (Ed.), *A Primatologia no Brasil*. Brasilia: Universidad Federal Districto Federal, pp. 159–180.

Rosenberger, A. L., Kinzey, W. G. 1976. Functional patterns of molar occlusion in platyrrhine primates. *American Journal of Physical Anthropology* 45(2): 281–298.

Rosenberger, A. L., Strier, K. B. 1989. Adaptive radiation of the ateline primates. *Journal of Human Evolution* 18: 717–750.

Rosenberger, A. L. 1992. Evolution of New World Monkeys. In Jones, S., Martin, R., Pilbeam, D. (Eds.), *The Cambridge Encyclopedia of Human Evolution*. Cambridge: Cambridge University Press, pp. 209–216.

Rosenberger, A. L., Tejedor, M. F. 2013. The misbegotten: long lineages, long branches and the inter-relationships of *Aotus*, *Callicebus* and the saki-uacaris. In Veiga, L. M., Barnett, A. A., Ferrari, S. F., Norconk, M. A. (Eds.), *Evolutionary Biology and Conservation of Titis, Sakis and Uacaris*. Cambridge: Cambridge University Press, pp. 13–22.

Rosenberger, A. L., Tejedor, M. F., Cooke, S. B., Pekkar, S. F. 2009. Platyrrhine ecophylogenetics in space and time. In Garber, P. A., Estrada, A., Bicca-Marques, J. C., Heymann, E. W., Strier, K. B. (Eds.), *South American Primates: Comparative Perspectives in the Study of Behavior, Ecology, and Conservation*. New York: Springer, pp. 69–113.

Schneider, H., Rosenberger, A. L. 1996. Molecules, morphology, and platyrrhine systematics. In Norconk, M. A., Rosenberger, A. L., Garber, P. A. (Eds.), *Adaptive Radiation of Neotropical Primates*. New York: Plenum, pp. 3–19.

Simpson, G. G. 1945. *The principles of classification and a classification of mammals. Bulletin of the American Museum of Natural History* 85: 1–350.

CHAPTER 5. How to Eat like a Monkey

Anthony, M. R. L., Kay, R. F. 1993. Tooth form and diet in ateline and alouattine primates: reflections on the comparative method. *American Journal of Science* 283(A): 356–382.

Bock, W. J., von Wahlert, G. 1965. Adaptation and the form-function complex. *Evolution* 19: 269–299.

Buffon, George-Louis Leclerc, Compte de. 1749–1788. *Histoire naturelle, générale et particulière*. Paris: Honoré Champion.

Carpenter, C. R. 1934. A field study of the behavior and social relations of howling monkeys. *Comparative Psychology Monographs* 10(2): 1–168.

Crompton, A. W., Hiiemae, K. 1969. How mammalian molar teeth work. *Discovery* 5(1): 23–34.

DiFiore, A., Link, A., Campbell, C. J. 2011. The atelines: behavioral and socioecological diversity in a New World monkey radiation. In Campbell, C. J., Fuentes, A., MacKinnon, K. C., Bearder, S. K., Stumpf, R. M. (Eds.), *Primates in Perspective*. New York: Oxford University Press, pp. 155–188.

DiFiore, A., Link, A., Dew, J. L. 2013. Diets of wild spider monkeys. In Campbell, C. J. (Ed.), *Spider Monkeys. Behavior, Ecology and Evolution of the Genus Ateles*. Cambridge: Cambridge University Press, pp. 81–137.

Eisenberg, J. F. 1981. *The Mammalian Radiations: An Analysis of Trends in Evolution, Adaptation, and Behavior*. Chicago: University of Chicago Press.

Garber, P. A. 1989. Role of spatial memory in primate foraging patterns: *Saguinus mystax* and *Saguinus fuscicollis*. *American Journal of Primatology* 19(4): 203–216.

Garber, P. A. 1988. Foraging decisions during nectar feeding by Tamarin Monkeys (*Saguinus mystax* and *Saguinus fuscicollis*, Callitrichidae, Primates) in Amazonian Peru. *Biotropica* 20: 100–106.

Hladik, C. M., Hladik, A. 1969. Rapports trophiques entre végétation et Primates dans la forêt de Barro Colorado (Panama). *La Terre et la Vie* 23: 25–117.

Janson, C. H. 1983. Adaptation of fruit morphology to dispersal agents in a Neotropical forest. *Science* 219: 187–189. https://doi.org/10.1126/science.219.4581.187

Kay, R. F. 1975. The functional adaptations of primate molar teeth. *American Journal of Physical Anthropology* 43: 195–216.

Kay, R. F. 1981. The nut-crackers—a new theory of the adaptations of the Ramapithecinae. *American Journal of Physical Anthropology* 55: 141–151. https://doi.org/10.1002/ajpa.1330550202

Kinzey, W. G. 1992. Dietary and dental adaptations in the Pitheciinae. *American Journal of Physical Anthropology* 88(4): 499–514.

Kinzey, W. G., Norconk, M. A. 1990. Hardness as a basis of fruit choice in two sympatric primates. *American Journal of Physical Anthropology* 81: 5–15.

Lambert, J. E., Chapman, C. A., Wrangham, R. W., Conklin-Brittain, N. L. 2004. Hardness of cercopithecine foods: implications for the critical function of enamel thickness in exploiting fallback foods. *American Journal of Physical Anthropology* 125: 363–368.

Lucas, P. W., Constantino, P. J., Chalk, J., Ziscovici, C., Wright, B. W., Fragaszy, D. M., Hill, D. A., Lee, J. J.-W., Chai, H., Darvell, B. W., Lee, P. K. D., Yuen, T. D. B. 2009. Indentation as a technique to assess the mechanical properties of fallback foods. *American Journal of Physical Anthropology* 140: 643–652. https://doi.org/10.1002/ajpa.21026

Marshall, A. J., Wrangham, R. W. 2007. Evolutionary consequences of fallback foods. *International Journal of Primatology* 28: 1219–1235. https://doi.org/10.1007/s10764-007-9218-5

Milton, K. 1980. *The Foraging Strategy of Howler Monkeys: A Study in Primate Economics.* New York: Columbia University Press.

Norconk, M. A., Veres, M. 2011. Physical properties of fruit and seeds ingested by primate seed predators with emphasis on sakis and bearded sakis. *Anatomical Record* 294(12): 2092–2111.

Oftedal, O. T. 1991. The nutritional consequences of foraging in primates: the relationship of nutrient intakes to nutrient requirements. *Philosophical Transactions of the Royal Society of London Series B* 334: 161–170.

Power, M. L. 2012. Nutritional and digestive challenges to being a gum-feeding primate. In Burrows, A. M., Nash, L. T. (Eds.), *The Evolution of Exudativory in Primates.* New York: Springer, pp. 25–44.

Rosenberger, A. L. 1978. Loss of incisor enamel in marmosets. *Journal of Mammalogy* 59: 207–208.

Rosenberger, A. L. 2010. Adaptive profile versus adaptive specialization: fossils and gumivory in early primate evolution. In Burrows, A. M., Nash, L. T. (Eds.), *The Evolution of Exudativory in Primates.* New York: Springer, pp. 273–295.

Rosenberger, A. L., Halenar, L. B., Cooke, S. B. 2011. The making of platyrrhine semifolivores: models for the evolution of folivory in primates. *Anatomical Record* 94: 2112–2130.

Rosenberger, A. L., Kinzey, W. G. 1976. Functional patterns of molar occlusion in platyrrhine primates. *American Journal of Physical Anthropology* 45: 281–298.

Rylands, A. B. 1984. Exudate-eating and tree-gouging by marmosets (Callitrichidae, Primates). In Chadwick, A. C., Sutton, S. L. (Eds.), *Tropical Rain Forest: The Leeds Symposium.* Leeds: Philosophical and Literary Society, pp. 155–168.

Smith, A. C. 2010. Influences on gum feeding in primates. In Burrows, A.M., Nash, L.T. (Eds.), *The Evolution of Exudativory in Primates.* New York: Springer, pp. 109–121.

Terborgh, J. 1983. *Five New World Primates.* Princeton: Princeton University Press.

CHAPTER 6. Arboreal Acrobats

Bergeson, D. J. 1998. Patterns of suspensory feeding in *Alouatta palliata, Ateles geoffroyi,* and *Cebus capucinus.* In Strasser, E., Fleagle, J. G., Rosenberger, A. L., McHenry, H. M. (Eds.), *Primate Locomotion: Recent Advances.* Boston: Springer, pp. 45–60. https://doi.org/10.1007/978-1-4899-0092-0_3

Biegert, J. 1961. Volarhaut der Hände und Füsse. *Primatologia* 2(1): 1–326.

Blainville, H.-M. D. 1839–1864. *Ostéographie, ou, Description iconographique comparée du squelette et du système dentaire des Mammifères récents et fossiles: pour servir de base à la zoologie et à la géologie. Atlas—Tome Premier. Primates.* Paris: J. B. Baillière et fils.

Dampier, W. 1700. *Voyage and descriptions. Volume 2.* London: James Knapton.

Garber, P. A. 2011. Primate locomotor positional behavior and ecology. In Campbell, C. J., Fuentes, A., MacKinnon, K. C., Bearder, S. K., Stumpf, R. M. (Eds.), *Primates in Perspective.* New York: Oxford University Press, pp. 548–563.

Gregory, W. K. 1910. The orders of mammals. *Bulletin of the American Museum of Natural History* 27: 1–102.

Napier, J. R., Napier, P. H. 1967. *A Handbook of Living Primates. Morphology, Ecology and Behaviour of Nonhuman Primates.* New York: Academic Press.

Organ, J. M. 2010. Structure and function of platyrrhine caudal vertebrae. *Anatomical Record* 293: 730–745.

Osborn, H. F. 1902. The law of adaptive radiation. *American Naturalist* 36: 353–363.

Richards, P. W. 1952. *The Tropical Rain Forest: An Ecological Study.* New York: Cambridge University Press.

Rosenberger, A. L. 1983. Tale of tails: parallelism and prehensility. *American Journal of Physical Anthropology* 60: 103–107.

Schmitt, D., Rose, M. D., Turnquist, J. E., Lemelin, P. 2005. Role of the prehensile tail during ateline locomotion: experimental and osteological evidence. *American Journal of Physical Anthropology* 126: 435–446.

Schultz, A. H. 1956. Postembryonic age changes. *Primatologia 1*: 887–964.

Slack, J. H. 1862. Monograph of the prehensile-tailed Quadrumana. *Proceedings of the Academy of Natural Sciences of Philadelphia* 14(1862): 507–519.

CHAPTER 7. Many Kinds of Platyrrhine Brains

Armstrong, E., Shea, M. A. 1997. Brains of New World and Old Word monkeys. In Kinzey, W. G. (Ed.), *New World Primates: Ecology, Evolution, and Behavior*. Abingdon-on-Thames: Routledge.

Beck, B. B. 1980. *Animal Tool Behavior: The Use and Manufacture of Tools by Animals*. New York: Garland.

Bronowski, J. 1973. *The Ascent of Man*. Boston: Little Brown.

Hartwig, W., Rosenberger, A. L., Norconk, M. A., Owl, M. Y. 2011. Relative brain size, gut size, and evolution in New World monkeys. *Anatomical Record* 294: 2207–2221.

Isler, K., Kirk, C. E., Miller, J., Albrecht, G. A., Gelvin, B. R., Martin, R. D. 2008. Endocranial volumes of primate species: scaling analyses using a comprehensive and reliable data set. *Journal of Human Evolution* 55: 967–978.

Janson, C. H. 2007. Experimental evidence for route integration and strategic planning in wild capuchin monkeys. *Animal Cognition* 10: 341–356.

Jerison, H. J. 1973. *Evolution of the Brain and Intelligence*. Chicago: University of Chicago Press.

Martin, R. D. 1990. *Primate Origins and Evolution*. Princeton: Princeton University Press.

Padberg, J., Cooke, D. F., Cerkevich, C. M., Kaas, J. H., Krubitzer, L. 2019. Cortical connections of area 2 and posterior parietal area 5 in macaque monkeys. *Journal of Comparative Neurology* 527: 718–737.

Radinsky, L. 1972. Endocasts and the study of primate brain evolution. In Tuttle, R. (Ed.), *The Functional and Evolutionary Biology of Primates*. Chicago: Aldine, pp. 175–194.

CHAPTER 8. The Variety and Means of Social Organization

Abbott, D. H., Barrett, J., George, L. M. 1993. Comparative aspects of the social suppression of reproduction in female marmosets and tamarins. In A. B. Rylands (Ed.), *Marmosets and Tamarins: Systematics, Ecology and Behaviour*. Oxford: Oxford University Press, pp. 123–151.

Agoramoorthy, G., Rudran, R. 1995. Infanticide by adult and subadult males in free-ranging Red Howler Monkeys, *Alouatta seniculus*, in Venezuela. *Ethology:* 99: 75–88.

Campbell, C. J, Gibson, K. N. 2008. Spider monkey reproduction and sexual behavior. In Campbell, C. J. (Ed.), *Spider Monkeys*. Cambridge: Cambridge University Press, pp. 266–287.

Du Mond, F. V., Hutchinson, T. C. 1967. Squirrel monkey reproduction: the "fatted" male phenomenon and seasonal spermatogenesis. *Science* 158: 1067–1070.

Dunn, J. C., Halenar, L. B., Davies, T. G., Cristobal-Azkarate, J., Reby, D., Sykes, D., Dengg, S., Fitch, W. T., Knapp, L. A. 2015. Evolutionary trade-off between vocal tract and testes dimensions in Howler Monkeys. *Current Biology* 25: 2839–2844.

Epple, G., Belcher, A., Küderling, I., Zeller, U., Scolnick, L., Greenfield, K. L., Smith, A. B., III. 1993. Making sense out of scents: species differences in scent glands, scent marking behavior and scent mark composition in the Callitrichidae. In A.B. Rylands (Ed.), *Marmosets and Tamarins: Systematics, Ecology and Behaviour*. Oxford: Oxford University Press, pp: 123–151.

Fernandez-Duque, E. 2011. Rensch's rule, Bergmann's effect and adult sexual dimorphism in wild monogamous owl monkeys (*Aotus azarai*) of Argentina. *American Journal of Physical Anthropology* 146: 38–48.

Fernandez-Duque, E., DiFiore, A., Huck, M. 2012. The behavior, ecology, and social evolution of New World monkeys. In Mitani, J. C., Call, J., Kappeler, J. B. (Eds.), *Evolution of Primate Societies*. Chicago: University of Chicago Press, pp. 43–64.

Fragaszy, D. M., Visalberghi, E., Fedigan, L. M. 2004. *The Complete Capuchin: The Biology of the Genus Cebus*. Cambridge: Cambridge University Press.

Freese, C. H., Oppenheimeier, J. R. 1981. The Capuchin Monkey, genus *Cebus*. In Coimbra-Filho, A., Mittermeier, R. A. (Eds.), *Ecology and Behavior of Neotropical Primates*. Rio de Janeiro: Academia Brasileira de Ciências, pp. 331–390.

Kay, R. F., Plavcan, J. M., Glander, K. E., Wright, P. C. 1988. Sexual selection and canine dimorphism in New World monkeys. *American Joutrnal of Physical Anthropology* 77: 385–397.

Kinzey, W. G. 1972. Canine teeth of the monkey, *Callicebus moloch*: Lack of sexual dimorphism. *Primates* 13: 365–369.

Merritt, D., Jr. 1980. Captive reproduction and husbandry of the Douroucouli, *Aotus trivirgatus*, and the Titi Monkey, *Callicebus* spp. *International Zoo Yearbook* 20:52–59.

Moynihan, M. 1976. *The New World Primates*. Princeton: Princeton University Press.

Moynihan, M. H. 1966. Communication in the Titi Monkey *Callicebus*. *Journal of Zoology, London* 150: 77–127.

Moynihan, M. H. 1964. Some behavior patterns of platyrrhine monkeys: I. The Night Monkey (*Aotus trivirgatus*). *Smithsonian Miscellaneous Collections* 146: 1–84.

Porter, L. M. 2007. *The Behavioral Ecology of Callimicos and Tamarins in Northwestern Bolivia*. Upper Saddle River, NJ: Pearson/Prentice Hall.

Robinson, J. H., Wright, P. C., Kinzey, W. G. 1987. Monogamous cebids and their relatives: intergroup calls and spacing. In Smuts, B. B., Cheney, D. L., Seyfarth, R. M., Wrangham, R. W., Struhsaker, T. (Eds.), *Primate Societies*. Chicago: University of Chicago Press, pp. 44–53.

Ross, C. N., French, J. A., Ortí, G. 2007. Germ-line chimerism and paternal care in marmosets (*Callithrix kuhlii*). *Proceedings of the National Academy of Sciences USA* 104: 6278–6282.

Rylands, A. B. (Ed.). 1993. *Marmosets and Tamarins: Systematics, Behaviour, and Ecology*. Oxford: Science Publications.

Rylands, A. B. 1996. Habitat and the evolution of social and reproductive behavior in Callitrichidae. *American Journal of Primatology* 38: 5–18.

Sekulic, R. 1982. The function of howling in Red Howler Monkeys (*Alouatta seniculus*). *Behaviour* 81: 38–54.

Snowdon, C.T., 2013. Language parallels in New World primates. In Helekar, S. A. (Ed.), *Animal Models of Speech and Language Disorders*. New York: Springer, pp. 241–261.

Strier, K. B. 1999. *Faces in the Forest: The Endangered Muriqui Monkeys of Brazil*. Cambridge, MA: Harvard University Press.

Stumpf, R. M. 2011. Chimpanzees and bonobos: Inter- and intra-species diversity. In Campbell, C. J., Fuentes, A., MacKinnon, K. C., Bearder, S. K., Stumpf, R. M. (Eds.), *Primates in Perspective*. Oxford: Oxford University Press, pp. 340–356.

Wolovich, C. K., Evans, S. 2007. Sociosexual behavior and chemical communication of *Aotus nancymaae*. *International Journal of Primatology* 28: 1299–1313.

Wyatt, T. D. 2003. *Pheromones and Animal Behaviour, Communication by Smell and Taste*. Cambridge: Cambridge University Press

CHAPTER 9. 20 Million Years

Ameghino, F. 1891. Los monos fósiles del Eoceno de la República Argentina. *Revista Argentina de Historia Natural* 1: 383–397.

Aristide, L., Rosenberger, A. L., Tejedor, M. F., Perez, S. I. 2015. Modeling lineage and phenotypic diversification in the New World monkey (Platyrrhini, Primates) radiation. *Molecular Phylogenetics and Evolution* 82: 375–385.

Aristide, L., Strauss, A., Halenar-Price, L. B., Gilissen, E., Cruz, F. W., Cartelle, C., Rosenberger, A. L., Lopes, R. T., Dos Reis, S. F., Perez, S. I. 2019. Cranial and endocranial diversity in extant and fossil atelids (Platyrrhini: Atelidae): a geometric morphometric study. *American Journal of Physical Anthropology* 169: 322–331.

Beck, B. B. 2019. *Unwitting Travelers: A history of primate reintroduction*. Berlin: Salt Water Media.

Bloch, J. I., Woodruff, E. D., Wood, A. R., Rincon, A. F., Harrington, A. R., Morgan, G. S., Foster, D. A., Montes, C., Jaramillo, C. A., Jud, N. A., Jones, D. S., MacFadden, B. J. 2016. First North American fossil monkey and early Miocene tropical biotic interchange. *Nature* 533: 243.

Bluntschli, H. 1931. *Homunculus patagonicus* und die ihm zugereihten fossilfunde aus den Santa Cruz Schichten Patagoniens. Eine morphologische revision an hand der originaltücke in der sammlung Ameghino zu La Plata. *Gegenbaurs Morphologisches Jehrbuch* 67: 811–892

Bond, M., Tejedor, M. F., Campbell, K. E., Jr., Chornogubusky, L., Novo, N., Goin, F. 2015. Eocene primates of South America and the African origins of New World monkeys. *Nature* 520: 538–541.

Cartelle, C., Hartwig, W. C. 1996. A new extinct primate among the Pleistocene megafauna of Bahia, Brazil. *Proceedings of the National Academy of Sciences USA* 93: 6405–6409.

Cooke, S. B., 2011. Paleodiet of extinct platyrrhines with emphasis on the Caribbean forms: three-dimensional geometric morphometrics of mandibular second molars. *The Anatomical Record* 294: 2073–2091.

Cooke, S. B., Gladman, J., Halenar, L. B., Klukkert, Z. S., Rosenberger, A. L. 2016. The paleobiology of the recently extinct platyrrhines of Brazil and the Caribbean. In Ruiz, M. (Ed.), *Molecular Population Genetics, Evolutionary Biology and Biological Conservation of Neotropical Primates*. New York: Nova Publishers, pp. 41–89.

Cooke, S. B., Mychajliw, A. M., Southon, J., MacPhee, R.D.E. 2017. The extinction of *Xenothrix mcgregori*, Jamaica's last monkey. *Journal of Mammalogy* 98: 937–949.

Cooke, S.B., Rosenberger, A.L., Turvey, S. 2011. An extinct monkey from Haiti and the origins of the Greater Antillean primates. *Proceedings of the National Academy of Sciences USA* 108: 2699–2704.

Delson, E., Rosenberger, A. L. 1984. Are there any anthropoid primate "living fossils"? In Eldredge, N., Stanley, S. (Eds.), *Casebook on Living Fossils*. New York: Fischer, pp. 50–61.

Eldredge, N., Gould, S. J. 1977. Punctuated equilibria: the tempo and mode of evolution reconsidered. *Paleobiology* 3(2): 115–151.

Fleagle, J. G., Bown, T. M. 1983. New primate fossils from late Oligocene (Colhuehuapian) localities of Chubut Province, Argentina. *Folia Primatologica* 41: 240–266.

Fleagle, J. G., Kay, R. F., Anthony, M. R. L. 1997. In Kay, R. F., Madden, R. H., Cifelli, R. L., Flynn, J. J. (Eds.), *Vertebrate Paleontology in the Neotropics. The Miocene Fauna of La Venta Colombia*. Washington, DC: Smithsonian Institution Press, pp. 473–495.

Fleagle, J. G., Powers, D. W., Conroy, G. C., Watters, J. P. 1987. New fossil platyrrhines from Santa Cruz Province, Argentina. *Folia Primatologica* 48: 65–77.

Fleagle, J. G., Tejedor, M. F. 2002. Early platyrrhines of southern South America. In Hartwig, W. C. (Ed.), *The Primate Fossil Record*. Cambridge: Cambridge University Press, pp. 161–173.

Halenar, L. B. 2011. Reconstructing the locomotor repertoire of *Protopithecus brasiliensis* I: Body size. *The Anatomical Record* 294: 2024–2047.

Halenar, L. B. 2011. Reconstructing the locomotor repertoire of *Protopithecus brasiliensis* II: Forelimb morphology. *The Anatomical Record* 294: 2048–2063.

Halenar, L. B., Cooke, S. B., Rosenberger, A. L., Rímoli, R. 2017. New cranium of the endemic Caribbean platyrrhine, *Antillothrix bernensis*, from La Altagracia Province, Dominican Republic. *Journal of Human Evolution* 106: 133–153.

Halenar, L. B, Rosenberger, A. L. 2013. A closer look at the "*Protopithecus*" fossil assemblages: new genus and species from Bahia, Brazil. *Journal of Human Evolution* 65: 374–390.

Hartwig, W. C., Cartelle, C. 1996. A complete skeleton of the giant South American primate *Protopithecus. Nature* 381: 307–311.

Hartwig W. C., Meldrum, J. D. 2002. Miocene platyrrhines of the northern Neotropics. In Hartwig, W. C. (Ed.), *The Primate Fossil Record*. Cambridge: Cambridge University Press, pp. 175–188..

Hershkovitz, P. 1970. Notes on Tertiary platyrrhine monkeys and description of a new genus from the late Miocene of Colombia. *Folia Primatologica* 12: 1–37.

Hershkovitz, P. 1982. Supposed squirrel monkey affinities of the late Oligocene *Dolichocebus gaimanensis. Nature* 298: 201–202.

Kay, R. F., Fleagle, J. G. 2010. Stem taxa, homoplasy, long lineages and the phylogenetic position of *Dolichocebus. Journal of Human Evolution* 59: 218–222.

Kay, R. F., Fleagle, J. G., Mitchell, T. R. T., Colbert, M., Bown, T., Powers, D. W. 2008. The anatomy of *Dolichocebus gaimanensis*, a stem platyrrhine monkey from Argentina. *Journal of Human Evolution* 54: 323–382.

Kay, R. F., Gonzales, L. A., Salenbien, W., Martinez, J.-N., Cooke, S. B., Valdivia, L. A., Rigsby, C., Baker, P.A. 2019. *Parvimico materdei* gen. et sp. nov.: A new platyrrhine from the Early Miocene of the Amazon Basin, Peru. *Journal of Human Evolution* 134: 1–16.

Kay, R. F., Hunt, K. D., Beeker, C. D., Conrad, G. W., Johnson, C. C., Keller, J. 2011. Preliminary notes on a newly discovered skull of the extinct monkey *Antillothrix* from Hispaniola and the origin of the Greater Antillean monkeys. *Journal of Human Evolution* 60: 124–128.

Luchterhand, K., Kay, R. F., Madden, R. H. 1986. *Mohanamico hershkovitzi*, gen. et sp. nov., un primate Miocéne moyen d'Amérique de Sud. *Comptes rendus de l'Académie des Sciences Paris* 303, series II, 19: 1753–1758.

MacPhee, R. D. E., Horovitz, I. 2004. New craniodental remains of the Quaternary Jamaican monkey *Xenothrix mcgregori* (Xenotrichini, Callicebinae, Pitheciidae), with a reconsideration of the *Aotus* hypothesis. *American Museum Novitates* 3434:1–51.

Macphee, R. D. E., Horovitz, I., Arredondo, O., Vasquez, O. J. 1995. A new genus for the extinct Hispaniolan monkey *Saimiri bernensis* Rimoli, 1977, with notes on its systematic position. *American Museum Novitates* 3134: 1–21.

MacPhee, R. D. E., Iturralde-Vinent, M. A., Gaffney, E. S. 2003. Domo de Zaza, an early Miocene vertebrate locality in south-central Cuba, with notes on the tectonic evolution of Puerto Rico and the Mona Passage. *American Museum Novitates* 3394: 1–42.

MacPhee, R. D. E., Meldrum, J. 2006. Postcranial remains of the extinct monkeys of the Greater Antilles, with evidence for semiterrestriality in *Paralouatta*. *American Museum Novitates* 3516: 1–65.

Madden, R. H., Savage, D. E., Fields, R. W. 1997. A history of vertebrate paleontology in the Magdalena Valley. In Kay, R. F., Madden, R. H., Cifelli, R. L., Flynn, J. J. (Eds.). *Vertebrate Paleontology in the Neotropics. The Miocene Fauna of La Venta Colombia*. Washington, DC: Smithsonian Institution Press, pp. 1–11.

Maier, W. 1983. Morphology of the interorbital region of *Saimiri sciureus*. *Folia Primatologica* 41(3–4): 277–303.

Marivaux, L., Adnet, S., Altamirano-Sierra, A. J., Boivin, M., Pujos, F., Ramdarshan, A., Salas-Gismondi, R., Tejada-Lara, J. V., Antoine, P.-O. 2016. Neotropics provide insights into the emergence of New World monkeys: new dental evidence from the late Oligocene of Peruvian Amazonia. *Journal of Human Evolution* 97: 159–175.

Miller, G. S., 1929. Mammals eaten by Indians, owls, and Spaniards in the coast region of the Dominican Republic. *Smithsonian Miscellaneous Collections* 82: 1–16.

Perelman, P., Johnson, W. E., Roos, C., Seuánez, H. N., Horvath, J. E., Moreira, M. A. M., Kessing, B. Pontius, J., Roelke, M., Rumpler, Y., Schneider, M. P. C., Silva, A., O'Brien, S. J., Pecon-Slattery, J. 2011. A molecular phylogeny of living primates. *PLOS Genetics* 3: e1001342.

Perez, I., Rosenberger, A. L. 2014. The status of platyrrhine phylogeny: a meta-analysis and quantitative appraisal of topological hypotheses. *Journal of Human Evolution* 76: 177–187.

Perez, S. I., Tejedor, M. F., Novo, N., Aristide, L. 2013. Divergence times and the evolutionary radiation of New World monkeys (Platyrrhini, Primates): an analysis of fossil and molecular data. *PLOS ONE* 8: e68029.

Rímoli, R., 1977. A new species of monkey (Cebidae: Saimirinae: Saimiri) from Hispaniola (original title in Spanish). *Cuadernos de CENDIA, Universidad Autonoma de Santa Domingo* 242: 5–14.

Rivero, M., Arredondo, O., 1991. *Paralouatta varonai*, a new Quaternary platyrrhine from Cuba. *Journal of Human Evolution* 21: 1–11.

Rocatti, G., Aristide, L., Rosenberger, A. L., Perez, S. I. 2017. Early evolutionary diversification of mandible morphology in the New World monkeys (Primate, Platyrrhini). *Journal of Human Evolution* 113: 24–37.

Rosenberger, A. L. 1977. *Xenothrix* and ceboid phylogeny. *Journal of Human Evolution*. 6: 541–561.

Rosenberger, A. L. 1979. Cranial anatomy and implications of *Dolichocebus*, a late Oligocene ceboid primate. *Nature* 279: 416–418.

Rosenberger A. L. 1982. Reply to: Hershkovitz P. (1982) Supposed squirrel monkey affinities of the late Oligocene *Dolichocebus gaimanensis*. *Nature* 298: 202.

Rosenberger, A. L. 2010. Platyrrhines, PAUP, parallelism, and the long lineage hypothesis: a reply to Kay et al. (2008). *Journal of Human Evolution* 59: 214–217.

Rosenberger, A. L. 2019. *Dolichocebus gaimanensis* is not a stem platyrrhine. *Folia primatologica* 90: 1–17.

Rosenberger, A. L., Cooke, S., Halenar, L. B., Tejedor, M. F., Hartwig, W. C., Novo, N. M., Muñoz-Saba, Y., 2015. Fossil alouattines and the origins of *Alouatta*: craniodental diversity and interrelationships. In Kowalewski, M. M., Garber, P. A., Cortés-Ortiz, L., Urbani, B., Youlatos, D. (Eds). *Howler Monkeys: Adaptive Radiation, Systematics, and Morphology*. New York: Springer, pp. 21–54.

Rosenberger, A. L., Hartwig, W. C., Takai, M., Setoguchi, T., Shigehara, N. 1991. Dental variability in *Saimiri* and the taxonomic status of *Neosaimiri fieldsi*, an early squirrel monkey from La Venta, Colombia. *International Journal of Primatology* 12: 291–301.

Rosenberger, A. L., Hartwig, W. C., Wolff, R. 1991. *Szalatavus attricuspis*, an early platyrrhine primate from Salla, Bolivia. *Folia Primatologica* 56: 221–233.

Rosenberger A. L., Klukkert, S. Z., Cooke, S. B., Rimoli, R. 2013. Rethinking *Antilllothrix*: the mandible and its implications. *American Journal of Primatology* 75: 825–836.

Rosenberger, A. L., Pickering, R., Green, H., Cooke, S. B., Tallman, M., Morrow, A., Rímoli, R. 2015. 1.32 ± 0.11 Ma age for underwater remains constrain antiquity and longevity of the Dominican primate *Antillothrix bernensis*. *Journal of Human Evolution* 88: 85–96.

Rosenberger, A. L., Setoguchi, T., Hartwig, W. C. 1991. *Laventiana annectens*, new fossil evidence for the origins of callitrichine New World monkeys. *Proceedings of the National Academy of Sciences USA* 88: 2137–2140.

Rosenberger, A. L., Setoguchi, T., Shigehara, N. 1990. The fossil record of callitrichine primates. *Journal of Human Evolution* 19: 209–236.

Rosenberger, A. L., Tejedor, M. F. 2013. The misbegotten: long lineages, long branches and the interrelationships of *Aotus*, *Callicebus* and the saki-uacaris. In Barnett, A., Veiga, L., Ferrari, S. F., Norconk, M. A. (Eds.), *Evolutionary Biology and Conservation of Titis, Sakis and Uacaris*. Cambridge: Cambridge University Press, pp. 13–22.

Rosenberger, A. L., Tejedor, M. F., Cooke, S. B., Halenar, L. B., Pekkar, S. 2009. Platyrrhine ecophylogenetics, past and present. In Garber, P. A., Estrada, A., Bicca-Marques, J. C., Heymann, E. W., Strier, K. B. (Eds.), *South American Primates: Comparative Perspectives in the Study of Behavior, Ecology, and Conservation*. New York: Springer, pp. 69–112.

Schneider, H., Sampaio, I. 2015. The systematics and evolution of New World primates—A review. *Molecular Phylogenetics and Evolution* 82(B): 348–357.

Setoguchi, T., Rosenberger, A. L. 1985. Miocene marmosets: first fossil evidence. *International Journal of Primatology* 6: 615–625.

Setoguchi, T., Rosenberger, A. L. 1987. A fossil owl monkey from the La Venta, Colombia. *Nature* 326: 692–694.

Stirton, R. A. 1951. Ceboid monkeys from the Miocene of Colombia. *Bulletin of the University of California Publications in Geological Sciences* 28: 315–356.

Stirton, R. A., Savage, D. E. 1951. A new monkey from the La Venta Miocene of Colombia. *Compilacion De Los Estudios Geologicos Oficiales en Colombia* 7: 345–356.

Takai, M., Anaya, F., Suzuki, H., Shigahara, N., Setoguchi, T. 2001. A new platyrrhine from the Middle Miocene of La Venta, Colombia, and the phyletic position of Callicebinae. *Anthropological Science, Tokyo* 109(4): 289–307.

Tejedor, M., Rosenberger, A. L. 2008. A new type for *Homunuculus patagonicus*, Ameghno, 1891. *PaleoAnthropology* 2008: 68–82.

Tejedor, M. F., Tauber, A. A., Rosenberger, A. L., Swisher, C. C., Palacios, M. E. 2006. New primate genus from the Miocene of Argentina. *Proccedings of the National Academy of Sciences USA* 103: 5437–5441.

Wilkinson, R. D., Steiper, M. E., Soligo, C., Martin, R. D., Yang, Z., Tavaré, S. 2011. Dating primate divergences through an integrated analysis of palaeontological and molecular data. *Systematic Biology* 60: 16–31.

Williams, E. E., Koopman, K. F. 1952. West Indian fossil monkeys. *American Museum Novitates* 1546: 1–16.

Woods, R., Turvey, S. T., Brace, S., MacPhee, R. D. E., Barnes, I. 2018. Ancient DNA of the extinct Jamaican monkey *Xenothrix* reveals extreme insular change within a morphologically conservative radiation. *Proceedings of the National Academy of Sciences USA* 115: 12769–12774.

CHAPTER 10. **South America Was Once an Island**

Anonymous. 2019. Jean-Jacques Savin: Frenchman completes Atlantic crossing in barrel. *BBC News* https://www.bbc.com/news/world-europe-48222707.

Beard, K. C. 2008. The oldest North American primate and mammalian biogeography during the Paleocene–Eocene Thermal Maximum. *Proceedings of the National Academy of Sciences USA* 105: 3815–3818.

Birkiatis, L. 2014. The De Geer, Thulean and Beringia routes: key concepts for understanding early Cenozoic biogeography. *Journal of Biogeography* 41: 1036–1054.

Buffon, Georges-Louis Leclerc, Compte de. 1797. *Buffon's Natural history, containing a theory of the earth, a general history of man, of the brute creation, and of vegetables, minerals, &c. &c.* London: Symonds.

Ciochon, R. L., Chiarelli, A. B. (Eds.). 1980. *Evolutionary Biology of the New World Monkeys and Continental Drift.* Boston: Springer.

Eberle, J. J., Greenwood, D. R. 2012. Life at the top of the greenhouse Eocene world—a review of the Eocene flora and vertebrate fauna from Canada's High Arctic. *Geological Society of America Bulletin* 124: 3–23.

Fleagle, J. G., Gilbert, C. C. 2008. *Elwyn Simons: A Search for Origins.* Springer: New York.

Heegan, R. J. 2015. *Floating Islands. An Activity Book.* http://www.unm.edu/~rheggen/FloatingIslands.htm

Hofstetter, R. 1972. Relationships, origins, and history of the ceboid monkeys and caviomorph rodents: a modern reinterpretation. In Dobzhansky, T., Hecht, M. K., Steere, W. C. (Eds.), *Evolutionary Biology, Vol. 6.* New York: Appleton-Century-Crofts, pp. 323–347.

Houle, A. 1999. The origin of platyrrhines: An evaluation of the Antarctic scenario and the floating island model. *American Journal of Physical Anthropology* 109: 541–559.

Iturralde-Vinent, M., MacPhee, R. D. E. 1999. Paleogeography of the Caribbean region: implications for Cenozoic biogeography. *Bulletin of the American Museum of Natural History* 238: 1–95.

Kay, R. F. 2015. New World monkey origins. *Science* 347: 1068–1069.

Kirk, E. C., Williams, B. A. 2011. New adapiform primate of Old World affinities from the Devil's Graveyard Formation of Texas. *Journal of Human Evolution* 61:156–68.

Lavocat, R., 1980. The implications of rodent paleontology and biogeography to the geographical sources and origin of platyrrhine primates. In Ciochon, R. L., Chiarelli, A. B. (Eds.), *Evolutionary Biology of the New World Monkeys and Continental Drift.* Boston: Springer, pp. 93–102.

Lehmann, T., Schaal, S. F. K. (Eds.). 2012. Messel and the terrestrial Eocene—Proceedings of the 22nd Senckenberg Conference. *Palaeobiodiversity and Palaeoenvironments* 92: 397–402.

Manchester, S. R., Wilde, V., Collinson, M. E. 2007. Fossil cashew nuts from the Eocene of Europe: biogeographic links between Africa and South America. *International Journal of Plant Sciences* 168: 1199–1206.

Matthew, W. D. 1915. Climate and evolution. *Annals of the New York Academy of Science* 24: 171–318.

Rose, K. D., 2012. The importance of Messel for interpreting Eocene Holarctic mammalian faunas. *Palaeobiodiversity and Palaeoenvironments* 92: 631–647. https://doi.org/10.1007/s12549-012-0090-8

Rosenberger, A. L. 2006. Protoanthropoidea (Primates, Simiiformes): a new primate higher taxon and a solution to the *Rooneyia* problem. *Journal of Mammalian Evolution* 13: 139–146.

Simons, E. L. 1976. The fossil record of primate phylogeny. In Goodman, M., Tashian, R. E. (Eds.), *Molecular Anthropology.* New York: Plenum, pp. 35–62.

Smith, T., Rose, K. D., Gingerich, P. D. 2006. Rapid Asia–Europe–North America geographic dispersal of earliest Eocene primate *Teilhardina* during the Paleocene–Eocene Thermal Maximum. *Proceedings of the National Academy of Sciences USA* 103: 11223–11227.

van Duzer, C. 2004. *Floating islands: A Global Bibliography*. Los Altos Hills, CA: Cantor.

Wallace, A. R. 1880. *Island Life*. London, Macmillan.

Webb, S. D., Opdyke, N. D. 1995. Global climatic influence on Cenozoic land mammal faunas. In Kennett, J. P., Stanley, S. M. (Eds.), *Effects of Past Global Change on Life*. Washington, DC: National Research Council, pp. 184–208.

Weil, M. 2014. Anthony Smith: Writer and adventurer who in his eighties sailed a raft made of plastic from the Canaries to the Bahamas. *The Independent* https://www.independent.co.uk/news /obituaries/anthony-smith-writer-and-adventurer-who-in-his-eighties-sailed-a-raft-made-of-plastic -from-the-9692165.html.

CHAPTER 11. After 20 Million Years of Existence, New World Monkeys Face Extinction

Anonymous. 2017. *Centro de Primatologia do Rio de Janeiro*. Instituto Estadual do Ambiente, Rio de Janeiro.

Bridgwater, D. D. (Ed.). 1972. *Saving the Lion Marmoset*. Wheeling, WV: Wild Animal Propagation Trust.

Candisani, L., Leitão, M., Strier, K., Mittermeier, R. A., Abranches, S. 2017. *Muriqui—Kings of the Forest*. Sao Paulo: Cámara Brasilera do Livro.

Coimbra-Filho, A. F., Mittermeier, R. A. 1972. Taxonomy of the genus *Leontopithecus* Lesson, 1840. In Bridgewater, D. D. (Ed.), *Saving the Lion Marmoset*. Wheeling, WV: Wild Animal Propagation Trust, pp. 7–22.

Darwin, C. 1832. *The Beagle Diary. 29th February & 1st March, 1832*. http://darwinbeagle.blogspot.com /2007/03/29th-feb-1st-mar-1832.html

Dean, W. 1997. *With Broadax and Firebrand: The Destruction of the Brazilian Atlantic Forest*. Berkeley: University of California Press.

Estrada, A., Garber, P. A., Rylands, A. B., Roos, C., Fernandez-Duque, E., Di Fiore, A., Nekaris, K.A.-I., Nijman, V., Heymann, E. W., Lambert, J. E., Rovero, F., Barelli, C., Setchell, J. M., Gillespie, T. R., Mittermeier, R. A., Arregoitia, L. V., de Guinea, M., Gouveia, S., Dobrovolski, R., Shanee, S., Shanee, N., Boyle, S. A., Fuentes, A., MacKinnon, K. C., Amato, K. R., Meyer, A. L. S., Wich, S., Sussman, R. W., Pan, R., Kone, I., Li, B. 2017. Impending extinction crisis of the world's primates: why primates matter. *Science Advances* 3. https://doi.org/10.1126/sciadv.1600946

Ferraz, D. da S., Tabacow, F., Mittermeier, R. A., Melo, F., Boubli, J., Jerusalinsky, L., Talebi, M. 2019. *Brachyteles hypoxanthus*. *The IUCN Red List of Threatened Species* 2019: e.T2994A17927482..

Gibbens, S. 2018. Tiny gold monkeys and pumas are getting their own highway. https://www.nationalgeo graphic.com.au/animals/tiny-gold-monkeys-and-pumas-are-getting-their-own-highway.aspx

Kleiman, D., Rylands, A. B. (Eds.). 2002. *Lion Tamarins: Biology and Conservation*. Washington, DC: Smithsonian Institution Press.

Mittermeier, R. A., Rylands, A. B., Wilson, D. E. (Eds.). 2013. *Handbook of the Mammals of the World. Volume 3. Primates*. Barcelona: Lynx Edicions.

Schwitzer, C., Mittermeier, R. A., Rylands, A. B., Chiozza, F., Williamson, E. A., Macfie, E. J., Wallis, J., Cotton, A. 2017. *Primates in Peril: The World's 25 Most Endangered Primates, 2016–2018*. https:// portals.iucn.org/library/sites/library/files/documents/2017-059.pdf

Strier, K. B. 1999. *Faces in the Forest: The Endangered Muriqui Monkeys of Brazil*. Cambridge, MA: Harvard University Press

INDEX